MICROBIAL DIVERSITY OF FISHES

A COLOUR HANDBOOK

MICROBIAL DIVERSITY OF FISHES

A COLOUR HANDBOOK

K. Dhevendaram

2024

Daya Publishing House®

A Division of

Astral International Pvt. Ltd.

New Delhi – 110 002

First Published, 2013

Reprinted, 2024

ISBN: 978-81-7035-786-5 (HB)

Disclaimer:

Every possible effort has been made to ensure that the information contained in this book is accurate at the time of going to press, and the publisher and author cannot accept responsibility for any errors or omissions, however caused. No responsibility for loss or damage occasioned to any person acting, or refraining from action, as a result of the material in this publication can be accepted by the editor, the publisher or the author. The Publisher is not associated with any product or vendor mentioned in the book. The contents of this work are intended to further general scientific research, understanding and discussion only. Readers should consult with a specialist where appropriate.

Every effort has been made to trace the owners of copyright material used in this book, if any. The author and the publisher will be grateful for any omission brought to their notice for acknowledgment in the future editions of the book.

Published by : **Daya Publishing House®**
A Division of
Astral International Pvt. Ltd.
– ISO 9001:2015 Certified Company –
4736/23, Ansari Road, Darya Ganj,
New Delhi-110 002
Ph. 011-43549197, 23278134
E-mail: info@astralint.com
Website: www.astralint.com

Printed at : Replika Press Pvt. Ltd.

— Dedicated to —

Late Prof. N. Balakrishnan Nair

Foreword

For human society, both coastal areas and Inland water bodies are an important interface between terrestrial land and water based activities. Indian continent has greater potentials of fishery resources. Today, our understanding of the productive eco-biology of aquatic ecosystems is still poor. Both capture fisheries and aquaculture are applied sciences drawing concepts, expertise and insight from biological, ecological and technological sciences. Can we design production system that mimic all processes occurring in nature, while providing the aquatic organisms with all necessary micro- and macro-nutrients? It is possible through proper management practices of micro flora of fish ponds to improve the gut micro flora of fishes. Eco-friendly integrated farming's pave the way for heal their ecosystem. Unsaturated fatty acids and fish meal for protein sources from fishes have both nutritive and medicinal values. Based on the above facts and scientific information, this book entitled *"Microbial Diversity of Fishes"* is a comprehensive approach on microbial diversity, fish nutrition, fish health management, antibiotic potentials, immunostimulants, processing technology and bio-fertilizers in fish ponds. For bringing out such a valuable book with advanced knowledge on technique know how of microbial activities in fisheries, I really appreciate Prof. K. Dhevendaran, SASTRA University, Thanjavur and it is an added feather to his cape of scientific tempo. For fishery technologists-a-wealth of technological, biological, ecological and economic information, are available in this text book to certain extent. Any advancement in this direction marks the milestone to develop and disseminate scientific technological knowledge to the Post Graduate students and also to research scholars.

I wish him all success in his endeavor.

Prof. R. Sethuraman

Vice -Chancellor,
SASTRA University,
Thanjavur, Tamil Nadu

Preface

Microbial diversity and their metabolic activity in fishes are such a rapidly evolving fields. It is a comprehensive approach on microbial domain in fish for the author and the publisher to plan to update it annually. The text book is linked to the very wheels of vehicle for modern approach and information dissemination. The investigation on intestinal and surface micro flora in fish has strong effect on the nutritional status of animals. Moreover, the nutritive and medicinal values of fin fish and shellfish have been recognized for long time. It is ubiquitous in society. It impacts the environment, nutrition, industry and health of animal and human.

In view of the foregoing, an introductory text book on microbial domain should, to some extent, integrate all areas of application. Another aspect of importance that should be addressed in an introductory fishery microbiology text book is the matter relating to the acceptance of the processes and products by the beneficiary of research in microbial activity of aquatic biota. This text book may be used as a preliminary text or supplementary book for a post graduate course.

I take this opportunity to thank our honorable Vice-Chancellor, Prof. R. Sethuraman for his unbounded encouragement and enthusiastic support. But for his moral guidance, it is highly impossible to bring out this important scientific text book for the benefits of post graduate students and research scholars. I am immensely indebted to Dr. Swaminathan, S., Dean, Sponsored Research and Dr. K. Sridharan, Dean, School of Chemical and Biotechnology for their advice, guidance and support.

I am very much obliged to scientists and technologists for constructive criticism.

The last but not the least I would extend my deep sense of appreciation to my beloved wife, P. Ragupathy Devi, M.A., M.Ed., and loving children, D. Vasanthi, B.E., M.S. (USA) and D. Diana, B.Tech. (Aravind Eye Hospital, Madurai) who stood behind me to complete the text book in time.

Prof. (Dr.) K. Dhevendaran

Contents

Chapter 1
General Introduction

The nature and the association of microorganisms over the surface of the body, inside the digestive system as well as in the reproductive system of fish, shellfish and prawn are thus warranting. There is an increasing interest in fish as food and as pets promoted the awareness of aquaculture as an industry. This would enjoy the same success as poultry farming as achieved in India. With increasing advancement in Science and Technology on industrialization and urbanization on the banks of the rivers and on the coastal belt of the oceans and estuaries, there is every possibility for the concomitant impact of anthropogenic pollutants on the very survival of the microbes as well as the physiological and biochemical activities of the host organisms like fish, shellfish and prawn and other biota.

The interactions of the microorganisms on the surface of seaweeds and in the fish and shellfish are quite unique and diversified. Microorganisms in the strict sense deal with bacteria, fungi, streptomycetes, blue green algae, virus, etc. The relationship between microorganisms and host animals may be for their mutual benefit. Such a symbiosis exists in a good number of organisms like fish and shellfish preferably the silver bellies such as leognathid fish (Figures 1.1 and 1.2).

The reason for this sort of association for the supply of food to the microorganisms the components of the micro biota has revealed a wide variety of functions to the host organisms.

Epiphytic Microflora

Furthermore, the interaction of microorganisms on the surface of a huge number of plants and animals is for the temporary settlement or at times for producing a layer swarming growth (Figure 1.3).

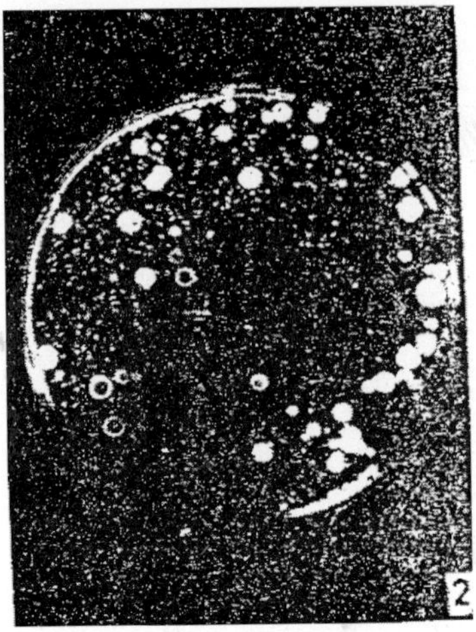

Figure 1.1

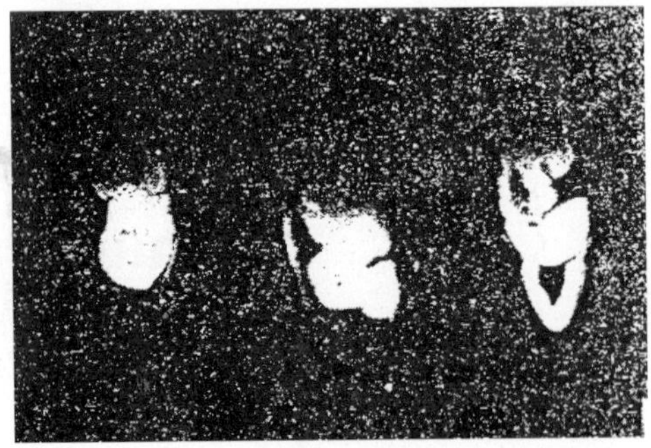

Figure 1.2

Invariably right from the cyanobacteria to aquatic mammals a number of creatures serve as substrate for settlement of microbes. The cyanobacteria colonized microbial cells where the rod shaped or coccal shaped bacterial colonies are arranged over the cell wall (Dhevendaran *et. al.* 2004). However, the vigorously growing algae eliminate the attachment of the microorganisms. It may be to some degree due to bacteriostatic or bactericidal substances exudated by the plants. Microbes are also found on the surface of the skin of the varieties of fish and shellfish. Dhevendaran and Maya (1998) observed mainly gram negative non-sporing rods and gram positive rods and coccals along the Arabian Sea (Figure 1.3). Obviously there were only minor taxonomic differences between bacterial population of fish and shellfish. *Vibrio* spp., *Pseudomonas* spp, *Micrococcus* spp., *Flavobacterium* spp., *Bacillus* spp., *Aeromonas* spp. and members of the family Enterobacteriaceae were on the surface of the fish and shellfish. The excreta of animals and the torn off tissues of animals could serve as nutrients for the microbial flora on the surface (Figure 1.4).

Commensals Inside Animals

The complex microbial diversity of the intestinal tract provides both nutritional benefit and protection against the pathogens. It is vital in modulating interactions with outside environment and inside development of beneficial immune responses. In the intestine of cultivable species of fish such as *Etroplus suratensis, E. maculatus, Oreochromis mossambicus, Labeo rohita, Carassius auratus* and *Cybrinus carbio* and in prawn, *Macrobrachium idella* the microbial diversity are almost similar as mentioned above (Dhevendaran and Manju 1997, 2002). These investigations have demonstrated that the intestinal micro biota has a strong effect on the nutritional status of animals which will be discussed in the later chapters. It became evident that the kind of food the fish or shellfish consumed has considerable influence on their indigenous intestinal micro flora. At a prolonged starvation bacteria could still be demonstrated in the empty intestine (Mercy Abraham 1989). Some authors (Dhevendaran *et al.*,

Figure 1.3

Red-Tail Butterfly Fish (*Chaetodon collare*)

Sponges (*Mycale mytilorum* (Annandale)

Figure 1.3–*Contd...*

Plants and Lakes

Seaweeds

Figure 1.4: Faecal Micro Flora of Fish

1978 and 2004; Maya *et al.*, 1990) believed that the gut micro flora have come from food of fish or shellfish. Whatever may be the source the end result may be the breakdown of the complex organic matter by the microbes to facilitate the utilization of the host organisms. Further more, Annie Mathew (1995) demonstrated the association of the antagonistic streptomycetes in the gut of the fish and shellfish and their involvement for the decomposition of the organic matter and the antagonistic property against the pathogens either in the gut or in the engulfed food materials (Figure 1.5).

This sort of probiotics is usually through the members of the healthy intestinal streptomycetes. Though certain strains of *Escherichia coli* and Coli forms are found in the intestine of the fish in the estuarine environment, it is (Figure 1.6) quite obvious that their occurrence indicates the sewage loaded water. (Dhevendaran and Sally, 2000).

Competition for Nutrients

Numerous creatures in water or in sediments feed on microorganisms. Some may live almost exclusively on bacteria and streptomycetes which are of high value protein feed (Manju and Dhevendaran, 1997; 2002). Certain filter feeders like *Mytilus californicus* feed on pure bacterial diet (ZoBell and Landon, 1937).

It is recognized that the indigenous micro biota has a pronounced effect on the structure, function and the metabolism of digestive tract of aquatic animals. Biochemical and physiological functions revealed that they serve as a source of nutrients, vitamins, enzymes and the microbial breakdown of complex organic esters.

Figure 1.5

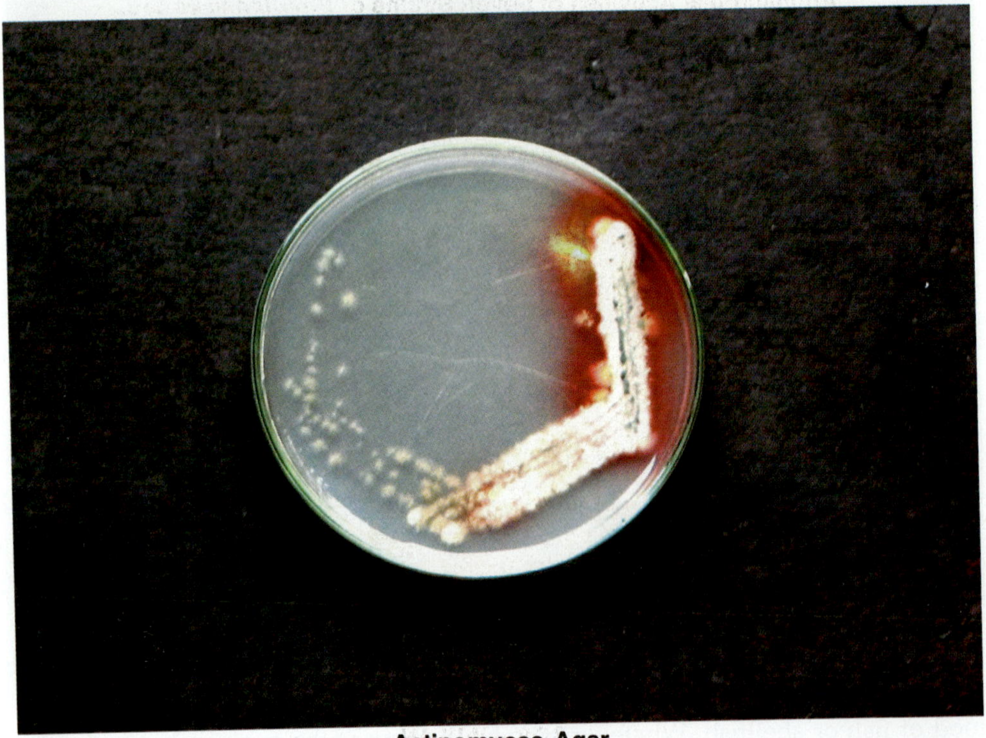

Actinomyces Agar

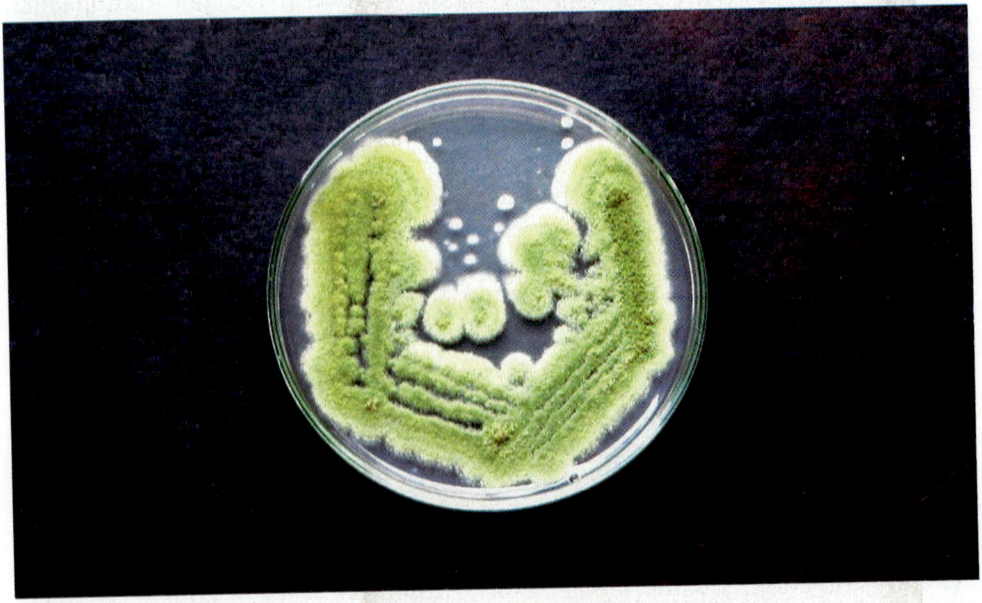

Potato Dextrose Agar

Figure 1.5–*Contd...*

Antimicrobial Activities of Seven Strains of *Streptomyces* spp., against *Aeromonus hydrophila*

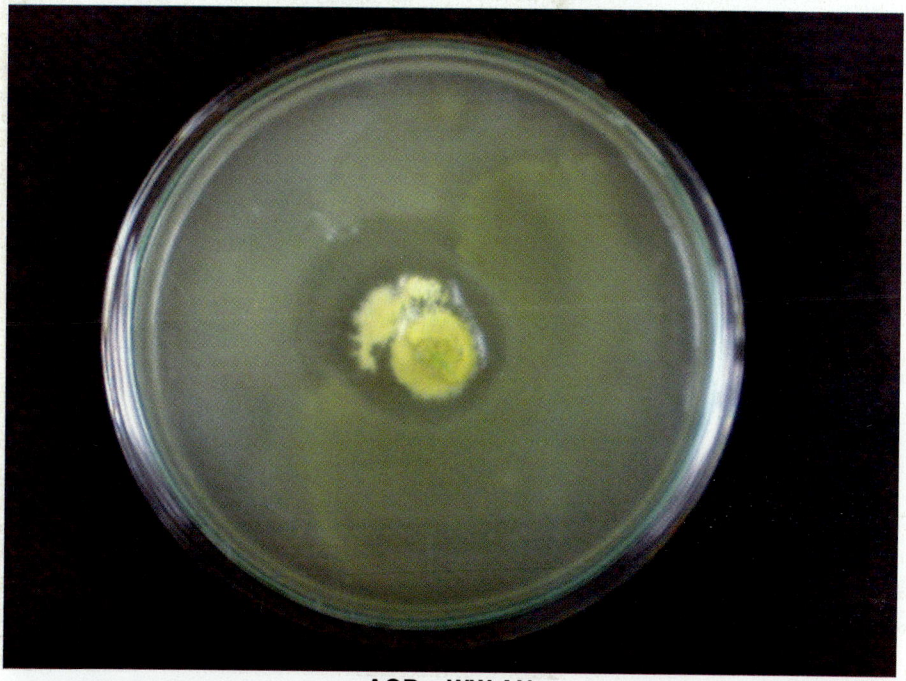

AQB – WW AN

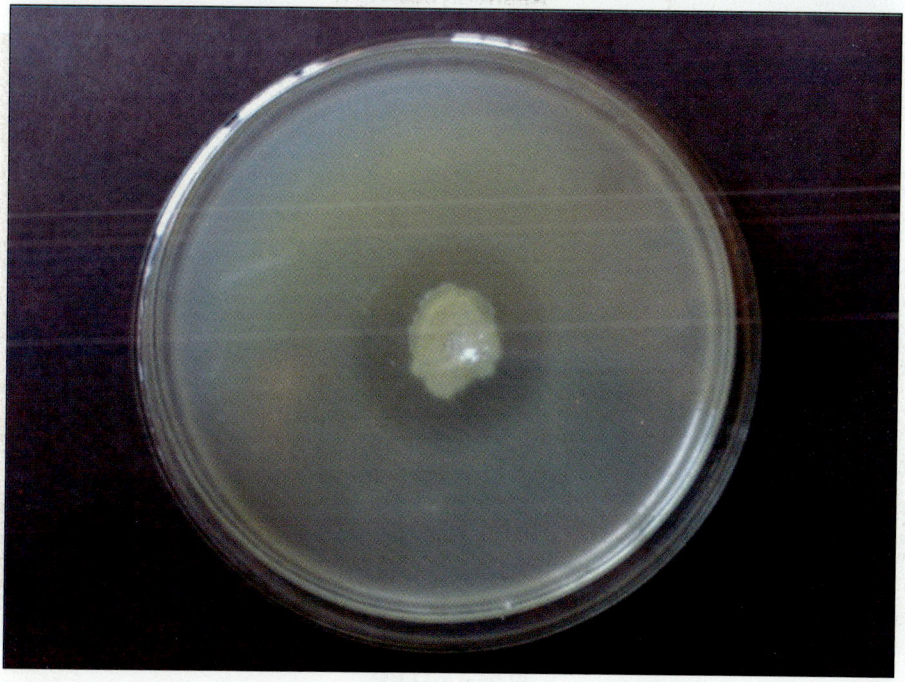

AQB – WY AN

Figure 1.5–*Contd...*

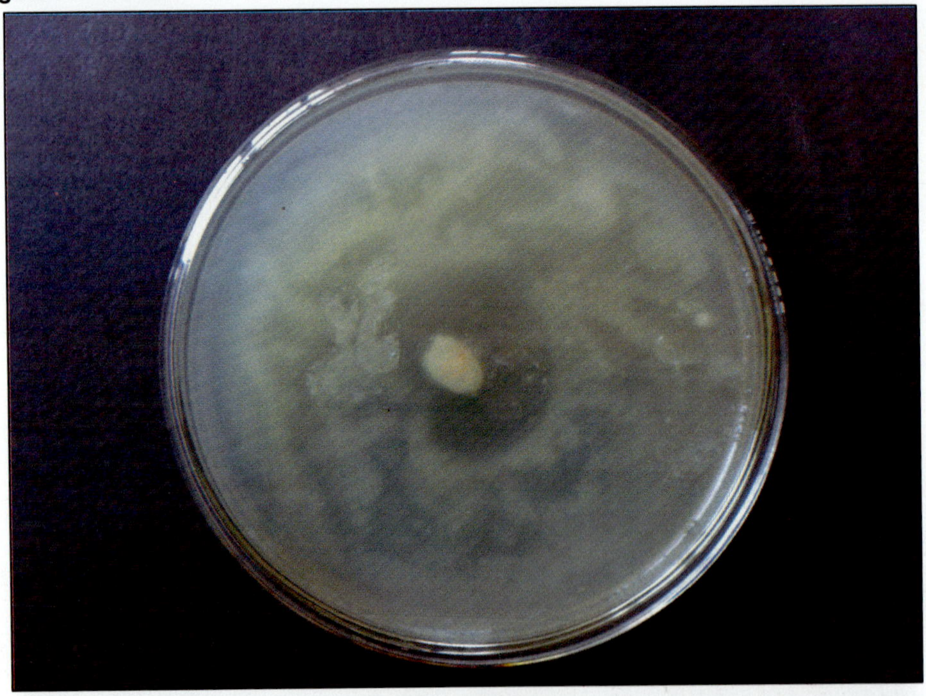

AQB – WB AN

AQB – WO EM

Figure 1.5–*Contd...*

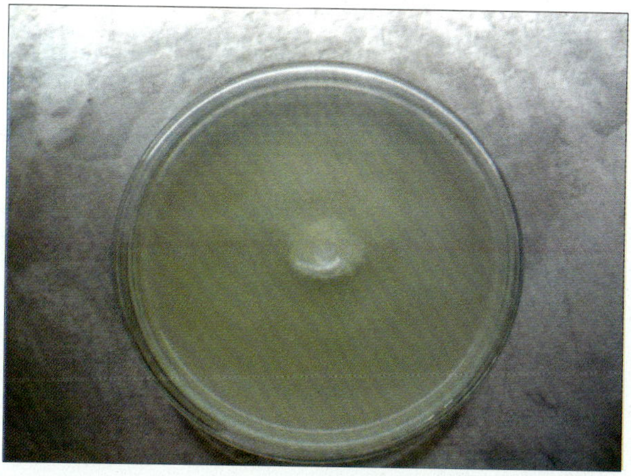

AQB – WP EM

Figure 1.6

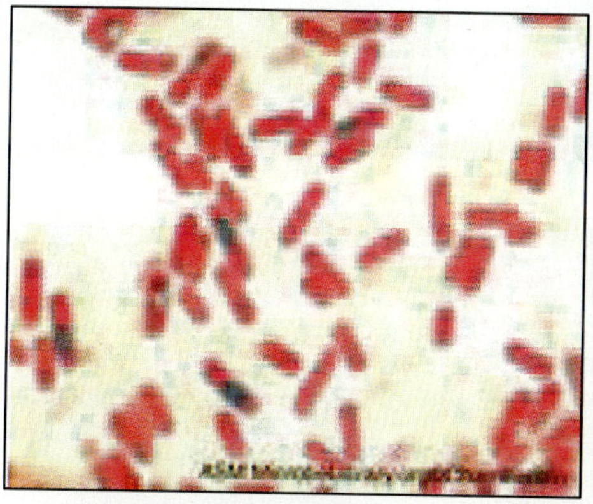

Bacillus

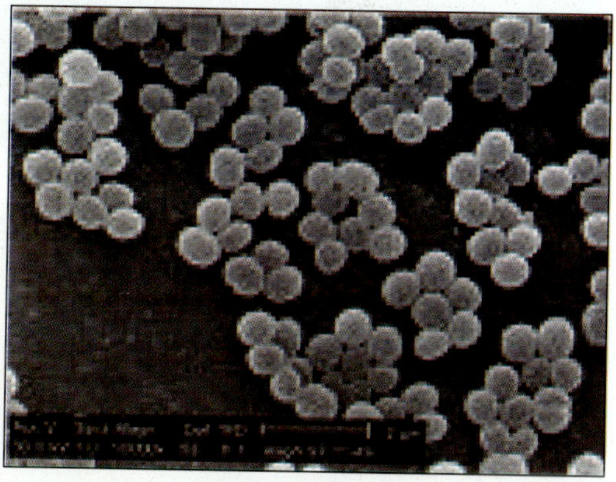

Staphylococcus

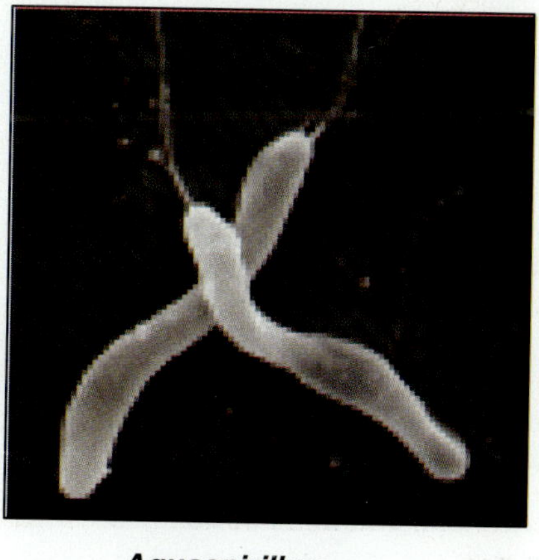

Aquaspirillum

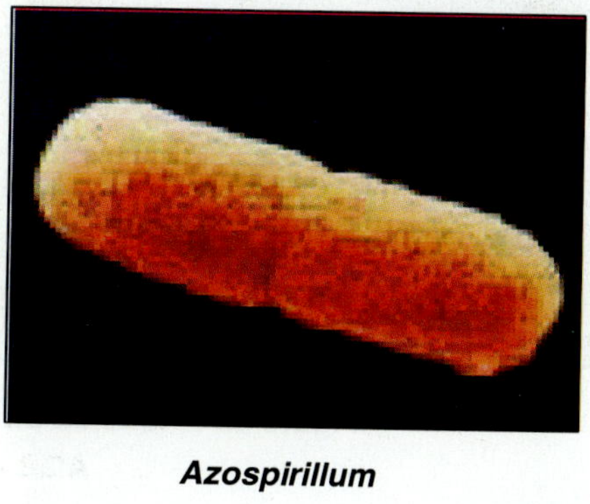

Azospirillum

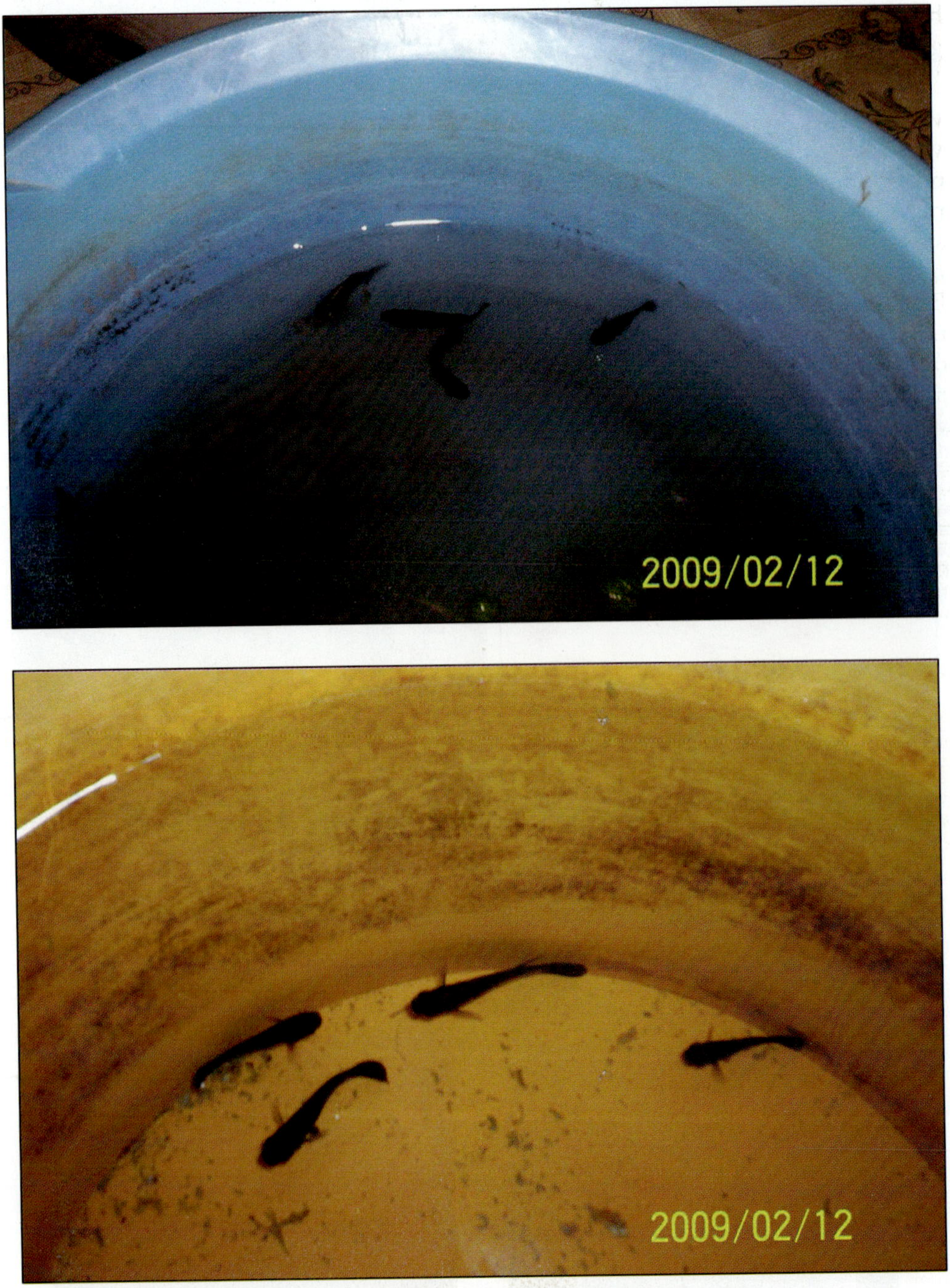

Figure 1.7: Fish Fed on the Artificial and Life Feeds

The nutritive and medicinal values of fish and shellfish have been recognized for a long time. Fish protein comprises all the amino acids in the balanced level for human consumption. In aquaculture operation, the production may be increased greatly by supplementing natural food with artificial ingredients (Figure 1.7).

Fish meal is being used as a major protein source in the pelleted feed all over the world. Since the fish meal is being costly the magnitude of fish meal replacement varied in different studies. The diversified microbial population preferably the nutritional grouping of microbial population plays an important role in the fish nutrition. Hence, the focused themes deserved to be studied on the following aspects such as Microbial biodiversity, Gastro-intestinal microflora, Nutritional grouping of microflora, Pigments of the microflora and photosynthetic bacteria, Microbial diseases, Bioactive compound and Bioremedial studies of microbes, Antibiogram studies of microbes against pathogens, Immunostimulant of microbes, Ornamental colouration of fish through microbial pigmentation, Growth hormones from microbial cells to juveniles, Microbial spoilage, Microbial probiotics, Microbial degradation of fishery wastes and Microbial biofertilizers in the fish and prawn culture ponds and Single Cell Proteins.

Chapter 2

Survey of Microorganism in Fish, Sponges and Photosynthetic Bacteria in Polluted Retting Zone

Introduction

The present investigation was attempted to understand, the selectivity and proliferation of various *Streptomyces* spp. in the microenvironments of gut regions of fishes and sponge tissue. The primary isolation of *Streptomyces* done under selective media like. Actinomyces agar, Glycerol Asparagine agar and Kuster's agar. This is the first time by choosing selective media instead of basal media and which were enriched with amino acid (Alanine), vitamin (Retinol), and yeast extract, sediment extract. Although the Nutritional classification of terrestrial bacteria has been established (Lochhead, 1952; Rouatt 1968) a systematic approach to understand, the original nutritional grouping of microorganisms in the normal gut of marine fish or sponges has not so far made. This proves that in some fishes or sponges enriched media is needed to get activate the inactive genes to get expressed to produce more potent strains showing high antibacterial activity, pigment synthesis and growth promoting substances. As far as the photosynthetic bacterial isolation from retting zones is concerned they create unique ecological environments that host diversified assemblages of fauna and flora. The retting husk itself is more important as a substrate for microbial biodiversity and nutrient regeneration in addition to the biodegradation of organic compounds like phenols. The benthic communities and microorganisms could survive despite strong osmotic potentials. The diversity of photosynthetic bacteria was vast in these areas. The microbial ecology and diversity of anoxygenic phototrophs have been investigated using a variety of techniques. While the traditional approach of liquid enrichment or establishing a most-probably number series leads to the isolation of a pure isolate, enrichment culture bias must be taken into consideration.

Despite the wide interest in bacteria for practical and scientific reasons, very little has been done on their organic nutritional requirements, although their inorganic requirements have received considerable attention (Robinson, 1950; Schoop, 1935). Various complex media utilizing fish infusions (Gibbons, 1936; Harrison and Kennedy, 1922; Lochhead, 1934), beef and yeast extracts or milk (Lochhead, 1934), and related substances (Robinson, 1950; Schoop, 1935) have been used for purposes of isolation and description, but few attempts have been made to culture these organisms on a chemically defined medium. Studies of the nutritional requirements of soil bacteria isolated on a nonselective basis (Lochhead and Chase, 1943) have shown the presence of a well-defined group of organisms dependent on soil extract for maximum growth. Some of these bacteria show no growth; others at best, submaximal growth in media containing sugars and inorganic salts, supplemented by yeast extract or combinations of amino acids and vitamins (not including B12).

Microbes are everywhere. They populate the air, the water and the soil, and have even evolved intimate relationships with plants and animals. Without microbes, life on earth would cease. This is due mainly to the essential roles microbes play in the systems that support life on earth, such as nutrient cycling and photosynthesis. Further, the physiology, nutrition and protection of plants and animals (including humans) are dependent on various relationships with microbes. Within the wide

Figure 2.1: Collection Sites

Vizhinjam

Figure 2.1–*Contd...*

Veli

Akkulam Lake

Figure 2.1–*Contd...*

Collection Site–Murukkumpuzha Retting Zone

continuum of aqueous environments ranging from fresh water to salt lakes, tremendous varieties of microbial species exist and interact. The interactions of microorganisms with aquatic biota are unique and diverse. The microbial population in our environment is both large and complex. An important aspect of microbiology in aquaculture operations is the phenomenon of gastro-intestinal microflora and their role in fish nutrition. The study of gastrointestinal (GI) microbial ecology involves investigation of the organisms present (abundance and diversity), their activity (usually determined *in vitro*, but ideally *in vivo* activity should be measured), and their relationship with each other and the host animal (synergistic and competitive interactions). This report will focus on the relationships between microbes and fish and also sponges. Retting ground in the backwater system covers the estuarine complex in the coastal zone. It is an interface between sea and land in tropical countries, having the conditions of changing salinity, tidal waves with muddy anaerobic sediments. It is the chesspool of stagnant water of foul smell with hydrogen sulfide. The organisms therein were well adapted to the changing environmental conditions due to progress of retting, inflex of freshwater and seawater by the tidal currents. The phototrophic purple and green bacteria are found in nearly all-aquatic environments particularly in retting zone. The photosynthetic or phototropic bacteria are a group of independent, self supporting microbes.

Fish are continually bathed in an aqueous suspension of microorganisms. Therefore, the external surface is in frequent contact with the microorganisms. The bacteria may enter the mouth either through water or through food uptake and pass down the alimentary canal. To isolate the nutritional grouping of bacteria, the isolates were classified into seven different groups based on their nutritional requirements (Lochhead, 1952). Bacteria may be enumerated from ingested food (Margolis, 1953; Srivasthava and Floodgate, 1966) or may be indigenous to the fish (Sera and Ishida, 1972; Maya *et al.*, 1990; Annie Mathew, 1995; Sarada,1996; Dhevendaran and Maya, 1998). Possibly some of them would be retained as part of resident microflora, whereas others may be destroyed by the digestive enzymatic action of fish or pass through the gut and be eliminated via the faeces (Mary *et al.*, 1975). In the intestinal tract, bacterial numbers appear to be much higher than the population in the surrounding water, indicating the presence of favourable ecological niche for microorganisms within the intestinal tract (Maya *et al.*, 1990; Dhevendaran and Maya, 1998).

The ornamental fishes were collected from Vizhinjam, Veli and Akkulam Lake. The marine ornamental fish: *Epinephelus diacanthus* (spotted grouper), *Tendania anhelans* (sponges) were collected from Vizhinjam. Estuarine fishes like *Oreochromis mossambicus, Etrophilus suratensis* were collected from Veli and Akkulam Lakes. The fresh water ornamental fishes like *Cyprinis carpio, Ospherenemous gourami, Carassius auratus* were collected from CARE (Centre for Aquaculture and Research Extension) belongs to Dept. of Aquatic Biology and Fisheries, University of Kerala. The selection of these three different places as well as fishes were done because to study the distribution of *Streptomyces* in the gut of fishes as well as strain potency. An attempt to understand the distribution pattern of Micro-organism in the micro-environment of gut regions of fishes was initiated.

Table 2.1: Microbial Population in Nutritional Grouped Fish Gut of *Epinephelus diacanthus* (Spotted Grouper)

Media Composition	*Number of Colonies CFU × 10^4g Dry Weight*		
	Bacteria	*Fungi*	*Streptomyces*
KA	5	–	2
KA+AA	1	–	3
KA+V	2	1	5
KA+AA+V	1	–	7
KA+Y	–	–	2
KA+Y+S	–	1	3
KA+S	1	–	3

KA: Kuster's agar; AA: Amino acid; V: Vitamin; Y: Yeast extract; S: Sediment extract.

Table 2.2: Microbial Population in Nutritional Grouped Fish Gut of *Tendania anhelans* (Sponges)

Media Composition	*Number of Colonies CFU × 10^4g Dry Weight*		
	Bacteria	*Fungi*	*Streptomyces*
KA	2	–	3
KA+AA	–	–	15
KA+V	–	–	5
KA+AA+V	1	–	31
KA+Y	3	1	2
KA+Y+S	–	1	5
KA+S	3	–	1

KA: Kuster's agar; AA: Amino acid; V: Vitamin; Y: Yeast extract; S: Sediment extract.

Table 2.3: Microbial Population in Nutritional Grouped Fish Gut of *Oreochromis mossambicus* (Tilapia)

Media Composition	*Number of Colonies CFU × 10^4g Dry Weight*		
	Bacteria	*Fungi*	*Streptomyces*
A	4	–	5
A + AA	1	–	4
A + S	3	–	7
A + AA + V	–	–	5
A + Y	1	–	4
A + Y + S	–	–	7
A + Y1	–	–	3

A: Actinomycetes agar; AA: Amino acid; V: Vitamin; Y: Yeast extract; S: Sediment extract.

Table 2.4: Microbial Population in Nutritional Grouped Fish Gut of *Etrophilus suratensis* (Pearl Spot)

Media Composition	*Number of Colonies CFU × 10^4g Dry Weight*		
	Bacteria	*Fungi*	*Streptomyces*
A	3	–	9
A + V	1	–	7
A + AA + V	5	1	6
A + Y	1	1	9
A + Y + S	–	–	7
A + S	–	1	6
A +AA	1	–	5

A: Actinomycetes agar; AA: Amino acid; V: Vitamin; Y: Yeast extract; S: Sediment extract.

Table 2.5: Microbial Population in Nutritional Grouped Fish Gut of *Cyprinis carpio* (Common Carp)

Media Composition	*Number of Colonies CFU × 10^4g Dry Weight*		
	Bacteria	*Fungi*	*Streptomyces*
GA	3	–	4
GA + V	1	–	7
GA + S	1	–	5
GA + Y	5	–	4
GA + Y + S	2	–	8
GA + AA + V	4	–	9
GA + G	1	–	4

GA: Glycerol asparagine agar; AA: Amino acid; V: Vitamin; Y: Yeast extract; S: sediment extract.

Table 2.6: Microbial Population in Nutritional Grouped Fish Gut of *Ospherenemous gourami*

Media Composition	*Number of Colonies CFU × 10^4g Dry Weight*		
	Bacteria	*Fungi*	*Streptomyces*
KA	1	1	4
KA+AA	3	–	5
KA+V	5	–	10
KA+AA+V	5	–	2
KA+Y	2	1	1
KA+Y+S	1	–	2
KA+S	2	1	4

KA: Kuster's agar; AA: Amino acid; V: Vitamin; Y: Yeast extract; S: Sediment extract.

Table 2.7: Microbial Population in Nutritional Grouped Fish Gut of *Carassius auratus* (Gold Fish)

Media Composition	*Number of Colonies CFU × 10⁴g Dry Weight*		
	Bacteria	*Fungi*	*Streptomyces*
KA	5	–	5
KA+AA	1	–	8
KA+V	3	–	3
KA+AA+V	4	–	2
KA+Y	1	1	4
KA+Y+S	5	1	1
KA+S	2	–	8

KA: Kuster's agar; AA: Amino acid; V: Vitamin; Y: Yeast extract; S: Sediment extract.

The sediment samples were colleted from retting areas of Murukkumpuzha along Kerala coast for isolation of photosynthetic bacteria. Retting zone in the backwater system covers estuarine complex in the coastal zone. The plants and animals therein are well adapted to the chess pool of stagnant water of foul smell with hydrogen sulfide and phenol. This is due to the progressive retting, influx of fresh water and seawater by the tidal currents. The retting husk itself is more important as a substrate for microbial biodiversity and nutrient regeneration in addition to the biodegradation of organic compounds like phenols.

In the present work, the microbial population (Bacterial, fungal and *Streptomyces* population) from gut of fishes (*Oreochromis mossambicus, Etrophilus suratensis, Cyprinis carpio, Ospherenemous gourami, Epinephelus diacanthus, Carassius auratus,* and sponge,*Tendania anhelans* collected from Vizhinjam, Veli and Akkulam Lakes were carried out. The present investigation was an attempt to understand the distribution pattern of Micro-organism in the micro-environment of gut regions of fishes. The microbial populations in nutritional grouped fish gut were assessed. The results indicated that maximal occurrence of *Streptomyces* spp and minimum occurrence of fungus were recorded (Tables 2.1–2.7). There are around 271 *Streptomyces* spp, 70 bacterial species and 12 fungal strains were isolated. The dominance of *Streptomyces* is due to selective media preferred instead of basal media and also these *Streptomyces* spp., produce antibiotics that prevent the occurrence of either bacteria or fungus. The reason may be that the selective media enriched with Vitamins, Amino acids, Yeast extract and sediment extract sources trigger or activate the particular genes for the expression of much more bioactivity and pigmentation than native one.

The photosynthetic organism was isolated from the samples collected from retting areas of Murukkumpuzha, Kerala coast by using the classical Winogradsky column (Figure 2.2). Appearance of green and purple colour colonies in Winogradsky column which were sub cultured using enrichment medium (Figure 2.3). They were enriched with addition of sulphide, Vitamin B_{12} and trace of biotin.

Figure 2.2: Winogradsky Column

Green sulfur bacteria

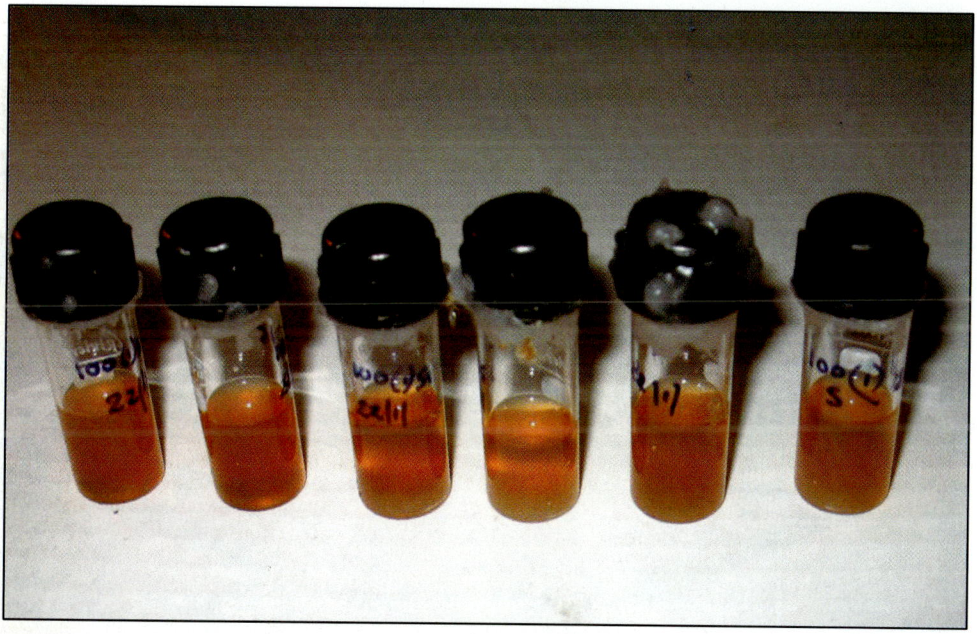

Purple sulfur bacteria

Figure 2.3: Photosynthetic Bacteria Isolated from Soil Samples of Retting Areas of Murukkumpuzha along Kerala Coast

The dominant occurrence of *Streptomyces* is mainly due to selective media as basal media and enriched with amino acid, vitamin, sediment and yeast extract. These *Streptomyces* synthesize secondary metabolites which prevents the occurrence of bacteria and fungus. The photosynthetic bacterial occurrence in retting sediments mainly due to presence of hydrogen sulphide as sulphur source and enriched media with little Vitamin B_{12} source need for culturing.

Chapter 3
Microbial Nutrition of Fishes

Introduction

The nutritive and medicinal values of fish have been recognized for a very long time. Fresh fish flesh provides an excellent source of protein for human diet. This protein has relatively of high digestibility, biological and growth promoting values. Fish proteins comprise all the essential amino acids in desirable strength for human consumption. Fish flesh therefore becomes a valuable supplement to human diet. Besides protein, it also contributes minerals, iodine, vitamins, fat, etc. It is easily digestible and provides a palatable taste. Further, fish oil has got high medicinal value. The rising protein demands of an expanding world population will not be met by merely traditional capture fisheries. At this juncture aquaculture has drawn the increasing attention from many quarters as a modern culture practice in both developed and less developed countries.

Aquaculture is the rational cultivation of aquatic plants and animals in confined water bodies, where rearing is not left to the mercy of nature. In fish culture operations the production may be increased greatly by merely supplementing the natural food along with some artificial ingredients. Fish meal is being used as the major source of protein in pelleted feeds the world over. However, fish meal is expensive besides being increasingly scarce. Therefore, there is an urgent need to develop low cost nutritionally balanced diets for aquaculture. Fish nutritionists have been trying to replace fishmeal partially or completely with plant protein sources to bring down the cost of feed. The magnitude of fishmeal replacement varied in different studies.

Figure 3.1

Alternative sources of protein for partial or complete replacement of fishmeal have been investigated by Abel *et. al*. (1984) and by Hossain and Jauncey (1989). Several unconventional sources of plant protein like grasses and weeds of terrestrial origin have been used for supplementation in fish feeds. Utility of aquatic weeds in fish feeds has also been developed. Besides leaf powder, the semi-digested form, commonly known as silage, can also be used to replace fishmeal in artificial diet (Figure 3.1).

In the context of increasing interest in ecologically sound integrated farming systems, multipurpose plants such as *Azolla* are becoming important. *Azolla* is a freshwater fern belonging to the family Azollaceae. It has a symbiotic relationship with the nitrogen fixing cyanobacterium, *Anabaena azollae*. From the aquaculture point of view, the herbivorous and omnivorous fish including carp, feed on *Azolla*. When both *Azolla* and formulated diets are given, they prefer *Azolla*. It was found that surface feeders (*Catla*), column feeders (rohu), bottom feeders (mrigal), herbivorous weed eaters (grass carp), omnivorous bottom feeders (common carp and tilapia) showed different rates of preference to *Azolla* (Daniel, 1994).Besides weeds (plants) and *Azolla*, the single-cell protein is also a valuable substance to replace fishmeal in artificial diets.

Single-cell protein is a broad term applied to a wide range of algae, fungi including yeasts, actinomycetes and bacteria, which are produced on large scale by fermentation processes for use as animal feed.

In view of the potential of SCP for protein production, it is perhaps not surprising that considerable attention has been focused on the use of this protein as a dietary replacement for fishmeal within compounded fish feeds. The nitrogen fixing bacteria play an important role in fish nutrition (Figure 3.2).

Figure 3.2

The inclusion of nitrogen fixing bacteria in the diet increased the overall nitrogen budget of the feed and thereby contributing a role in the growth of the fish (Saradha, 1996). The bacteria, especially the photosynthetic bacteria and actinomycetes can be used as substitutes for fishmeal in artificial ration. They tremendously increase the conversion efficiency and growth potentials of fish and prawns (Manju and Dhevendaran, 1997).

Azolla Feed

Azolla spp. are consumed by various fish and could improve the availability of nitrogen in semi-intensive agro-pisciculture systems (Cassani, 1981; Vincke and Micha, 1985; Barbier *et. al.*, 1985; Antoine *et. al.*, 1986). Jauncey (1982) dealt with the effects of varying dietary protein utilization and body composition of juvenile *Oreochromis mossambicus*. Saradha (1996) conducted experiments with *Azolla* as an ingredient in the formulated feed of *Tilapia* and observed the influence on gut flora and nutritional grouping of bacteria in relation to concentration of *Azolla* incorporated (Figure 3.3).

Figure 3.3: Azolla

Silage Based Diet

Considerable work has been devoted to the quantification of plant production in aquatic ecosystems utilizing total dry matter or ash-free dry weight as a measure of plant biomass. The role of microorganisms in the decomposition of the submerged plant *Potamogeton pectinatus* L. was studied by many scientists (Howard Williams *et. al.*, 1978; Howard Williams and Davies, 1979; Robb *et. al.*, 1979). Differentiation of plant material is the fragmentation of litter through animal feeding activities together with environmental and microbial influence, whereas decomposition is the degradation of plant litter into its constituent elements by animal digestion and microbial activities (Hanlon, 1982). The microbial decomposed plant materials could be used as fish feed, which are commonly known as silage. Attempts have been made on this line to develop a quality fish feed that could be produced from simple plant material by way of microbial processing. Incorporation of many types of aquatic weeds and algae in dry pelleted feeds for different species of fish has been tried by many researchers.

Diet with Single Cell Protein (SCP)

Many review articles on microorganisms as sources of food have appeared in the past two decades to give a fairly good idea of the current status of the protein problem in the world (Grey, 1970). In an experiment with *Mugil cephalus*, Odum (1970) observed that the fine particles eaten by the fish were covered with bacteria and protozoa. Kobayashi *et. al.* (1971) explained the cultivation of photosynthetic bacteria in municipal sewage sludges. Several workers have suggested that photosynthetic bacteria may be a potential feed source (Figure 3.4) for aquaculture systems or live stock (Wai, 1972; Thanii and Simard, 1973; Shipman, 1974; Shipman *et. al.*, 1975; Manju and Dhevendaran 1997).

Biomass harvested from photosynthetic cultivation in sewage and animal waste could be used as supplement for various livestock feeds (Coe and Turk, 1973). The importance of carotenoides in shrimp diets to the pigmentation of shrimp flesh has

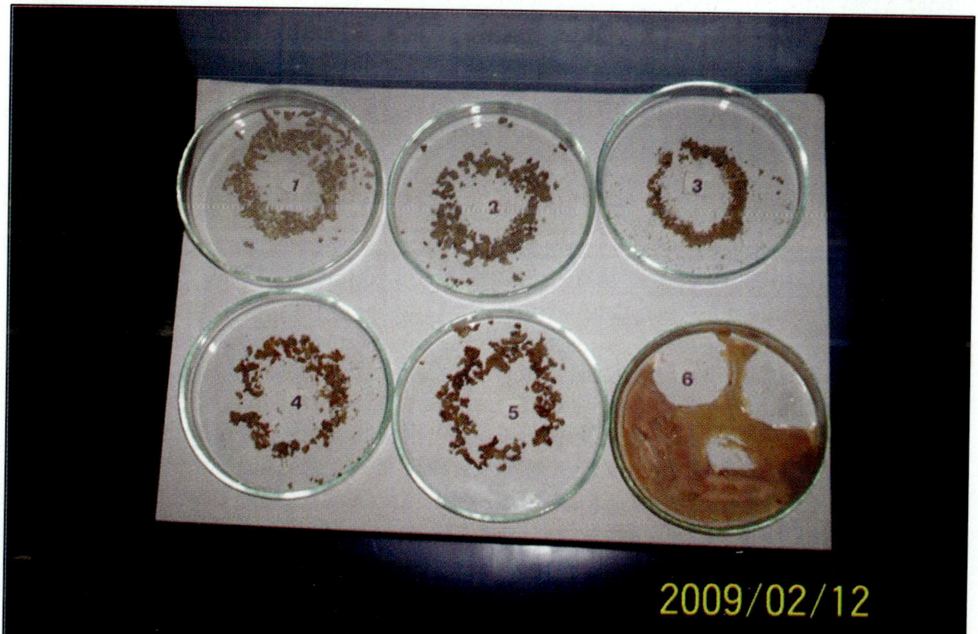

Figure 3.4: Artificial Feed with Microbes

been demonstrated by some earlier investigators (Joseph and Williams, 1975; Sandifer and Joseph, 1976).

Heterotrophic bacteria play a major role in culture systems. They are highly significant because of their role in the maintenance of water quality, nutrient recycling and the nutrition of fish. The methanol bacterial SCP can replace up to 75 per cent of the fishmeal in a salmonid production diet (Tacon and Jackson, 1985).

Gut Microflora

An important aspect of microbiology in aquaculture operations is the phenomenon of gastro-intestinal microflora and their role in fish nutrition. Bacteria may be enumerated from ingested food (Margolis, 1953; Srivastava and Floodgate, 1966) or may be indigenous to the fish (Sera *et. al.*, 1972; Maya *et. al.*, 1990; Annie Mathew *et. al.*, 1994; Maya *et. al.*, 1995; Saradha, 1996; Dhevendaran and Maya, 1998).Many researchers have observed that microorganisms are the main source of food for deposit-feeding animals (Newell, 1970; Fenchel, 1971). Dhevendaran and Maya(1998) reviewed the gut microflora of various fishes in different biotopes and stated that a relationship might exist between the commensal flora of fish and environment.

It is possible that the gastro-intestinal microflora may play a major role in the nutrition, growth and disease susceptibility of the fish (Figure 3.5). Quantitative studies on bacteria and algae in the food of the mullet, *Mugil cephalus* L. has been carried out by Moriarty (1976). In 1989, he also reported that some aquatic animals interacted with the bacteria in their environment to enhance their growth. Further, Mercy Abraham (1989) made a detailed study on the nutritional grouping of gut microflora in the starved and normal fed fish, *Labeo rohita*. Bacteria floccule was successfully used as diet for filter feeding fishes by Zhu Xuebao (1989). It is interesting to note that Maya *et. al.* (1990) observed the nutritional grouping of gut flora in *Therapon jarbua*. Subsequently Rekha *et al.* (1992) studied the influence of artificial feed on the gut microflora of *Mystus gulio*.

Cellulase

Cellulase was once expected to be an important source of energy for plant-eating fish. This opinion has received some reverses although it still deserves considerably more investigation. The cellulase activity that has been reported for certain species may have been contributed by the gut microflora acquired from the food. Cellulase activity in certain fish was related to the amount of dead plant material, which must have been intensively colonized by bacteria before ingestion (Prejs and Blaszczk, 1977). No microflora, however, has yet been demonstrated to be uniquely adapted to the fish gut (Sera *et al.*, 1974; Trust and Sparrow, 1974), and specialized to produce the enzyme required to degrade cellulase. Gerking (1984) studied the assimilation and maintenance ratio of a herbivorous fish, *Sarpa salpa*, feeding on a green alga, *Ulva lactuta* and described about the cellulase activity of aerobic gut microflora. Dhevendaran *et. al.* (1992) dealt with the microbial enzymes and cellulolytic activity in *Aeromonas* of retting ground. The bacterium, *Cellulococcus* sp. isolated from the gut of *Oreochromis mossambicus* showed cellulase activity during its growth phase

Figure 3.5

Figure 3.5–*Contd...*

(Saradha, 1996). She reported that cellulolytic activity of bacteria in the gut of fish is an important phenomenon for easy digestibility of cellulase molecules in the dietary *Azolla*.

Experiments in *Azolla* Feeds

Growth Studies

The results of the analysis relating to the growth of the fish *O. mossambicus* fed with different types of *Azolla* diets are shown in Tables 3.5 and 3.6. Table 3.5 depicts the mean percentage increase in length of the fish after four months. From the analysis it is clear that all *Azolla* diets did not show significant variation in growth in terms of length. Duncan's multiple range test showed that mean percentage increase in length of fish fed with diets D_2 D_5 and D_6 differed significantly from those fed with other three diets, namely D_1, D_3 and D_4 ($P < 0.01$). The diet D_5 fed fish showed maximum percentage increase in length (141.4cm) but did not differ significantly from the group fed with D_2 (140.8cm). The fish fed with D_6 and D_3 did not differ from each other but, lower than D_2, D_5 and D_6. The fish fed with D_1 showed least percentage increase in length (131cm). Table 3.7 summarizes the food conversion efficiency and conversion ratio.

Nutritional Grouping of Gut Microflora

In order to understand the indigenous gut microflora specific to the nutritional grouping of the cultured fish, *O. mossambicus*, we have used seven types of media. The results are shown in Table 3.8. The results show the distribution of bacteria, fungi and actinomycetes in the gut of the cultured fish. The appearance of the colony forming units (CFU) were noticed and counted. All seven media enhanced the proliferation of bacterial population, but it was not so in the case of fungi and actinomycetes. Maximum count of bacterial population ($78 \times 10^3/g$) was observed in fish fed with diet containing TNAU – 1 hybrid (D_6), where the plating was done in the basal medium supplemented with yeast and sediment extract. Minimum number of bacterial population was observed in fish fed with *A. filiculoides* ($18 \times 10^3/g$), where the medium used for plating was the basal medium. An interesting thing to be noted is that the bacterial colonies were maximum in basal medium supplemented with yeast extract and sediment extract and minimum in basal medium, irrespective of the diet provided. The maximum bacterial count in other treatments were $74 \times 10^3/g$ (D_5), $58 \times 10^3/g$ (D_4), $56 \times 10^3/g$ (D_3), $58 \times 10^3/g$ (D_2) and $64 \times 10^3/g$ (D_1).

Table 3.5: Means, Standard errors and F-ratio of Percentage Increase in Length of *O. mossambicus* Fed with Different *Azolla* Diets

Diet	*Mean±Standard Error*	*F-ratio*
D_1	$131.0 \pm 0.8028^{a,b}$	
D_2	140.8 ± 1.9877^{d}	
D_3	$134.9 \pm 2.3826^{b,c}$	9.5094**
D_4	127.7 ± 1.6469^{a}	
D_5	141.4 ± 1.8809^{d}	
D_6	$136.4 \pm 1.3350^{c,d}$	

** Significant P < 0.01 (Due to Duncan's test).

Means with the same subscripts do not differ significantly.

Table 3.6: Means, Standard Errors and F-ratio of Percentage Increase in Weight of *O. mossambicus* Fed with Different *Azolla* Diets

Diet	*Mean±Standard Error*	*F-ratio*
D_1	485.30 ± 8.3879^{a}	
D_2	$503.00 \pm 6.8832^{a,b}$	
D_3	507.30 ± 3.7120^{b}	11.4050**
D_4	$502.70 \pm 4.1608^{a,b}$	
D_5	520.40 ± 9.3538^{b}	
D_6	555.00 ± 7.7817^{c}	

** Significant P < 0.01 (Due to Duncan's test).

Means with the same subscripts do not differ significantly.

Cellulase Activity

Screening of bacteria for cellulase activity shows the cellulolytic activity at different incubation periods. The fish fed with diet containing *A. microphylla* was selected for the study. The potent culture *Cellulococcus* AQB – BM_1 was selected for further study. The cellulase activity of bacteria varied at different intervals during

Table 3.7: Weight Gain, Food Conversion Ratio and Conversion Efficiency of *O. mossambicus* Fed with Different *Azolla* Diets

Parameters		Diets						F-Ratio
		D_1	D_2	D_3	D_4	D_5	D_6	
Initial weight (g)	Mean	0.98	1.12	1.06	1.08	0.98	0.99	
	S.D.	0.0122	0.0837	0.0548	0.0447	0.0110	0.0110	
Final weight (g)	Mean	1. 79	2.12	2.16	1.88	1.81	2.04	
	S.D.	0.0045	0.0837	0.0548	0.0164	0.0110	0.2081	
Mean weight (g)	Mean	1.38	1.62	1.61	1.48	1.40	1.52	
	S.D.	0.0082	0.0671	0.0548	0.0204	0.0000	0.1095	
Production (g) (trough/30 days)	Mean	0.81	1.00	1.09	0.80	0.82	1. 056	
	S.D.	0.0084	0.1000	0.0179	0.0536	0.0219	0.1972	
Feed given (g)		0.25	0.25	0.25	0.25	0.25	0.25	
Unconsumed feed (g)		0.05	0.075	0.10	0.10	0.10	0.10	
Feed consumed (g)	Mean	0.198	0.150	0.190	0.150	0.136	0.146	
	S.D.	0.0045	0.0100	0.0137	0.0022	0.0055	0.0055	
Conversion efficiency	Mean	67.52[a]	95.00[b]	96.80[b]	90.00[b]	98.8[b]	119.80[c]	11.9004**
	S.D.	0.9391	17.4642	6.5727	5.7879	1.6432	18.0748	
Conversion ratio	Mean	1.48[c]	1.07[b]	1.03[b]	1.10[b]	1.01[b]	0.85[a]	18.5047**
	S.D.	0.0141	0.2051	0.0767	0.0654	0.0110	0.1369	

** P< 0.01

Means with same subscripts do not differ significantly.

Table 3.8: Nutritional Grouping of Total Heterotrophic Microorganisms in the Gut of *O. mossambicus* Fed with Different Types of *Azolla* Feed

Nutritional Grouping Media	*Microorganisms (x 10^3/g Dry Weight)*																	
	D_1			*D_2*			*D_3*			*D_4*			*D_5*			*D_6*		
	B	*F*	*A*	*B*	*F*	*A*	*B*	*F*	*A*	*B*	*F*	*A*	*B*	*F*	*A*	*B*	*F*	*A*
Basal medium	26	–	–	18	–	1	20	–	–	29	–	–	41	–	–	36	–	–
Basal medium + Vit. B_{12}	31	1	–	20	–	–	25	–	–	31	1	–	44	–	–	39	–	–
Basal medium + 2 Amino acids	34	1	–	28	–	1	29	3	2	40	–	–	38	–	1	48	1	–
Basal medium + Vit. B_{12} + 2 Amino acids	40	–	2	30	1	1	38	–	–	26	–	–	40	1	–	60	–	1
Basal medium + Yeast extract	52	–	–	43	–	–	44	–	–	44	–	–	51	1	1	71	–	1
Basal medium + Sediment extract	58	–	–	49	–	–	51	–	2	55	–	–	59	–	–	58	–	–
Basal medium + Yeast extract + Sediment extract	64	–	–	58	–	2	56	–	3	58	–	4	74	–	2	78	–	2

growth. The growth of bacteria rapidly increased from 2 hr (OD – 0.370) to 10 hr (OD – 790) intervals. *Cellulococcus* AQB – BM_1 showed elevated cellulase activity at all intervals.

In the context of increasing interest in ecologically important integrated farming systems, multipurpose plant such as *Azolla* is becoming very much important. It is a well- known fact that the herbivorous and omnivorous fish utilize aquatic weeds apart from their regular palletized formulated diets. In view of the potential use of *Azolla* as a cheaper and high quality protein source for fish in recent years attempts have been made to utilize *Azolla* as a dietary replacement for fish meal within compounded fish feeds.

Natural and artificial feeds exerted significant influence over the indigenous gut microflora of the fish. Knowledge of enteric bacterial flora of commercially important cultivable fish was thus found to be indispensable because of two reasons. They are:

1. The control of microbial fish diseases and
2. The preparation and handing of proper fish feeds.

Since the nutritional factor is of basic significance in determining the bacterial inter-relationships, study of the nutritional properties of the different bacterial population is considered essential. In the present study a systematic approach has been made to assess the nutritional requirements of bacteria in the gut of *O. mossambicus* fed on different types of *Azolla* diets.

It is speculated that soil extract of the environment from where the fish was collected, contained growth promoting factors and other essential nutritive compounds. This may be the reason for the appearance of higher bacterial populations in the basal medium supplemented with yeast and sediment extracts. Fathima (1973) noticed the same pattern of results with the microbes in the gut of the Indian Mackerel, *Rastrelliger kanagurta*. Similarly Mary and others (1975) observed that yeast extract and sediment extract were found to favour the growth of all the isolates tested. Further Mercy Abraham (1989) also confirmed with the similar results in the nutritional grouping of total heterotrophic microorganisms in the gut of *Labeo rohila* fed with artificial feeds. Saradha (1996) proved this fact that the bacterial isolates from the gut of *O. mossambicus* showed a greater preference towards basal medium enriched with yeast and sediment extracts.

Experiments with Aquatic Macrophyte Feed (Silage and Powder)

Growth Studies

The percentage increase in length exhibited in O. *mossambicus* fed with different types of aquatic macrophyte diets. The results revealed that the test diets show a significant difference in growth in terms of length over the control diet (D_7) ($P<0.01$). But, the difference in growth among the macrophyte diets themselves was not statistically significant. Maximum growth was observed in group fed with diet containing *Eichhornia* silage (D_8). In this treatment the percentage increase obtained

Chapter 5
Microbial Probiotics in Fishes

Introduction

Probiotics generally includes bacteria, cyanobacteria, microalgae, fungi, etc. Some Chinese researchers translate it into English as "normal microbiota" or "effective microbiota". It includes photosynthetic bacteria, *Lactobacillus*, Actinomycetes, Nitrobacteria, Denitrifying bacteria, *Bifido bacterium*, Yeast etc. (Hary *et al.*, 1997). The emphasis in disease management should be on prevention, which is likely to be more cost-effective than care. This may lead to less reliance on the use of chemicals, which largely treat the symptoms of the problem and not the cause (Planas and Cuhna, 1999). So far, the conventional approaches, such as the use of disinfectants and antimicrobial drugs have had limited success in the prevention or cure of aquatic diseases (Subasinghee, 1997). In agriculture the value of probiotics notably Gram positive bacteria such as *Lactobacillus*, has come to be appreciated as an alternative to be antibiotic in disease control strategies (Fuller and Turvey, 1971; Parker, 1974; Roach, and Tannock, 1980; Fuller, 1978; Smoragiewicz *et al.*, 1993). In English literature, probiotic bacteria are generally called the bacteria which can improve the water quality of aquaculture, and (or) inhibit the pathogens in water thereby increasing production. "Probiotics", "Probiont", "Probiotic bacteria" or " Beneficial bacteria" are the terms synomously used for probiotic bacteria. Parker (1974) introduced the modern concept of probiotics more than 30 years ago.

Antogonism against Pathogens

Antagonism to pathogens is a characteristic of a good aquatic biocontrol agent. Antogonism may be mediated not only by antibiotics, but also by many other inhibiting substances like organic acids, hydrogen peroxide and siderophores (Gatesoupe 1999). These compounds provided by the biocontrol agents highly dependent on experimental conditions that are different *in vitro* and *in vivo* conditions.

The importance of microbial communities in aquaculture systems and pond productivity cannot be over stressed Bioremediation agents serve to modify or manipulate the communities in water and sediment such that they reduce or eliminate selected pathogenic microbes and generally improve growth and survival of targeted species. These are various ways through which bioremediation agents could be used in aquaculture systems. Austin *et al.* (1995) reported a probiotic strain of *Vibrio alginolyticus*, which did not cause any harmful effects in salmonids. By using the cross-streaking method, the probiotic was observed to inhibit the fish pathogens. Maeda and Nagami (1989) reported some aspects of the biocontrolling method in aquaculture. In their study bacterial strains possessing *Vibrio* static activity that improved the growth of prawn and crab larvae were observed.

Chapter 6
Probiotics and Efficacy in Aquaculture

Introduction

The word *"Probiotic"* is derived from the Greek meaning *"For Life"* and had several meanings over the years. It was believed that beneficial organisms could balance the intestinal environment, by preventing the growth of pathogenic bacteria and as a consequence, improve health and prolong life. The original work of Metchnikoff (1908) and his associates was using the "Bulgarium *Bacillus*". It is closely related to *"Lactobacillus"* starter (bacterium used for fermentation of milk) of yoghurt and to this date, *Lactobacillus* has remained the most commonly used probiotic organisms. Currently available probiotic microbes are *L. delbrueckii* sub. sp. *bulgaricus, L. acidophilus, L. casei, L. fermentum, L. plantanum, L. brevis, L. cellobiosus, L. lactis* and *L. reuteri.*

The first use of *Streptococcus* was in the form of soured milk and yoghurt. There are three species of this genus employed for probiotic and are *S. lactis, S. cremoris* and *S. intermedius.* Besides this there are other bacteria, fungi and yeast. Some of them are *Leuconostoc, Pediococcus, Propionibacterium* and *Bacillus.* They may be administered by direct insertion into the mouth or by inclusion in the food/feed or water. Experiments have also been done by administration into newly hatched chicks by spraying into the surrounding atmosphere. An effective probiotic is required for favourable environmental conditions.

Attributes of Probiotics

Viability

In order to impart the beneficial effects, the organisms must remain alive in the host intestine.

Colonization

Barriers in colonization: Due to physiological orientation of the digestive system, the various barriers that probiotics are challenged with are:

1. Saliva in mouth (enzymes such as lactoperoxidase) and
2. Other digestive enzymes
3. Gastric juices in stomach
4. Bile salts
5. Peristalsis
6. Phagocytosis and immune response and
7. Competition with native microbial community

Stabilization

Carbohydrates which are not digested by human beings can promote the growth of selected indigenous probiotic bacteria, especially Bifidobacteria in the intestinal tract.

Functionality of Probiotics

This is the ability of probiotic to survive in Gastrointestinal (GI) tract, adherence to the intestinal epithelial line, production of antimicrobial compounds towards pathogen, ability to stabilize the intestinal microflora, anti – genotoxic property, a short generation time, a good shelf – life in a food or powdered preparation and non – pathogenic properties. Yoghurt protects against diarrhoea, constipation, heart disease, colon cancer and lactase deficiency as well as providing effective therapy for sunburn and vagina thrush. All these are attributable to Lactic Acid Bacteria (LAB), which are the common probiotic organisms in use (Table 6.1).

Table 6.1: Desirable Properties of Probiotic Lactic Acid Bacteria (LAB)

Probiotic Strain Characteristics	*Functional and Technological Property*
1. Human origin	Species – dependent health effect and maintained viability applicability to fermented food.
2. Acid and bile stability	Survival in intestine.
3. Adherence to human intestinal cells	Exclusion of pathogens immune modulation.
4. Colonization of human intestinal tract	Multiplication in the intestinal tract temporarily.
5. Production of antimicrobial substance	Pathogens inactivation and normalization of gutflora.
6. Antagonism against carcinogenic and pathogenic bacteria	Prevention of dental decay and prevention of pathogen.
7. Safety in food and clinical use	Accurate genus, species and strain identification documented safely.
8. Shelf-life and stability during processing and storage	All the above should be maintained during processing and storage.

Growth Promotion

LAB is responsible for fermentation in GI tract. They act on carbohydrate complex like starch and cellulase and split them into simpler compounds for absorption besides synthesizing Vitamin K and B complex. The presence of Bifidobacteria in breastfed milk and they serve as probiotics for better growth and development of babies and is also attributed to the presence of N – acetylglucosamine. An overall increase in weight due to infection control and increased digestibility of nutrients. *Bifidobacterium globosum* increases the growth rate and feed intake in weaning pig.

Immunostimulation/Immunomodulation

Probiotic bacteria offer new dietary alternatives of immunotherapy. LAB administered orally stimulated the immune system in non – specific way. *Lactobacillus* spp. serve as adjuvant to an oral vaccine to rotavirus in children. *Lactobacillus casei* acts as immunomodulator altering anti – tumor response.

Antagonism

Some probiotics have been seen to produce certain antibiotics like acidolin, acidophylin, lactobacilin and lactocidin, effective against many pathogenic bacteria like *Estherichia, Salmonella, Streptococcus, Shigella, Proteus, Klebsiella, Pseudomonas, Bacillus* and *Vibrio*.

Antibioses for Host Benefits

Intestine microflora of higher animals contributes to resistance against invading pathogens. LAB bacteriocins have been reported to inhibit *Clostridium botulinum* spores. *P. pentosaceous, Lactobacillus* and *Pediococcus* are probiotic against entropathogens in poultry.

Why Probiotic – Feeding?

Beneficial role of probiotics is for treatment of diarrhoea, the alleviation of GI side-effect of antibiotic treatment, the relief of constipation and balancing and stabilization of intestinal integrity. Probiotics once colorized in the GI tract, can affect the flora in three ways:

1. By providing the preformed metabolites.
2. By continuous provision of viable cells which do not colonize but survive and metabolize *in-situ*, and
3. By providing viable cells which colonize the gut and produce useful metabolites, *in-situ*.

Probiotics in Aquaculture

In aquaculture, probiotics have been tried in cultivation of shrimp larvae. *Vibrio alginolyticus* inoculated in shrimp culture with an aim to suppress pathogenic vibrios like *V. parahaemolyticus, Vibrio harveyi* and *V. splendidus*.

Reports on the presence of LAB in the intestinal micro biota of fish, suggest that there exist LAB as indigenous gut microflora. *Carbonbacterium piscicola* and *C. divergens* which produce bactericin active against *Listeria monocytogenes.* Probiotics have been successfully used in aquaculture to enhance both internal and external microbial environments. In aquaculture, the current trend is to replace antibiotics by probiotics for ecological considerations. Intensive shrimp larval rearing technology promote better bacterial equilibrium by using probiotics and biological filters. Generally, the best approach is to isolate the probiotic bacteria from the culture and enhance their natural effects, favouring their growth by adding some nutrients or distributing large quantities of them in the culture. Besides enzymes, bacteria can bring some nutrients to the host, *e.g*: Vitamin B_{12}. The probiotics are likely to release antibiotics but their work is at a very low concentration.

The concept of 'prebiotics'is defined as a non-digestible food ingredient that beneficially affects the host by selectively stimulating the growth and/or the activity of one or a limited number of bacteria in the colon and thus improves the host health. Cow's milk is the natural prebiotic. It contains lactose, lipids, proteins, salts and traces of vitamins, enzymes, etc. It has also been proposed to call it "Symbiotic", the combination, of a pro- and pre-biotic in the same food products.

Commercialization

A bacterial strain for use as a probiotic should preferably be derived from the gastrointestinal tract of the same animal species, to which it is planned to be fed. The efficiency of the chosen probiotic strains can be recognized by testing strains in field trials, assessing *in vitro* for ability of the bacteria to adhere and grow in the gut and for anti-bacterial activity and by taking good colonizers and by genetic manipulation to induce it to over express the antibacterial substance.

Chapter 7
Microbial Association with Fishery Products

Introduction

Fish and Fishery constitute an important food component for a large section of world population, more so in developing countries, where fish forms a cheap source of protein. In the last two decades there has been an increase in awareness about the nutritional and health benefits of fish consumption. The low fat content of some fish and the presence of polyunsaturated fatty acids in red meal fishes, which are known to reduce the risks of cornory heart disease, have increased the dietary and health significance of seafood consumption. In U.S.A. seafood consumption has increased from an average of 4.5kg per person in 1960 to about 7.0kg per person in 2002. Total fishery production in 2001 was reported to be 130.2 million tones, of which 37.9 (30 per cent) million tones was from aquaculture, 90 per cent of which come from Asian countries. On the other hand, seafood is also known to have been responsible for a significant percentage of food borne diseases. With increased fish consumption there is also an increase in the number of food-borne illness associated with Fishery products. In the United States, 10-19 per cent of food borne illness involved. Seafood as a vehicle and between 1993 and 1997, 6-8 per cent of the food borne illness involved consumption of fish and shellfish.

Microorganisms Associated with Fish

In the case of live fin fish, microorganisms may be associated with gill, gut and skin. Filter feeding shellfish like mollusks concentrate different types of microorganisms present in the environment around them. Some of the microorganisms of human health concern may be native to the aquatic environment. Such as the members of the genus *Vibrio*, but others like *Salmonella, E.coli* and many viruses are introduced into the aquatic environment through human activities such as contamination by domestic sewage.

Microorganism Associated with Shellfish

Faecal contamination of shellfish growing areas in the major route of entry for human pathogenic viruses into shellfish. Shellfish harvested from areas with higher fecal coli forms or *E.coli* counts are to be used only alter depuration relaying or heat processing. However, human volunteers studies in Australia and natural outbreaks studies in UK and U.S.A. suggest that depuration may fail to eliminate enteric viruses from contaminated shellfish and that compliance with *E.coli* or fecal coli forms never standard nor does guarantee absence of viruses.

Marine Biodiversity

Marine environment is more productive and vast than terrestrial or freshwater environment. Varieties of algae, sponges, fishes, microorganisms, etc. are present in marine ecosystem. The animal diversity in tropical sea is always enhancing. This ranges from the primitive invertebrates to highly evolved mammals. However, diversity in its true sense is observed with the fishes. Apart from the very diverse groups of food fishes, the coral reefs of the tropical waters provide a very good abode for a variety of ornamental fishes. Marine organisms provide many of the pharmaceutically active compounds.

The antibiotic and similar compounds are from perhaps the most interesting, most dynamically increasing and one of the practically most important groups of the natural products. Natural products are the carbon compounds isolated from diverse living things. These compounds may derive from primary or rather secondary metabolism of living organisms. The primary metabolites are common in all biological systems. The secondary metabolites are, however, low molecular, chemically and taxonomically extremely diverse compounds with obscure function, characteristic mainly to some specific, distinct types of organisms.

Microbial Association with Fish and Shellfish

Microorganisms can be defined as life forms that can not be seen with the unaided eye. The broad definition encompasses an extensive and diverse assemblage of organisms which exhibit widely different morphological, ecological, physiological characteristics. They are represented by groups like viruses, bacteria, fungi, diatoms, algae, protozoan, etc. The microbial diversity encompasses vast number of species and the number of species of these groups known to science is only a tip of the iceberg.

Marine Microbial Diversity

Nearly 75 per cent of the earth is covered with seawater. Microbial world in the marine ecosystem is complex and diverse in nature. The study of Marine microbiology includes the forms that live from the coastal to offshore regions, specialized ecosystems like black smokers of hot thermal vents at depths, tropical coral reef ecosystems and specialized ecosystems like estuaries, lagoons, saltpans, backwaters and mangroves. They are nutritionally versatile, as that of terrestrial counterparts ranging from phototrophy to chemolithotrophy and chemoheterotrophy.

Prawn and Shrimps

Mangrove habitats are rich in shrimp and prawn resources. Commercial prawn fishery yields are greater in the coasts with luxuriant mangrove important nutrient for food webs. During the decomposition of mangrove liter, a large amount of nutrients are released, and detritus food is formed. This detritus food contributes to the prawn and shrimp fishery production. The one welled detritus transported to offshore can also provide food and habitat for juvenile shrimps outside mangrove waterways. Mangroves are of primary importance to fisheries in many ways, plants of species make use of mangrove ecosystems in various manner, some being obligate *i.e.* spending their entire life in this ecosystem or a crucial part of it (*e.g.* penaed prawn) using it as shelter or as a source of food; others acultive *i.e.* being able to survive and reproduce even in absence of mangroves but yet showing preference for the habitats and nutrition provided therein.

Fin-Fish

Fish are cold-blooded aquatic vertebrates possessing gills and fins. There are 24,600 species of known fishes which comprise almost half the number of total vertebrates. About 95 per cent of fishes are teleosts. Mangrove habitats usually contain a rich ichthyofauna. The mangrove water provides ideal niche for the juvenile fish owing to less water motion, soft substratum.

Ornamental Fishes

Ornamental fishes have always fascinated the mankind for their brilliant and attractive colouration, tiny size and peculiar movements. The "marine living jewels" is no way exceptional to these remarks. Ornamental fishes are receiving increased attention in view of their demand in export trade. The annual world trade in ornamental fishes is of the order of U.S. $4.5billion of which nearly half is constituted by marine ornamental. On a global basis, although coral reefs cover less than one quarter of 1 per cent of the marine environment, they are considered to be most biologically rich and productive ecosystem on the entire earth. Therefore coral reef is often known as the rain forest of the sea. Coral reefs support over 4000 species of ornamental fishes.

Chapter 8

Complement System of Fish

Introduction

The two parts of the vertebrate immune system are the acquired and innate components. Acquired immunity first appeared about 450million years ago in bony and cartilaginous fishes. It plays an important role in protection of host against recurrent infections by generating memory cells and specific soluble and membrane-bound receptors, such as immunoglobulins (Ig) and T cell receptors (TCR), to remove specific pathogens. On the other hand, innate immunity act as first line of defence and ward off many microbial attacks or keep them in check until acquired immunity develops.

Complement is one of the important components of innate immune system, consists of about 35 soluble and membrane-bound proteins and cell receptors that plays a critical role in host defence by interacting with components of both the innate and acquired immunity. The functions of the complement are numerous but it's most well known for its capacity to kill pathogens by creating pores in their surface membranes.

The study of fish complement, which began in the 1960's in cartilaginous fish, and in 1970's in teleosts, fellowing the methodology developed for studying human or marine complement. All the literatures deal with comparative biology of complement examining phylogenetic evolution in fish with relation to mammals, biological activities and antimicrobial activities. The characterization of fish complement is still incomplete and is not nearly as well described as the mammalian complement system.

Basic Functions of Complement

The biological activities of this system affect both innate and acquired immunity.

After initial activation, the various complement components interact in a regulatory cascade to carry out following major functions:

1. Immune clearance; removing immune complexes from the circulation
2. Lysis of bacteria, viruses and cells.
3. Opsonization; Promoting phagocytosis of particulate antigens.
4. Binding to specific complement receptors on cells of the immune system, triggering specific cell functions, inflammation and secretion of immunoregulatory molecules.

Complement Components of the Fish

The components of the complement system, made up of proteins and glycoproteins, are synthesized mainly by hepatic cells, monocytes, macrophages, and epithelial cells of gastrointestinal and genitourinary tracts in mammals. These components constitute 5 per cent (by weight) of the serum globulin fraction. Most of those circulate in the serum as proenzymes being functionally inactive or zymogens, which are inactive until proteolytic cleavage exposing the active site. The complement reaction starts with an enzyme cascade.The complement system of cartilaginous fishes consists of classical and an alternative pathway.

Opsonization

Complement mediated opsonization has been studied in many fish species *viz.*, carp, salmonids, channel cat fish, Japanese eel, tilapia and yellow tail. The opsonization of pathogens either by complement or by antibody results in an enhanced phagocytosis and is mediated through ligand- receptor interactions between the surfaces of the two cells. The role of antibody or complement in opsonization process can be studied by comparing the effects of heat-inactivated or normal serum on the phagocytic activity of leucocytes. C3b is the major opsonin of the mammalian complement system.

The mechanism of opsonization in fish is still poorly understood and this activity in fish is primarily efficient against non-pathogenic bacteria. The nature of pathogen and host decides whether the pathogen will be opsonized by complement or antibody. C3 receptors are present on the macrophages of salmonids, tilapia and yellow tail and on the neutrophils of carp and channel catfish. C5a-like chemotactic factors have been noticed in trout and eel. Rainbow trout complement inactivates the toxicity of extracellular products of *A.salmonicida*.

In fish, the antigen is trapped in maleano macrophage centers present in spleen and kidney for immune clearance. The involvement of complement and it's receptors in the translocation of these complexes to the spleen of fish is yet to be known.

Immunocompetent Organs in Teleosts

The teleosts are masters of the aquatic environment and are diverse as well as highly specialized in the water as the higher tetrapods on land. They include the majority of food fishes. Recent works on the teleost immune system have used similarities with the defence system of higher vertebrates to identify various genes or their products. However, there exists a difference in structure and form of the immune system between fish and mammals as well as with in the teleost species. Fish lack bone marrow and lymph nodes. Instead, kidney is the major lymphoid organ in teleosts in addition to the thymus, spleen and mucosa-associated lymphoid tissues(MALT). The chapter discussed here and presents the occurrence, structure and functions of different immunocompetent organs in fish.

Lymphoid Organs

There is a great variation in the age of development in lymphoid organs. However, in freshwater fish the thymus is the first lymphoid organ to develop and the first to become lymphoid. Subsequently, kidney, spleen and gut-associated lymphoid tissue (GALT) develop. Very little is known on the aging of the immune system of fish. It is well known that the involution of thymus occurs with relation to age, season and hormonal status of fish. In some fish such as carp and shark, which are long-lived, the thymus does not appear to involute at all.

There are two types of lymphoid organs. They are primary lymphoid organs and secondary lymphoid organs.

Primary Lymphoid Organs

Primary lymphoid organs include thymus, head kidney that produce mature stem cells.

Thymus

It is a paired organ situated in the dorsolateral region of the gill chamber close to the opercular cavity. The histology of the thymus varies within the species. The thymus may starts developing as early as 24 hours after fertilization in tilapia species. Gradually the cells differentiate into thymocytes and epithelial cells. The thymus is the first organ to be lymphoid during histogenesis and acquires mature lymphocytes. Thymocytes appear in carp on day 4 post-fertilization. In Channel catfish, monocytes are first leucocytic cells to be appeared. In marine species, the lymphoid organs develop in the following order: head kidney, spleen and thymus. In general lymphocytes appear later in the development of marine fishes than in freshwater fishes.

Anterior Kidney

Like other vertebrates, the kidney is situated retroperitoneally, exterior to the dorsal wall or body cavity. The anatomy of the kidney varies from species to species. The anterior kidney is the lymphocyte precursor tissue and mostly the generation and maturation of 13 lymphocytes occur in this tissue.

Secondary Lymphoid Organs

The secondary lymphoid organs include kidney, spleen, epigonal organs and mucosa associated lymphoid tissue.

Kidney

The kidney of fish is provided with a low-pressure renal portal system. The sluggish blood supply ramifies into sinusoids with a haematopoietic stroma whose framework of reticular cells and phagocytes resembles that of the bone marrow of higher vertebrates. The head kidney also serves as a secondary lymphoid organ- a lymph node analogue, important in the induction and elaboration of immune responses.

Spleen

The spleen is the last organ to develop during histogenesis of the lymphoid organs of teleosts. It plays roles both in haematopoiesis and immune reactivity. The spleen in Indian major carps is multifocal and distributed in the mesenteric tissue, while it is spherical or elliptical and single in salmonids. The presence of exocrine pancreas inside spleenic tissue of Indian major carps and few other species is a common feature.

Mucosa-Associated Lymphoid Tissues (MALT)

Mucosal surfaces of gut, gills and skin are protected by both humoral and cellular defence mechanisms. The mucus secreted by goblet cells, forms a physical barrier by constant secretion and shedding. The mucus also contains number of defence molecules *viz.*, lysozyme, antibody and specific immunoglobulins, complements, etc.

The Immunological Defence Mechanisms of Fish Against Bacteria and Viruses

Fish are in intimate contact with their environment, which contain very high concentrations of bacteria and viruses. Many bacteria are saprophytic, some are pathogenic and both are very capable of digesting and degrading the fish's tissues. However, under normal conditions the fish maintains a healthy state by defending itself potential invaders by a complex system of innate and acquired defence mechanisms.

Chapter 9
Innate Defence Mechanisms Against Bacteria

Introduction

The innate defence mechanisms of fish against bacteria include production of broad-spectrum anti-microbial substances and acute-phage proteins, non-classical complement activation, release of cytokines, inflammation and phagocytosis. While these mechanisms provide potent defences against invasion of pathogenic bacteria which have evolved means of avoiding many of them.

Integumental Innate Defences

Many observations suggest the importance of innate defences in the integument of fish. For instance it is well known that rainbow trout is virtually resistant to immersion challenge by *Aeromonas salmonicida* while Atlantic salmon is very susceptible. However, there is little difference in the LD_{50} value when these fish species are injected with the bacteria.

Mucus

The protective role of skin mucus has been demonstrated by carefully removing mucus with a swab and then challenging turbot or ayu with *Vibrio anguillarum* and inducing increased mortality. Mucus also contains many substances with antibacterial activities and these may play important roles.

Antibacterial Peptides

These substances are low molecular weight peptides with the ability to disrupt bacterial membranes. They have been identified from mucus secretions of a number of fish species but little is yet known about their ability to kill fish pathogenic bacteria. Pathogenic strains of *Aeromonas salmonicida* are less susceptible to cecropin than non-pathogenic strains lacking the A-layer, but they are nevertheless, killed by higher concentrations.

Proteases

Trypsin-like proteases and cathepsin L and B proteases have been found in skin mucus of a number of fish species. The ability of these enzymes to lyse formalin-killed *V.anguillarum* has led to the suggestion that they may play a role in defence against bacteria but their action on live bacteria does not appear to have been studied yet.

Lectins

Once bacteria have made contact with their fish host, many pathogenic species can adhere to the mucus and the epithelial cells by surface molecules known as adhesins. Once the bacterium has attached to the host cell, the latter is induced to endocytose the bacterium that can grow and spread in the host to produce disease. An example of this is vibriosis.

Lysozyme

This enzyme can attack the peptidoglycan layer of bacterial cell walls causing them to lyse. Lysozyme has been found in fish mucus, serum and ova. Peritoneal macrophages and blood neutrophils contain lysozyme and the latter are thought to be the source of serum lysozyme. The production of lysozyme by Atlantic salmon macrophages *in vitro* is enhanced in the presence of yeast glucan or bacterial lipopolysaccharide. Fish lysozyme occurs in two forms and one of these appears to much more bactericidal than lysozyme of higher vertebrates. There are several reports of lysozyme isolated from fish serum and ova being bactericidal even for important fish pathogens like *A. salmonicida* and *A. hydrophila*.

Acquired Defence Mechanism Against Bacteria

Anti-adhesins

Bacteria must first adhere to the surface epithelium of the fish host and they possess a variety of structures in the cell wall to do so. Antibodies, which react with these adhesins, may help to prevent the bacteria gaining access. Fish produce antibodies in mucosal tissue like the gut, gill and antibodies have been detected in the mucus of gill and skin. The role of antibody in the prevention of bacteria adhering to the gut lining may not be present in fish, though it's feasible defence against bacterial skin and gill diseases.

Anti-toxins

Many bacterial pathogens produce potent toxins and antibodies, that neutralize them, may have a protective role. A typical strains of *Aeromonas salmonicida* (ssp. achromogenes) produce a potent lethal metallo-protease and antibody responses against toxin confer good protection against experimental challenge.

Anti-invasins

Some bacteria *e.g.*, *Photobacterium damsela* ssp. and *Piscicida* are capable of invading non-phagocytic host cells, where they can avoid many defence mechanisms. Antibodies, which block such invasion, can be protective by rendering the pathogen exposed to various other means of elimination such as killing by phagocytes.

Defence Mechanisms Against Viral Diseases

Viral diseases cause very significant losses in aquaculture. Some viruses mainly affect young fish *e.g.*, infectious pancreatic necrosis virus(IPNV) in salmonids, while others cause mortality in fish throughout their life *e.g.*, viral hemorrhagic septicemia virus (VHSV) and infectious salmon anemia virus (ISAV). Many fish viral pathogens produce a persistent carrier state in the host. These fish can shed virus into the environment and be a source of infection to other fish. Low levels of virus can often be isolated from the kidney of carrier fish and evidence indicates that some viruses *e.g.*, IPNV, can replicate in and be released from the kidney macrophages without being cytolytic.

The Immunological Defence Mechanisms of Fish Against Fungal Infections

Kingdom Fungi is an extremely large and diverse group and includes members with widely differing characteristics and uncertain taxonomic relationships. They share some characteristics, such as heterotrophy and a life cycle that usually incorporates either a thalloid or mycelial stage. A fungal pathogens of fish, such as *Dermocystidium* sp. and *Candida sake*, have a thalloid morphology which consists of a group of associated but unspecialized cells.

Commonly Encountered Fungal Diseases

As previously mentioned, the disease known as saprolegniasis is the first fungal disease of fish to be described. However, it did not attract much interest until 1877, when a large scale outbreak of a then unknown disease in Great Britain prompted a Parliamentary enquiry and attracted the attention of many of the biologists of the day including the great Thomas Huxley, who was among the first to recognize the *Saprolegnia ferax*, which he described as a fungus and is now classified as an oomycete, as the primary cause of the outbreak. A number of methods have been reported for inducing saprolegniasis by immersion challenge, although very little infection has been reported in fish not previously exposed to stress factors. Wounding or abrasion is commonly used and cortisol implants have also been shown to increase susceptibility. While most of the literature on saprolegniasis concerns its effect on salmonids. *S. parasitica* has been reported from a wide range of fish and amphibians. It is a disease frequently encountered by salmonid farmers and a major constraint hatchery production in Africa.

Fish Antibodies

Fish are phylogenetically the lowest group of vertebrates that possess the capability to synthesize antibody molecules in response to antigens. Many features of fish antibodies are similar to mammalian antibodies which regard to structure, function, etc. However, many distinct variations do exist with fish antibodies, which are discussed in the following section.

Antibody Structure in Fish

Fish in general produce a single high molecular weight polymeric antibody, which is considered as analogous to mammalian IgM. The basis monomeric structure consists of two identical heavy (H) chains and two identical light(L) chains as in mammals. However, in fish, the molecular weight of each heavy chain is of around 70KD and each light chain is of around 25KD. The degree of polymerization varies among different groups of fish.

Antibody Response in Fish

The antibody response against a particular antigen in fish is almost similar to mammals. They too exhibit the primary response when come in contact with the antigen for the first time and the secondary response in subsequent exposure to antigen. The characteristic features of secondary response such as the shorter lag phase and increased antibody concentration are also evident in fish, which clearly indicate the presence of specific memory cells. However, there is no class switching or affinity maturation of antibodies, as observed in mammals. Another characteristic feature of antibody response in fish is temperature dependency *i.e.*, fish produce a quicker and better antibody response at higher temperature in the physiologically tolerable range than when kept at lower temperature.

Parasites of Fish

Among disease causing protozoans, the most pathogenic are ectoparasitic protozoa, which affect the skin and gill epithelium. Next in order being the tissue parasites that result in growth reduction, dysfunction and deformation of vital organs of host fish leading to death (Lom 1970a 1970b). To date, as many as 2,400 species of fish infecting protozoa have been recorded, many of them being severe pathogens and several known as 'true pests' in fresh water and mariculture and breeding ornamental fish (Lom and Dykova 1992). Taxonomically, all 7 major groups of protozoa namely, sarcodines, mastigophorans, opalinids, apicomplexans, myxozoans, microsporans and ciliates have been described from fishes. Their pathogenic impact, however, varies. Not all of them are important in culture systems.

Among flagellates, 'costiasis' is a severe disease of young carp especially in rearing and spawning ponds. It is found mainly in carp breeding areas in West Bengal, Orissa and Andhra Pradesh. Dull spots on the sides of fish, which eventually become fused into continuous grey mucus film. Damaged tissue between rays of fins evident. Bare rays common. Fry become emaciated. Head appears large. Gills pale and discoloured.

Cryptobia sp is another bodonid that infects gill filaments of the catfish, *Mystus vittatus* in West Bengal (Mandal, 1984), Indian carp, *Catla catla* in Nagarjunasagar reservoir in Andhra Pradesh (Lakshminarayanan, 1988). A slender form, with broad anterior and pointed posterior ends.

Parasites of Shellfish

In shrimp culture, protozoan symbionts are often ignored although the ciliates often cause passive and chronic infections that spread by contamination. They generally exhibit a high degree of specificity and seasonality, infections being more predominant during monsoon period. In general severe pathogens of shrimps are due to four major protozoan groups namely gregarines, coccidians, microsporidians and ciliates.

Gregarines are the common inhabitants of intestinal track. They are relatively less pathogenic in adults but are known to reduce feed intake, growth rates and induce external fouling.

Symptoms

Infected individuals found emanciated and perceptible reduction in length, weight and feed intake noticed.

Chapter 10

Colour Changes and the Chromgenic Variations of the Fish by the Application of the Different Mycelial Coloured *Streptomyces* spp. to the Feed

Introduction

Carotenoids are a group of over 600 molecules which can be found in most life forms and fulfill diverse functions, ranging from their evolutionary role as photosynthetic of light-quenching pigments to anti-oxidants, precursors of vitamin A, or pigments involved in the visual attraction of animals such as flower pollinators or mating partners.

The basic structure of carotenoid is composed of eight isoprene units. The structural formula of all carotenoids comes from that lycopene, starting with different structural modifications. Among the diversity of physiological functions, the provitamin A activity of carotenoids with β-ionone rings like Astaxanthin is without doubt of central importance. Carotenoids also affect the endocrine system with respect to gonadal development and maturation, fertilization, hatching, viability and growth, particularly in fish and crustaceans. In addition to their role as a source of dietary Pro-vitamin A, carotenoids have been proposed to stimulate animal immune systems, increase stress tolerance, and enhance embryonic development. Other ascribed functions like reduction in the mortality rate during embryonic development, increase the ability tolerate harsh environmental conditions such as elevated water temperature, elevated ammonia levels, harmful effects of lights and perform a respiratory function under conditions of limited O_2. Another important function of carotenoids in conjuction with protein molecules (carotenoproteins of crustaceans), appear to improve the stabilization of proteins and their tertiary structure. It also involved in the cross-membrane calcium transfer, in the neuronal respiratory chain as oxygen reservoirs and in photo protection. The carotenoid are functioning as antioxidants or shields during haemolymph transport and in the eggs; protecting nutrients and embryonic tissues from oxidative damage. Czeczuga (1979) found that fish with a high level of carotenoids were more resistant to bacterial and fungal diseases than fish with low carotenoid levels. A conspicuous and far better understood role of carotenoids is their influence on pigmentation. The abnormally pigmented Atlantic halibut (*Hippoglossus hippoglossus*) are more susceptible to skin damage from ultraviolet radiation than normally pigmented fish. Production of protective pigments and growth promoting factors by *Streptomyces* have also been reported (Rambow and Rose; 1963). TLC, U.V-Vis Spectrophotometers and Column Chromatography techniques clearly prove that *Streptomyces* spp. have antioxidants and their property mainly due to the presence of antioxidants. Carotenoids known to influence sexual dichroism and protective colouration in fish. Carotenoid are located mainly in some autonomous organells, chloroplasts and chromoplasts. Each organelle uses a unique mechanism to request carotenoids (Barttley and Scontnik 1999).

Carotenoids in Microorganisms

Spirulina has several carotenoids, the major components being β-carotene, zeaxanthin echineone, myxo-xanthophylls and β-cryptoxanthin. Several isomers of β-carotene also exist. In some species of algae, fungi, higher plants and some bacterial species, various types of carotenoid pigments are present (Bulton 1995; Ronald, W.S.).

The Basidiomycetes yeast, *Phaffia rhodozyma*, possesses the ability to accumulate very large amounts of astaxanthin, exceeding amounts 0.1L of the yeast dry weight.The crustacean hypodermis and exoskeleton accounts from about 60 to as much as 90 per cent of the animals pigment content. Astaxanthin, the main pigments of crustaceans is a red C-40 dihydroxy di keto carotenoid. The bulk of pigments in the crustacean epidermal tissue however, are found in the ester form. American Flamingo contains astaxanthin, canthaxanthin and adnorubin whereas cock of the rock contains canthaxanthin and phoeniconene. Carotrnoids are found in the livers of the *Monitor varanus* comodenised and the tortoise, *Testudo graeca*.

Carotenoids, accumulated in the plumage of birds, the colour and patterns are important both for sexual attraction and visual communication (Brush 1981). Carotenoids may be important because of their role as Vitamin A precursors and free radical trapping agents. It has been suggested that dietary carotenoids influence longevity in humans as well as in animals (Cutler, 1989). Since fish do not synthesize ketocarotenoid pigment de novo, they must get them from their diet. The typical colour of salmonid muscle is due to carotenoids and so carotenoids may be involved in giving the yellow pigmentation. Salmonids like other animals are unable to synthesize carotenoids de novo and obtain from their diet, but there may be some was 137. Minimum value was obtained in fishes fed with diet D_7 (126.20). The percentage increase in length obtained in other diets namely D_9, D_{10} and D_{11} were 133.90, 134.40 and 134.50 respectively.

Nutritional Grouping of Gut Microflora

It has been studied and summarized the nutritional grouping of microorganisms, isolated from the gut of *O. mossambicus*, fed with different types of macrophyte diets. The results proved that microorganisms like bacteria, fungi and actinomycetes were distributed in the gut of the fish, but the distribution was not in uniform pattern in all the media used. An interesting thing to be noted was that, the silage harboured more bacteria. Maximum bacterial count (79 x 10^3/g) was observed in fish fed with diet containing *Pistia* silage (D_{10}), where the culture medium used was basal medium enriched with yeast and sediment extract. The distribution of baceria, fungi and actinomycetes were comparatively low in fish fed on control diet (D_7) than other test diets. Minimum bacterial count was found in fish fed with D_7, where the basal medium was supplemented with yeast extract only. Several species of aquatic macrophytes are used as livestock fodder but their high moisture content is a major constraint. The conversion of aquatic macrophytes into silage has been proposed as a method for reducing or eliminating the need for drying the plants.By using the macrophytes in fish feeds the cost can be reduced and the useless weeds could also have a role in fish nutrition. If the cellulase present in aquatic macrophyte can be degraded, it can be utilized as an important ingredient in supplementary feed for fish. In nature, cellulolytic enzyme secreted by microorganisms helps in the degradation of plant matter (Montencourt, 1983; Dhevendaran *et al.*, 1992).

The silage-based diets provided good results than the powder form of macrophytes. The growth in terms of weight was maximum in fish fed on diet containing *Pistia* silage, whereas maximum length was attained in fish fed on *Eichhornia* silage. Likewise, better conversion efficiency was exhibited by the fish fed with diet containing *Eichhornia* silage followed by diet containing *Pistia* silage. These findings corroborated with the findings of Das *et al.* (1994), where the maximum growth of *Labeo rohita* was found in *Eichhornia* based diet. Boyd (1968) observed that dehydrated samples of *Eichhornia* leave contain satisfactory levels of essential amino acids. The dried powder of *Eichhornia* also contains essential vitamins capable of promoting growth of the fish.

The gut micro flora was maximum in fish fed on diet containing *Pistia* silage. Both *Eichhornia* and *Pistia* silages incorporated diets provided higher number of bacterial count than the diets containing *Eichhornia* and *Pistia* powder. The microorganisms showed an affinity towards the basal medium supplemented with yeast and sediment extracts.The silage harboured many types of microorganisms which may be ingested by the fish in the culture systems. This may be the reason for obtaining higher number of microorganisms in the fish fed on silage – based diets. The microbial degradation of macrophytes can provide detritus and a reasonably good microflora as food for fish.The development of a low – cost fish feed, which can be produced in simple and practical ways, is important to both large and small aquaculture systems. A feed – production system based on simple, vegetative starter materials, has the potential of expanding natural feed production in remote rural areas, where resources are minimum and human protein deficiencies are often high.

Production and utilization of microbial proteins have been studied extensively during the last few decades. One of the problems involved in economically sound utilization of this approach is that usually dehydration and processing of the product are necessary. Among the unconventional protein sources, single cell proteins (SCP) of bacterial origin appear to be a promising substitute for fishmeal. This is one of the ways of reducing the feeding cost of the fish in the culture systems. In the present study, we used SCP individually and in mixtures for replacing 25 per cent of the fishmeal in compounded diets of *Oreochromis mossambicus*.

The highest increase in length was obtained in fish fed with photosynthetic bacteria, whereas highest weight was attained in fish fed with diet containing bacterial mixture. The photosynthetic bacteria (Figure 3.6) synthesise vitamins such as Riboflavin (B_2), Pyridoxine (B_6), Folic acid, Cobalamine (B_{12}), Ascorbic acid (C), and Cholecalciferol (D_3) (Kobayashi, 1970). This may be the reason for obtaining high growth percentage of fish fed with photosynthetic bacteria. They are able to synthesize the low-molecular weight precursors of cell macromolecules and vitamins. This may be the reason for better growth rate recorded in fish fed on single-cell protein diets, when compared to the control feed, which do not contain single-cell proteins.

There are reports relating to the importance of single – cell protein as a source of food for copepods. Caumette (1987) observed that, after a red tide, a dense development of a copepod community of *Tisbe* sp. which fed on photosynthetic bacteria.

Different authors have shown that bacteria play an important role in the nutrition of harpacticoides, and that certain bacteria strains were entirely sufficient for the development of certain copepods (Itoh, 1973; Rieper, 1978; Ustach, 1982). Rieper (1978) observed that a mixture of strains often supplies better results than the use of a pure strain. Certain bacterial strains exert a great attraction, whereas others have a non – attractive character (Rieper, 1982). Some bacteria may provide indispensable growth factors for copepods (Provasoli and Others, 1970).

In an experiment to study the population dynamics of copepods, *Tisbe holothuriae*, Guerin and Rieper – Kirchner (1991) employed successfully three strains of bacteria, namely *Micrococcus* sp., *Moraxella* like *Coccobacillus* and *Flavobacterium* sp. The results of the study were quite interesting and promising.

Regarding the nutritional grouping of gut microflora, we have noticed maximum number of bacterial population in fish fed with diet containing both bacteria and actinomycetes (Figure 3.7), where the medium used was the basal medium, supplemented with yeast and sediment extracts. This result coincides with the findings of Saradha (1996) where, the gut microflora showed more affinity towards basal medium enriched with yeast and sediment extracts. The microorganisms found in the gut region have a direct relationship with the growth of the fish (Shiranee and Others, 1993). It is probable that gastro – intestinal microflora may play a major role in the nutrition of the fish. It is further suggested that the microflora might even be the source of food as SCP for the fish (Trust and Sparrow, 1974; Seki, 1989; Saradha,

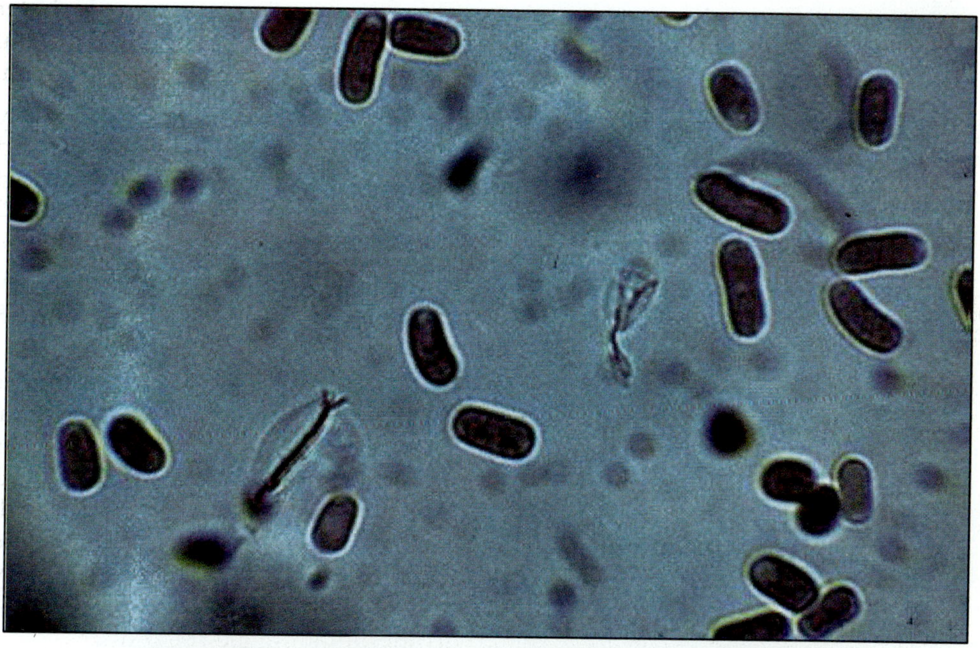

Purple Sulphur Bacteria (*Chromatium* spp.), Cell Size – 4.5 µm

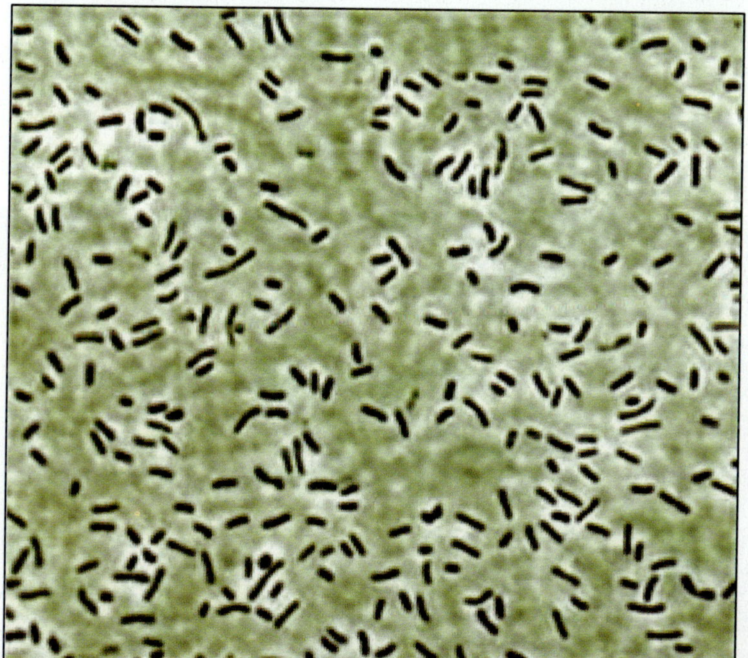

Purple Green Sulphur Bacteria (*Chlorobium* spp), Cell Size – 0.8 µm

Figure 3.6

1996; Manju and Dhevendaran, 1997).The protein content of the fish flesh was maximum in fish groups fed with photosynthetic bacteria, followed by fish fed with both bacteria and actinomycetes. The fish fed with control diet showed lowest value for protein. Similar results were reported by Manju and Dhevendaran (1997), where the SCP fed prawns showed a higher protein content in the muscle than that of control diet. This may be due to the fact that the microbial SCP contains more protein

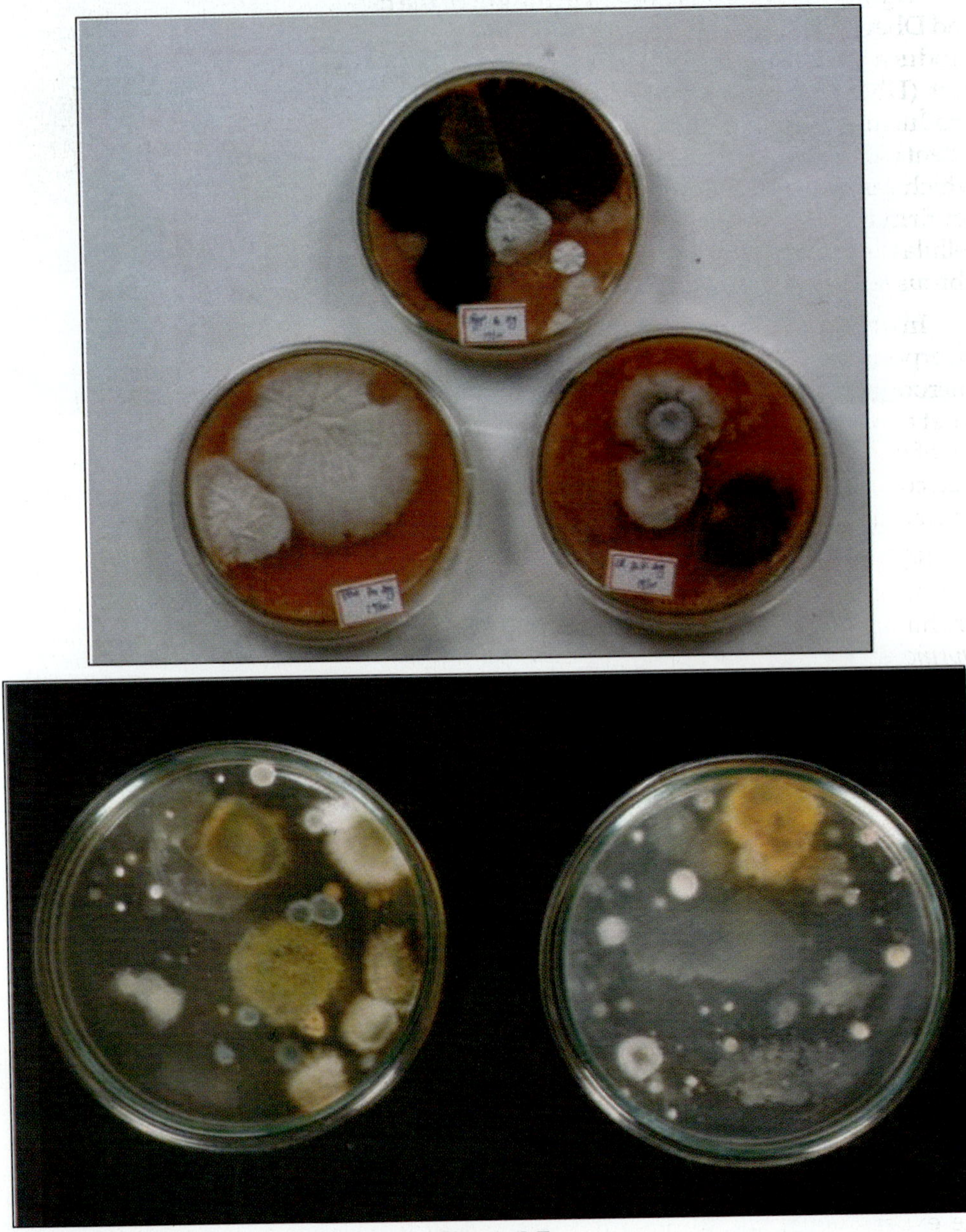

Figure 3.7

than fishmeal. A significant proportion of the nitrogen present within single – cell protein is in the form of non – aminoacid nitrogen containing compounds, particularly as purine and pyrimidine bases of nucleic acids. Some workers (Kaushik and Luquet, 1980) suggested that high nucleic acid inclusion in SCP may be deleterious to fish and shellfish. But, in this study we did not get any deleterious effects such as mortality of the fish.

The microorganisms present in the gut of the fish act as a source of food (Manju and Dhevendaran, 1997). Besides this, some bacteria present in the gut of certain fish produce the enzyme arylsulfatase, which helps in the hydrolysis of sulfates in the diet (Dhevendaran and Maya, 1998). The phosphobacteria and phosphatase-producing bacteria in the gut may contribute to the phosphate value to a certain extent (Alex and Others, 1993). The bacteria present in the gut may produce cellulase, which can degrade cellulose present in the diet of the fish (Saradha, 1996) thereby confirming the present observation. Dhevendaran *et al.* (1992) already observed cellulase enzyme activity in *Aeromonas* spp. isolated from retting ground containing fibrous coconut husk.

In the nutritional grouping of the fish, both *Eichhornia* and *Pistia* silage incorporated diets provided higher number of bacterial flora in the gut region. The microorganisms showed an affinity towards the basal medium supplemented with yeast and sediment extracts (Figure 3.8). The microbes harbour the silage, reach the gut of the fish, proliferate there, and act as secondary source of food for the fish. The feed containing *Pistia* fish has incorporated higher percentage of protein content in the muscle.

SCP used for the present study has bacterial mixture containing *Bacillus, Moraxella, Pseudomonas* and *Corynebacterium*, isolated from the retting grounds of Anchuthengu, Kerala. *Actinomyces* sp.was isolated from the estuarine clam, *Villorita cyprinoides* and photosynthetic bacteria, *Rhodospirillum rubrum* as Reference strain. The protein content of these used strains varied between 35 to 70 per cent and hence selected for replacing fishmeal as protein substitute. We could successfully replace 25 per cent fishmeal from the artificial diet of *O. mossambicus*.

The fish fed on SCP diets exhibited remarkable increase in growth and better conversion efficiency as compared to the control diet. Maximum growth in terms of percentage increase in weight was observed in fishes fed with diet containing bacterial mixture. The diet containing photosynthetic bacteria provided better conversion efficiency and conversion ratio. When the biochemical analysis was taken into consideration, the SCP diets fed fish contained more protein percentage rather than the control diet.The photosynthetic bacterial cells contain approximately 65 per cent protein and sufficient quantities of essential aminoacids. In addition, they contain various B- vitamins, ascorbic acid and Cholecalciferol (D3).They are capable of fixing nitrogen, which is of fundamental importance to the nitrogen economy of the soil. The use of photosynthetic bacteria for land animal feeding and even as biofertilizers is of potential factors affecting world food production (Kobayashi and Haque, 1971).

A total of sixteen diets (Table 3.9) were tested for the three sets of experiments. The experimental conditions provided for all experiments being the same, a

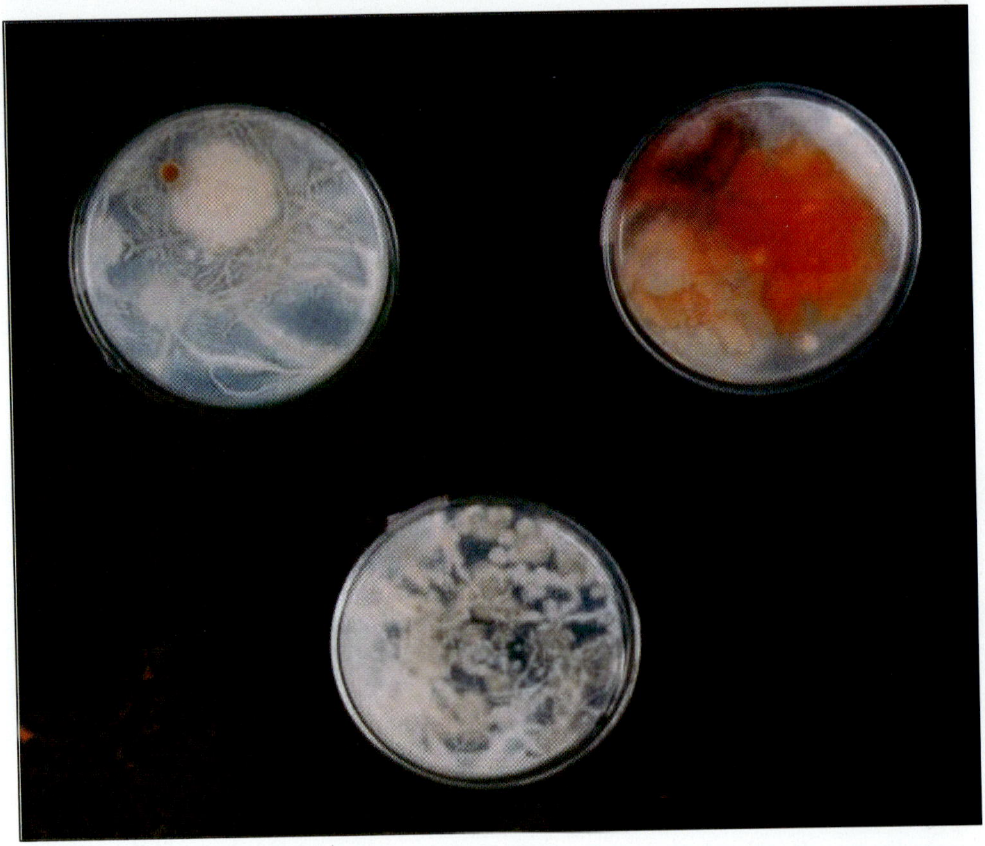

Figure 3.8

comparison can be done between different types of diets. Out of the sixteen diets, the TNAU -1 hybrid incorporated diet (D6) provided highest growth of *Oreochromis mossambicus* in terms of percentage increase in weight. The feed containing TNAU-1 hybrid and that with *Eichhornia* silage provided better conversion efficiency.With regard to the nutritional grouping, the diet containing both bacterial mixture and actinomycetes helped the fish to proliferate maximum number of gut microflora. When the proximate composition was considered, the diet containing photosynthetic bacteria fed fish exhibited highest protein content in the flesh. In general, it can be concluded that, all experimental diets, in one sense or the other proved their efficiencies as a fish meal substitute in the artificial diet of *O.mossambicus*. It can be seen that the *Azolla,* aquatic macrophytres and SCP do not have any harmful effect to the growth and health of the fish, since they themselves act as the feed for the growing fish. The increase in load of the gut flora reflects the progressive growth of the fish in natural fish ponds cultured with *E. suratensis* as observed by Shiranee and Others, 1993). Our findings strongly support the above experimental observations. The present study was carried out to elucidate the effects of replacing the fishmeal from the pelletised diet of *Oreochromis mossambicus* (Peters) by dried *Azolla* powder, aquatic macrophytes like *Eichhornia* and *Pistia* and various types of single-cell protein (SCP). The influence of experimental diets on the growth, conversion efficiency, nutritional grouping of gut microflora, cellulolytic activity and bio-chemical analysis was carried out.

Table 3.9: Feed Conversion Efficiency, Conversion Ratio and F-Ratio of Fish Fed on the 16 Experimental Diets

Diets	Conversion Efficiency	F-Ratio	Conversion Ratio	F-Ratio
D_1	67.52[a]		1.48[c]	
D_2	95.00[b]		1.07[b]	
D_3	96.80[b]	11.9004**	1.03[b]	18.5047**
D_4	90.00[b]		1.10[b]	
D_5	98.80[b]		1.01[b]	
D_6	119.80[e]		0.85[a]	
D_7	56.80[a]		I.77[c]	
D_8	117.20[c]		0.84[a]	
D_9	93.00[b]	93.8522Z**	1.07[b]	60.2227**
D_{10}	113.00[c]		0.87[a]	
D_{11}	92.40[b]		I.08[b]	
D_{12}	53.66[a]		1.862[d]	
D_{13}	69.00[b]		1 464[c]	
D_{14}	72.40[b,c]	52.4381**	I.38[b,c]	44.0804**
D_{15}	86.60[d]		1.140[a]	
D_{16}	74.82[c]		1.315[b]	

** Significant ($P < 0.01$)

The SCP diets did not adversely affect the biochemical composition of the fish. The protein content was highest in fish fed on diet containing photosynthetic bacteria, followed by that prepared with bacteria and actinomycetes. From the experiments it is well appreciablc that the new hybrid of *Azolla* has a promising future in fish nutrition. *Eichhornia* and *Pistia* which are considered to be harmful weeds by nature in the aquatic environment could be made use of by the microbial decomposition as silage to the feed of the fish. Photosynthetic bacteria, the proliferating bacteria from the retting grounds and potential antibiotics producing actinomycetes have been suggested to be the potential feed resources for the partial replacement of fishmeal in the artificial diet as they synthesize vitamins and the secondary metabolites that may enhance the growth of the fish.

Chapter 4

Application of *Streptomyces* to the Feed as Single Cell Protein for Growth of Ornamental Fish

Introduction

Single cell protein broadly refers to the microbial biomass or protein extract used as food or feed additive. If the SCP is suitable for human consumption, it is considered as food grade. SCP is regarded as feed grade, when it is used as animal feed supplement, but not suitable for human consumption. These *Streptomyces* which is already proved to synthesize growth promoters used as SCP source of high protein content. The microbial incorporated feed was prepared and supplemented to ornamental fishes like *Xiphophorus maculates*, *Xiphophorus helleri* and *Vieja synspilus* (Figure 4.1). This was done to evaluate the colouration rise and growth of the fishes.

Sound nutrition promotes reproductive success and healthy larval development and is essential in the production of high quality aquaculture products. To reduce dependence on live-food, counter measures such as shortening of the preweaning period (Person *et al.*, 1993) and development of complete artificial diets for precocious cofeeding (Holt, 1993) were undertaken with some success. However, the exclusive use of artificial dry diets as first feed was deemed unsuitable until Cahu *et al.* (1998) succeeded in developing a complete feed containing yeast and fish protein hydrolysate for the seabass larvae, totally replacing live-prey. Nevertheless, providing an acceptable first food with reliable availability, proper size and nutrient composition, remains a major challenge in aquaculture science. Single-cell proteins (SCPs) are playing a greater role in the evolution of aquaculture diets. With excellent nutrient profiles and capacity to be mass produced economically, SCPs have been added to

Figure 4.1: Fishes Selected for the Feeding Experiment

Xiphophorus helleri

Xiphophorus maculates

Figure 4.1–*Contd...*

Vieja synspilus

aquaculture diets as partial replacement for fishmeal (Coutteau and Lavens, 1989; Olvera-Novoa *et al.*, 2002; Li and Gatlin, 2003) and for PUFA-fortification of rotifer and *Artemia* (McEvoy *et al.*, 1996).

In recent years, the concept of 'functional foods' has developed in human and animal nutrition. Functional foods, which first emerged in Japan, have been defined as 'foods or dietary components that may provide a health benefit beyond basic nutrition' (IFIC, 2004). The presence of biologically active components of functional foods can have an impact on health benefits or other desirable physiological effects in addition to their nutritional function. The term 'functional foods' covers a broad range of products including, for example, DHA- and selenium-enriched eggs, selenium-enriched pork, stanol- and sterol-enriched margarine, etc. Also included under this category are dietary 'probiotics' and 'prebiotics'. A probiotic is defined, in the strict sense, as "a viable microbial dietary supplement that beneficially affects the host through its effects in the intestinal tract" (Roberfroid, 2000). It should be noted that the term has also been widely and incorrectly applied in aquaculture, and especially shrimp culture, to include the use of live microbes to beneficially alter the microbial balance in the culture system itself. Prebiotics, on the other hand, have been described as "non-digestible substances that provide a beneficial physiological effect on the host by selectively stimulating the favorable growth or activity of a limited number of indigenous bacteria" (Gibson and Roberfroid, 1995).

Nucleotides have been recognised as important elements in mammalian nutrition especially during periods of rapid growth or physiological stress as well as appearing to play a key role in efficient immune system function (Uauy, 1989; Barness, 1994; Van Buren, 1994). They can be synthesised directly or scavenged by salvage pathways in the body although it appears that exogenous dietary sources are preferentially used (Uauy, 1994). However, immune cells and intestinal cells cannot synthesise nucleotides and depend on nucleotides from other sources (Quan, 1992).

Synthesis and salvage of nucleotides are thought to be energy intensive in metabolic terms and dietary nucleotides may reduce the metabolic cost of *de novo* nucleotide synthesis. Dietary sources of nucleotides may also benefit rapidly dividing tissues, such as those of the immune system, especially under a challenge, and the term 'conditionally essential' has been used to describe their role in nutrition (Carver and Walker, 1995). In many biochemical processes, primary nucleotides such as 5'AMP, 5'CMP, 5'GMP, 5'IMP and 5'UMP are used to produce a number of intermediate metabolites through a series of enzymatic reactions. Supplementation of primary nucleotides in the diet provides a ready source of nucleotides for use in the synthesis of intermediate nucleotides when required (Mateo, 2005, personal communication).

Any ingredients of animal and plant origin containing cellular material are potential sources of nucleotides, usually in the form of nucleoproteins. The nucleotide content is particularly high in ingredients such as fish solubles, animal protein solubles, fish meal, legumes (adenine content is particularly high in black-eyed peas), yeast extracts and unicellular organisms such as yeasts and bacteria that are rich in RNA or DNA. The content, proportion and availability of nucleotides differ among ingredients. Muscle protein is a poor source of nucleotides as they are mainly in the form of actin-myosin protein. Oilseeds, such as soybeans, grains, fruits, vegetables and processed milk products are also poor sources of nucleotides (Barness, 1994; Devresse, 2000; Mateo, 2005). Among marine protein sources, anchovies and sardines, for example, have much higher guanine levels than squid, clams or mackerel. Availability and digestibility are also important issues. Whole yeast is much less digestible than yeast extract, possibly due to the need to digest the yeast cell wall and yeast extract having much higher levels of soluble protein. Fish and animal protein solubles are highly digestible but they leach easily, affecting overall availability (Devresse, 2000).

The term SCP is used to refer to single cell protein which is used to refer to microbial biomass used as food and feed additives. Either the isolated cell protein or the total cell material may be called SCP. Microorganisms have a very short generation time, and can thus provide a rapid mass increase. Microorganisms can be easily modified genetically to produce cells that bring about desirable results. The protein content is high. The production of SCP can be based upon raw materials readily available in large quantities. SCP production can be carried out in continuous culture and thus be independent of climatic changes. Gastrointestinal disturbances are common complaints in man following the consumption of algal and bacteria SCP, and these and other problems associated with the consumption of SCP have been reviewed elsewhere. When gram-negative bacteria are used as SCP sources for human use, the endotoxins must be removed or detoxified.

Right from the time immemorial plant and animal extracts are widely utilized as the medicines for mankind as well as animal kingdom. The ocean which covers over 70 per cent of the earth contributes significantly a good number of animals and plants harbouring potential bioactive compounds. Preferably microorganisms are ubiquitous and unique in their distribution and contribution of certain pharmaceutically important compounds. An outstanding example of how a traditional medical practice can lead to the isolation of a medically active substance provided by the work of mangrove plants. The chemistry of marine organisms preferably microbes differ from terrestrial counterpart in many respects. The halogenated compounds seldom found in marine organisms, because of chlorine and bromine compounds. Similarly, marine organisms are rich in sterols and are quite unique in the marine environment.

Streptomycetes are widely distributed in coastal environment and are possibly isolated from deeper oceans also. Ecological factors of marine environment are responsible fully for developing new strategy to the exploration of bioactive compounds for multipurpose. If they are isolated from marine ecosystem means that they are totally adopted to this environment eventhough their ancestors may have come from terrestrial origin. The ocean is an untapped resources and the occurrence and distribution of antagonistic streptomycetes have been carried out by Walker and Henrici (1934), Lakshmanapermalsamy and Vanajakumar (1978; 1981) from east coast and by Dhevendaran and Annie (1999 a&b) from south west coast of India.

In aquaculture operations, essential and expensive components of the feed are proteins, especially in the fishmeal. Since the supply of fishmeal has become uncertain, it is of great importance to replace the fishmeal to a minimum possible extent in fish rations. Among unconventional protein sources, single cell protein (SCP) of microbial origin appear to be a promising substitute for fishmeal, which can replace upto 25-50 per cent of fishmeal. Marine streptomycetes incorporated into the artificial diet on growth, can enhance food conversion efficiency and protein increment of the juvenile prawn, *Macrobrachium idella* and fish, *Oreochromus mossombicus* (Manju Dhevendaran, 1997, 2002).

Microbial incorporated feed (Figure 4.3) was prepared and supplemented to fishes like *Xiphophorus maculates, Xiphophorus helleri* and *Vieja synspilus*. On choosing this fresh water ornamental fish is because of commercial value and visible external colouration. The experimental step has five fishes each troughs (Figure 4.2) and the feed were prepared by square method (Figure 4.3). The results showed that the FCE values were increased when compared to control; FCR values were less when compared to control (Tables 4.1–4.6). The strains of *Streptomyces* spp that incorporated are rich in protein and they may synthesize the lower weight precursors like macromolecules and vitamins. This may be the reason for the better growth rate. The protein present in each strains was accessed by Lowery *et al*. (1951) method and it was found to be AQB-T-55 per cent, AQB-E-47 per cent, AQB-S-52 per cent, AQB-Gr-

Figure 4.2: Experimental Setup for Feed Trails Prepared with Combination of Normal Ingredients and Microbe Incorporated

Control

Feed 1 (AQB-G + AQB-C)

Figure 4.2–*Contd...*

Feed 2 (AQB-E + AQB-H)

Feed 3 (AQB-Gr + AQB-T)

Figure 4.2–*Contd...*

Feed 4 (AQB-S + AQB-G)

Table 4.1: Food Conversion Ratio and Conversion Efficiency of *X. maculates* with Formulated Diets

Parameters	*Diets*				
	Control	*Feed 1*	*Feed 2*	*Feed 3*	*Feed 4*
Weight of feed given	0.2	0.2	0.2	0.2	0.15
Weight of unfed	0.071	0.067	0.069	0.085	0.071
Weight of feed intake	0.129	0.133	0.14	0.115	0.079
Weight of excreta	0.049	0.042	0.030	0.042	0.047
Initial wt of fish	0.4	0.4	0.4	0.4	0.3
Final wt of fish	0.44	0.5	0.47	0.45	0.35
Weight gain	0.04	0.1	0.07	0.05	0.05
Food Conversion Ratio (FCR)	3.22	1.33	2	2.3	1.58
Food conversion efficiency (FCE)	31.0	75.1	50	43.4	63.3

Figure 4.3: Feed Prepared from *Streptomyces* Strains Isolated from Gut of Fishes

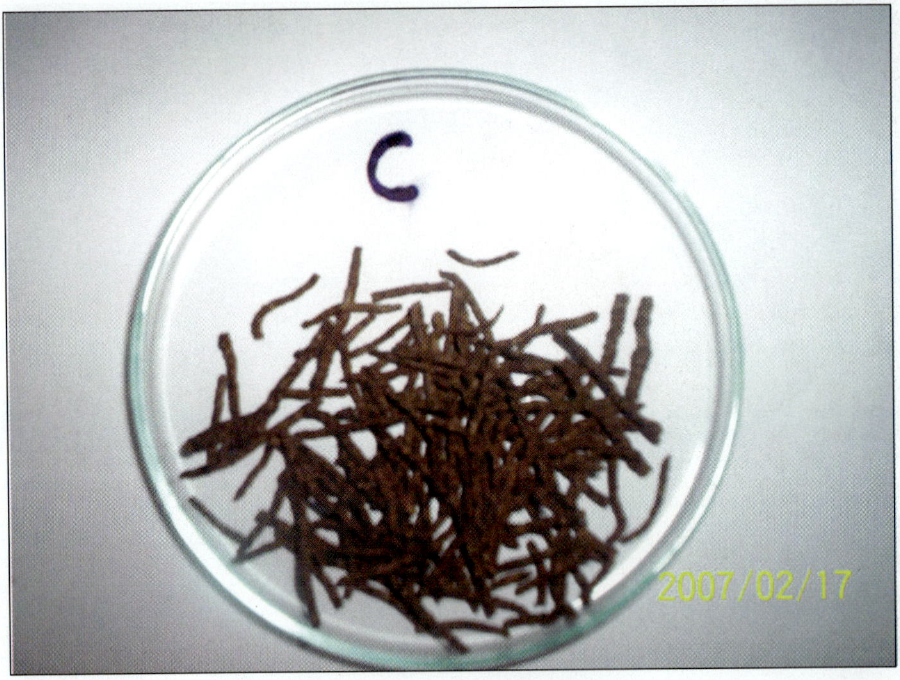

Control Feed – Artifical feed

Feed 1- Gourami + Carp (AQB-G + AQB-C),

Figure 4.3–*Contd...*

Feed 2- Etroplus + Hymavathy pond sediment sample (AQB-E + AQB-CA)

Feed 3- Grouper + Tilapia (AQB-Gr + AQB-T)

Figure 4.3–*Contd...*

Feed 4 - Sponges + Gouraml (AQB-S + AQB-G) + Colorants

Table 4.2: Ingredients Composition of Formulated Diets (25 per cent) Expressed in Grams

Ingredients	*Control*	*Feed 1*	*Feed 2*	*Feed 3*	*Feed 4*
Rice bran	20	20	20	20	20
Ground nut Oil cake (GOC)	35	35	35	35	35
Bengal gram	15	15	15	15	15
Fish meal	15	12	12	12	12
Tapioca	15	15	15	15	15
Streptomyces spp.	–	3	3	3	3

45 per cent, AQB-C-42 per cent, AQB-G-54 per cent, AQB-CA-57 per cent. The results corroborated with the finding (Manju and Dhevendran, 1997), in which the SCP fed fish provided better growth and conversion efficiencies. Supplementary feeding is a major expense in aquaculture operation. The commonly used convensional feed in India is a mixture of Rice bran and oil cake in equal proportions. But this feed is

Table 4.3: Ingredients Composition of Formulated Diets (25 per cent) Expressed in Grams

Ingredients	*Control*	*Feed 1*	*Feed 2*	*Feed 3*	*Feed 4*
Rice bran	15.5	15.7	15.4	15.5	15.9
Ground nut Oil cake (GOC)	34.8	35	35.8	34	35.1
Bengal gram	20	18	18.4	20	18.5
Fish meal	4.7	2.5	2.4	2.5	2.5
Tapioca	25	25	25	25	25
Streptomyces spp.	–	3	3	3	3

Table 4.4: Food Conversion Ratio and Conversion Efficiency of *X. helleris* with Formulated Diets

Parameters	*Diets*				
	Control	*Feed 1*	*Feed 2*	*Feed 3*	*Feed 4*
Weight of feed given	0.239	0.234	0.232	0238	0.231
Weight of unfed	0.062	0.049	0.052	0.051	0.064
Weight of feed intake	0.177	0.185	0.180	0.187	0.167
Weight of excreta	0.074	0.057	0.059	0.052	0.073
Initial wt of fish	1.594	1.572	1.548	1.590	1.582
Final wt of fish	1.684	1.701	1.658	1.717	1.681
Weight gain	0.09	0.129	0.11	0.127	0.102
Food Conversion Ratio (FCR)	1.966	1.43	1.64	1.47	1.63
Food conversion efficiency (FCE)	50.85	69.73	61.11	67.91	61.08

Table 4.5: Ingredients Composition of Formulated Diets (25 per cent) Expressed in Grams

Ingredients	*Control*	*Feed 1*	*Feed 2*	*Feed 3*	*Feed 4*
Rice bran	15	15	15	15.5	15.5
Ground nut Oil cake (GOC)	35	33.5	34.5	34	35.5
Bengal gram	20	19	19	20	18.5
Fish meal	5.0	2.5	2.5	2.5	2.5
Tapioca	25	27	26	25	25
Streptomyces spp.	–	3	3	3	3

nutritionally imbalanced to achieve the fast growth of the fish (Varghese *et al.*, 1976). The supply of fishmeal has become increasingly uncertain and the price has been raised rapidly. Thus the increase in the cost necessitated to look for cheaper alternate source with efficient growth promoters. It is better and cost efficient to supplement microbial incorporated feed.

Table 4.6: Food Conversion Ratio and Conversion Efficiency of Flower Horn Fishes (*Vieja synspilus*) with Formulated Diets

Parameters	*Diets*				
	Control	*Feed 1*	*Feed 2*	*Feed 3*	*Feed 4*
Initial wt of fish	5.0	4.0	3.9	4.3	3.9
Final wt of fish	5.3	4.5	4.3	4.9	4.3
Weight of feed given	1.25	1.0	0.97	1.07	0.97
Weight of unfed	0.50	0.30	0.37	0.43	0.23
Weight gain	0.3	0.5	0.4	0.6	0.5
Food Conversion Ratio (FCR)	2.5	1.4	1.5	1.06	1.85
Food conversion efficiency (FCE)	40	71.42	66.66	93.75	54.05
% Increase in Wt	6	12.5	10.25	13.95	10.25

Single cell protein refers to the dried microbial cells or total protein extracted from pure microbial cell culture which can be used as food supplement to animals as feed grade. Here, in the experiment the growth of fishes was assessed by supplementing microbial incorporated feed. Since FCR values were less when compared to FCE values. The percentage of growth rate was high when compared with control. The high protein content of *Streptomyces* may be one of the reasons which favours the growth of these ornamental fishes.

structural modification before deposition in the muscle. Carotenoids known to influence sexual dichroism and protective colouration in fish. It reduces the incidence of cancer and degenerative disease in man, probably due at least partly to their antioxidant role in protecting sensitive biological molecules from highly reactive oxygen species or free radicals.

The colouration of ornamental fishes gives the market value for the fish. If any external source may be biological or chemical affects the colouration or pigmentation of fishes, the market value of that fish is reduced or not sold. An attempt made on supplementation of microbial incorporated feed to fishes which are affected either biologically or chemically and recovery of colouration or pigmentation of the fishes was studied. The chemical treatment like bleaching the ornamental fishes (*Xiphophorus helleri*) by exposing them to sunlight and chlorine stress was done. The candidate microbe here used was *Streptomyces* strains incorporated to the feed which has high storage of pigments and growth factors. The retainment of colouration to the fishes again gives the fishes the market value.

Nutrition and health are two concepts intimately linked with one another and it is sometimes difficult to identify the boundary between nutrition and health or pathology. All nutritional studies will follow at least two parameters in order to quantify the requirement of a specific nutrient: growth rate and survival. When monitoring survival the nutritionist obviously agrees that nutrition does affect health status. However, most of the reviews on nutrition (National Research Council, 1993) report on the nutritional requirement without much emphasis given on the possible effect that macro or micro nutrients might have on health. Furthermore, nutrient requirements are almost always determined under standard, well defined and favourable conditions. However, these ideal conditions are not always prevailing in commercial farms; animal often being stressed, confronted with pathogens, and unfavourable environmental conditions. Under these conditions, the requirement for certain nutrients will almost certainly increase due to an increased activity of the immune system.

Manufacturing a feed for fish or shrimp involves several steps that all have a defined effect on the quality and performance of the final product. Chronologically, we can roughly classify them as follows: selection of raw materials, formulation and processing. Processing essentially shapes a particle and gives the ingredient mash the physical presentation that is required for handling and optimal feeding of the animal. If the production process is done according to basic rules, it will not significantly affect the quality of the diet. On the other hand, selection of raw materials and formulation do have a decisive influence on the performance of the diet. Imbalances in formulation or presence of toxins in the ingredients can severely affect fish health status and can lead to increase sensitivity to diseases. Although a large number of feed ingredients may be taken into account for feed formulation (Tacon, 1987, 1993, 1994) most of the feed formulations are simply based on a few "universal" ingredients: fish meal and fish oil, soya and wheat (or another starch source). In most cases, for "dense" formulas, fish meal and soya (*e.g.* defatted soya 44) will be the most cost effective protein source as can be demonstrated by the use of least-cost software (linear programming).Once the formulation is designed, selection of raw material is of prime importance to ensure the quality of the diet and its performance. Bad quality fish meal, inadequately toasted soya, cereals contaminated with mycotoxins can certainly cause reduced growth rate and finally can lead to increased mortalities.

Vitamin A is a fat soluble vitamin, important for maintaining the integrity of epithelial and mucosal surfaces. Vitamin A influences in homeotherms which is haematopoiesis of phagocytes and lymphocytes. Phagocytosis and intracellular killing are increased in animals given supplemental vitamin A. It restores normal humoral and cellular activity in steroid-treated animals. Retinoids and carotenoids increase proliferation of T helper cells, the induction of cytotoxic T cells and expression of IL-2 receptors on natural killer cells. Vitamin A is a relatively weak anti-oxidant (Bendich, 1992; Blazer, 1992; Lall and Olivier, 1993). It has been illustrated that astaxanthin has an anti-cancer effect and increases the number of communicating cells. Carotenoids (ß-carotene and canthaxanthin) are potent antioxidants, enhance proliferative responses of T and B lymphocytes to mitogens, increase cytotoxic T-cell and macrophage tumor-killing activities and stimulate the secretion of tumor necrosis factor and simultaneously lower the tumor burden. Canthaxanthin lacks vitamin A activity while β-carotene is a precursor of vitamin A. (Bendich, 1989, 1992; Burton, 1989; Blazer, 1992; Lall and Olivier, 1993; Chew, 1995). Different doses of astaxanthin (O, 50, 400 ppm) were evaluated in rainbow trout (Verlhac *et al.*, 1996). Astaxanthin improved phagocytosis and complement activity between 50 and 400 ppm, although not significantly. Tacon and Kurmaly (1996) reported that wild shrimp contain more astaxanthin (50-80 ppm) than cultured shrimp (10 ppm). Best growth rates have been obtained in *Fija ponicus* fed 200 ppm astaxanthin.

Females in many animal species frequently prefer to mate with the most elaborately ornamented males (Andersson, 1994), but how such displays reveal a male's worth is a contentious issue (Hill, 1999; Olson, and Owens, 1998; Muller *et al.*, 2000, Hudon, 1994). For a signal to honestly reveal an individual's quality, it must be costly to produce (Zahavi, 1975; Grafen, 1990). The expression of many ornamental traits depends on carotenoids, red and yellow pigments that animals cannot synthesize *de novo* and ultimately must obtain through their diet. Carotenoids are antioxidants and immunostimulants (Chew, 1996; Stahl and Sies, 1999) and it has recently been hypothesized that a trade-off exists in carotenoid allocation between maintaining health and ornamentation. Males in better condition should require fewer carotenoids for immune function and could therefore use more of this resource to enhance ornamental display, thereby advertising their superior health (Lozano, 1994; Von Schantz, *et al.*, 1999). However, whether variation in circulating carotenoid levels directly causes variation in both immune function and sexual attractiveness has not been determined in any species.

Dietary differences may actually explain the inter-population variation in both circulating carotenoids and skin pigmentation in different animal groups [*e.g.*, fish in Grether *et al.* (1999); birds in Negro *et al.* (2000); reptiles in Costantini *et al.* (2005b)]. All species possessing the *crtYc* and *crtYd* genes belong to the order Actinomycetales, a group of gram-positive bacteria with a high GC content of their DNA. Surprisingly, in *Streptomyces* also belonging to the Actinomycetales, a lycopene cyclase of the classical CrtY-type has been functionally identified [Gerhard Sandmann (2002)].

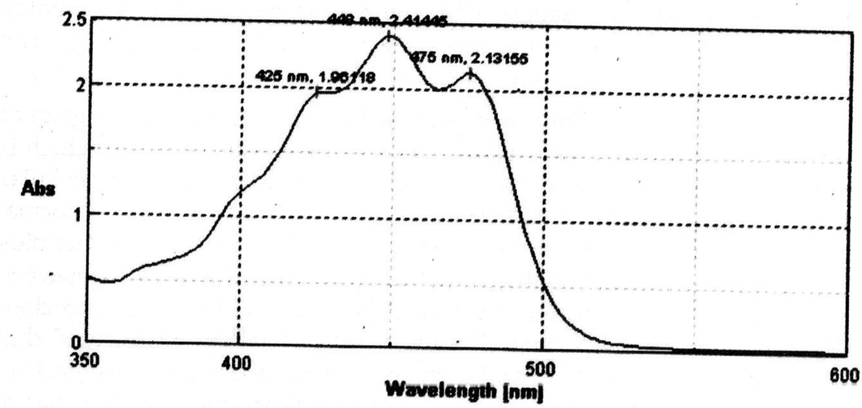

C01-RED SWORD TAIL (CONTROL)

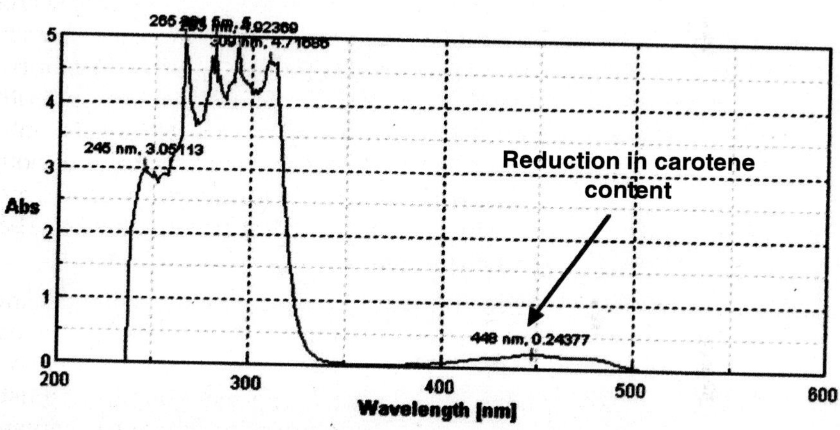

C02-SUNLIGHT EXPOSED (OUT DOOR)

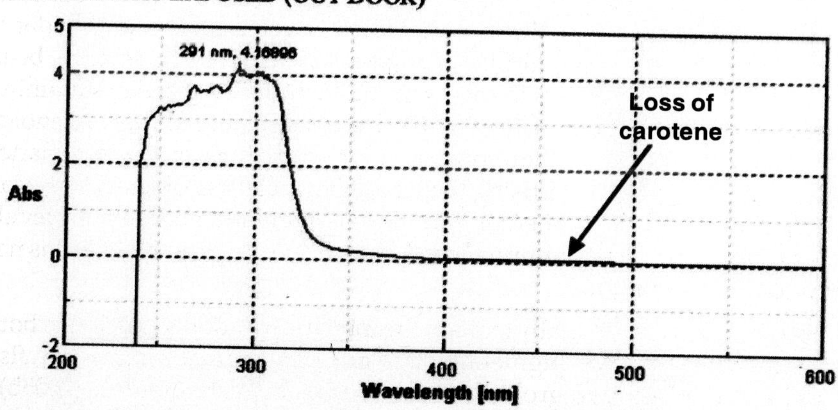

C04-CHLORINE INDOOR

Figure 10.1: U.V. Spectral Analysis Carotenoid Loss During the Stress Given to the Fishes

Figure 10. 2: Retainment of Carotenoid Pigments was Observed 12 Days After Feeding Trails Over

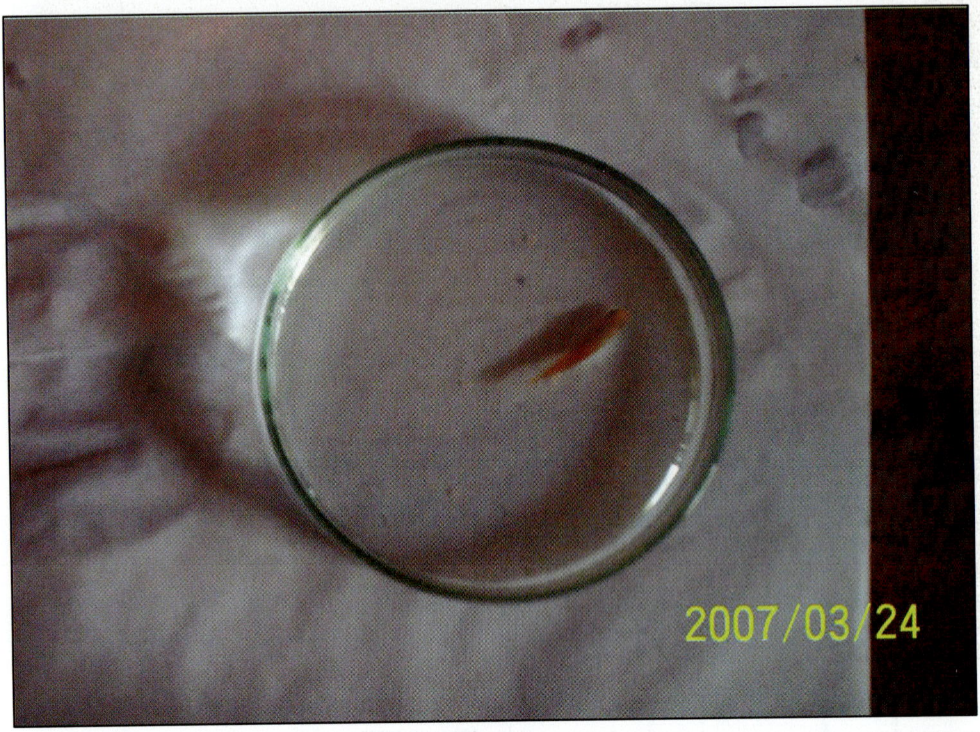

Control

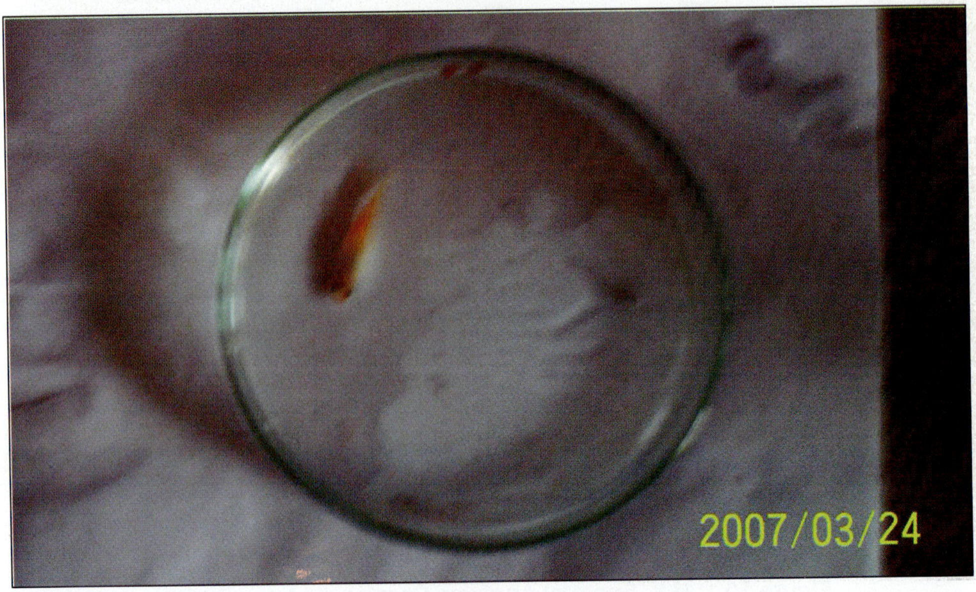

Feed 1

Figure 10.2–*Contd...*

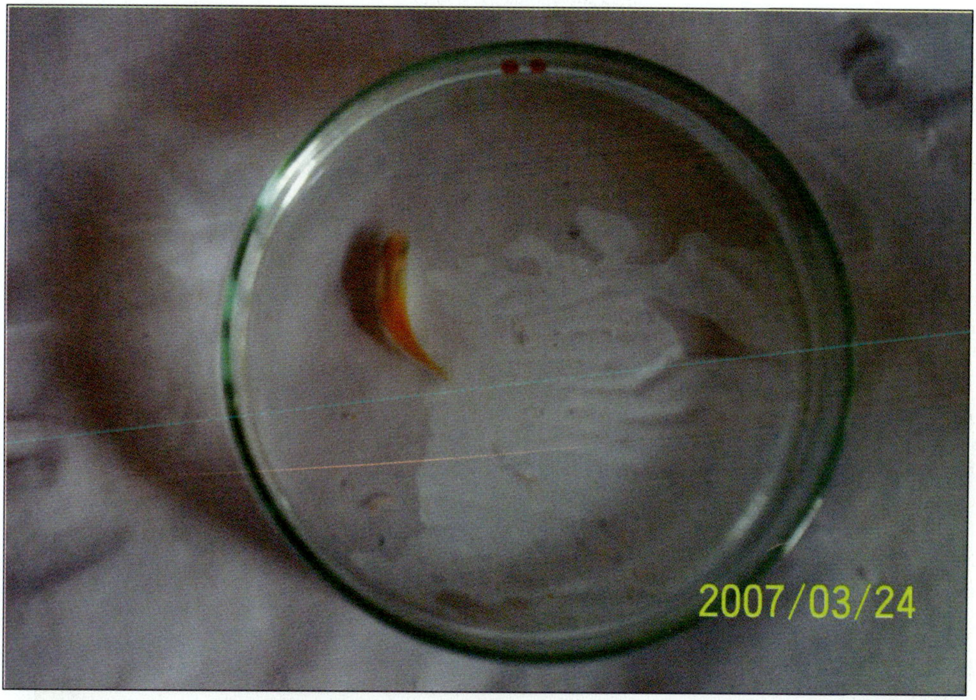

Feed 2

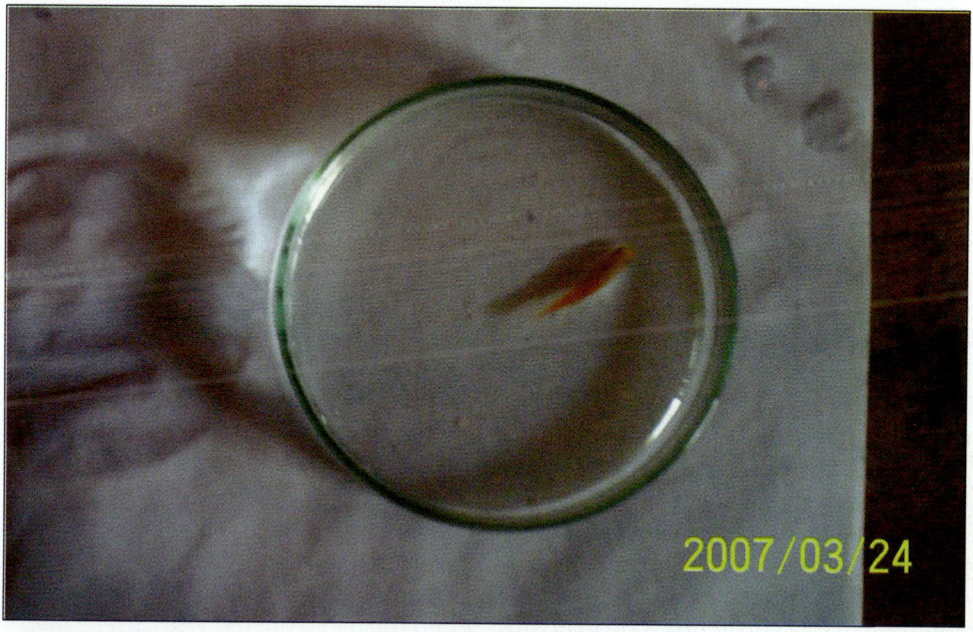

Feed 3

Figure 10.2–*Contd...*

Feed 4

During experiment the fading of fish done on natural sunlight and chlorine exposure and the carotenoid expression were done on UV- Vis Spectral analysis. The characterization of carotenoids are done (Figure 10.1). The complete loss of carotenoids occurred mainly in chlorine exposed fish and partial loss seen in sunlight exposed one. The formulated feeds were prepared by incorporating *Streptomyces* with Basal feed. The fishes that were faded under each type of exposure were given the formulated feed of which has *Streptomyces* incorporation. After a week of trial, the fish sample were collected and quantitative and qualitative analysis of carotenoids were done,where the maximal recovery of fishes of carotenoids were found (Figure 10.2). The recovery of carotenoids was proved through UV spectral analysis (Figure 10.3). The whole experimental set up is based on utilization of *Streptomyces* which has major storage of secondary metabolites particularly we concentrate on carotenoids. These *Streptomyces* which has previous record as antibiotic, antioxidant, anticancer, antiviral properties (Nishino, 1999) has also large synthesis of secondary metabolites. Among the proposed functions of carotenoids in aquaculture have been those of pigmentation, antioxidant functions, cellular protection from photodynamic damage, enhancement of growth and reproductive potentials. The results conclude the fast retaining of lost carotenoids on exposure, when the fish were fed with *Streptomyces* incorporated feed. These main reasons behind this are that the precursor in carotenoid

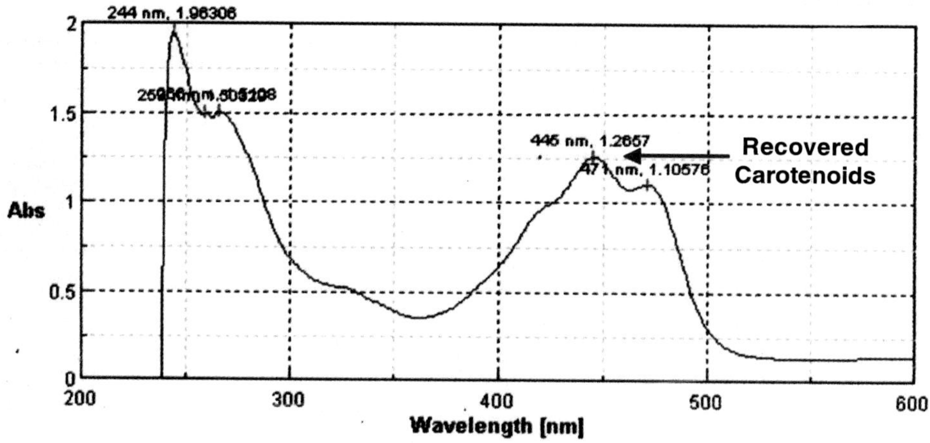

C1- FISH MEAL+STREPTOMYCES (BW) SUNLIGHT EXPOSED

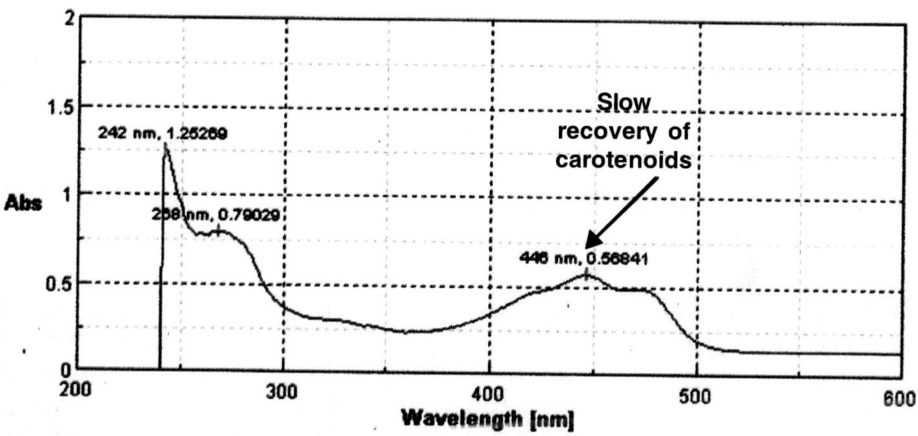

C9-FISH MEAL+STREPTOMYCES (BW)+CHLORINE

Figure 10.3: Recovered Carotenoids of the Fishes by Feeding *Streptomyces* Incoroporated Feed

metabolism is mainly synthesized by the *Streptomyces* spp. Normally, animals cannot synthesize the carotenoids by conversion of one form to another. The conversion of carotenoids of different forms are initiated by enzymes. These precursors are converted to a particular form, which is normally stored and expressed. As far as Red Sword tail (*Xiphophorus helleri*) is concerned the main stored carotenoids include Astaxanthin, Lycopene, β-carotene, that are completely retained when fed with *Streptomyces* incorporated feed.

The Microbial incorporated feed supplemented to fishes are the precursors for the retaiment of colouration or pigmentation. The whole experimental set up is based on utilization of *Streptomyces* which has major storage of secondary metabolites particularly we concentrate on carotenoids. These *Streptomyces* which have previous record as antibiotic, antioxidant, anticancer, antiviral properties have also large synthesis of secondary metabolites. The results conclude the fast retaining of lost colouration, when the fish were fed with *Streptomyces* incorporated feed. The experimental results proves that *Streptomycetes* can be a affective tool in Biotechnology as an Immunostimulant, feed preparation, as disease protectant, against photo oxidative stress, against harmful chemicals and also this can be a least cost effective antibiotic.

Chapter 11
Metazoan Parasites of Fishes

Introduction

The group Metazoa comprises all eukaryotes that consist of many cells functioning in a highly integrated fashion. In protozoans the individual cell has to perform all tasks needed for living, survival and reproduction. Metazoans have developed specialized organs that are fully adapted for different functions. In parasitic animals the systems described in the following section are most important, since the dysfunction of any one would lead to the inevitable death of the parasitic aggressor. Because there are many functional similarities, these main systems will be dealt with here in a comparative manner.Metazoan retains vestiges of their unicellular origin, as shown by their development from unicellular "eggs", some of which may develop even if they are not fertilized. They also have the ability to reconstruct their whole bodies from a single cell, as the sponges easily do, as well as the fertilized oocytes of vertebrates.

Musculature

Among metazoan parasites a variety of different types of muscles has developed. In general either smooth or striated muscle fibers occur but intermediate forms can also be found. Muscles of the smooth type are usually observed in parasites with a relatively soft body cover (e.g., platy helminthes, acanthocephalans), whereas parasites with a relatively stiff exoskeleton have strong, striated muscle bundles and additional smooth filaments along the intestine, etc. There are many basic differences in shape, arrangement and fine structure of muscles within the different systematic taxa (Platyhelminthes, Nematodes, Acanthocephalan, Pentastomida, Insects, Ticks, Crustacea, Leeches, *Hirudo medicinalis*).

Integument

The parasite-host interface is the place where nutrients are taken up and is the site of the attacks on the host's defense system. In the parasites considered here, two main types of body cover exist: An acellular filamentous cuticle (excreted by an underlying cellularly organized hypodermis) is found, for example, in nematodes, pentastomids and arthropods (ticks, mites, insects, crustaceans). A syncytial cytoplasmic tegument where giant nuclei or nuclear fragments may occur covers the surface of larval monogeneans, digeneans, cestodes and acanthocephalans, whereas the tegument (*i.e.* neodermis = new skin) of adult parasitic platyhelminths lacks nuclei. The surface of both types of outer body cover (cuticle or tegument) may be lined by more or less thick layer(s) of carbohydrates, mucopolysaccharides, or even membranous material depending on the species and the site of parasitism. This surface coat that fulfils its tasks in defense against environmental influences while covering "normal" and parasitic cells is thought to be the ancestor of all cuticular systems of the whole animal and plant world. Apparently during evolution the glycocalyx, while situated between body- and/or cell protrusions (e.g. microvilli, protuberances) became fortified by enclosure of fibers of collagen, chitin, cellulose and/or calcium carbonate components etc. Thus, the original protection system received a second function, *i.e.* the preservation of the body shape as a system belonging to the exoskeleton. However, the uptake of nutrients through this increasing outer surface remained possible using a variety of mechanisms and carrier systems. Thus, *e.g.*, schistosomes are able to take in huge amounts of glucose through their surface membranes and nematodes may be decimated by several drugs due to their cuticular uptake, too (chemotherapy, mode of action).

Intestine and Food Uptake

Among the parasitic metazoans two groups lack an intestine in all developmental stages: cestodes and acanthocephalans. In other groups only the larval stages may not develop an intestine (Example: microfilariae of filariae) or the intestine may lose its function during a developmental phase; for example, unsheathed larvae of hookworms and even some motile pupae of insects (Example: mosquitoes) do not feed. In other cases adult worms lose their intestinal functions by partial or total degeneration of the alimentary system (Example: *Dracunculus medinensis* and relatives; *Onchocerca volvulus*). The intestinal tract is blind-ending in monogeneans and digeneans, whereas in other groups an anus is present although it may become closed secondarily (Example: Philometridae, Dracunculidae).

The feeding but intestineless groups (which in general are parasites of the host's alimentary system) are obliged to take up whatever food their structures allow. The smooth-walled tapeworms are able to feed by endocytosis and additional active membrane transport, whereas the latter is probably the only possible method for the stiff-walled acanthocephalans. In general, parasites are adapted to different ways of feeding and to different foods. For detailed information on many groups please refer to the respective headwords.

Almost all groups of animals and plants contain at least a few parasites. Parasites are a group of organisms that may or may not cause illness in fishes, depending on a number of factors. Parasites are a natural occurrence, not contamination. They are as common in fish as insects are in fruits and vegetables. Some parasites are harmful to their hosts; others are not. A parasite that might be harmful if present in large numbers might not be harmful if just a few has parasitized the host. There are different types of hosts. A definitive host is the host harboring the sexual, adult stage of the parasite. An intermediate host is one harboring an asexual or larval stage of the parasite (Margolis, 1970). A parasite vector is an arthropod or other invertebrate that transmits the parasite from one vertebrate host to another. An infection is parasitism by an internal parasite, while an infestation is parasitism by an external one. A facultative parasite is able to live either free or as a parasite, while an obligatory parasite must be parasitic. The term life-cycle refers to the development of a parasite through its various stages. A reservoir host is a host other than the ones normally involved in the life-cycle. Interestingly, we still don't know exactly how intestinal parasites manage to survive without being themselves digested along with the other contents of the intestinal tract. As you can imagine, the inside of the gastrointestinal tract is a pretty inhospitable place to live, as it is full of acids and enzymes designed to digest anything passing through it.Parasites are typically divided into two groups: ectoparasites, which live on the outside of the host (including the gills, mouth, skin and fin surfaces), and endoparasites, which live in the tissues, blood and/or organs (including the gastrointestinal tract). The helminthes – monogeneans, digeneans and nematodes are common parasites of marine fishes (Sindermann, 1990).

Monogenetic flukes are flatworms that are usually found as ectoparasites of fish. They often have a hooked attachment organ and they have a simple, direct life cycle. They are just visible to the naked eye. They often feed on skin and gill tissue, but usually only cause problems if found in large numbers. *Gyrodactylus* lives on the skin, fins and gills of many species of fish.

Digeneans are endoparasitic flukes. They have two suckers, one at each end. They have an indirect life cycle, with fish acting as both intermediate hosts (carrying metacercaria larvae) and final hosts. Adult digeneans are usually found in the gastrointestinal tract; the larvae (which may be encysted) can be found throughout the body and cause problems if they invade such organs as the eyes (as does *Diplostomum*) or heart, in high numbers.

Tapeworms (cestodes) are endoparasites. Adult tapeworms live in the digestive tract. Tapeworms are most common in newly imported or wild-caught fish. They have an indirect life cycle, with fish serving as both intermediate and final hosts. There is a public health concern, as some cestodes can be infective to humans, causing zoonotic infections. *Diphyllobothrium* has a larval stage that can be fatal to fish and infective to humans. Signs of tapeworms may be a swollen abdomen, general unthriftiness, and wasting. Fish may have difficulty swimming. Treatment involves using praziquantel, which will remove tapeworms from the intestinal tract, but will not eliminate those residing within the body cavity.

Acanthocephala, also called spiny or thorny headed worms, are highly specialized for life in the digestive tract of the host. They have spines on the eversible proboscis at the anterior end of the body, by which the worm anchors to the tissues of the digestive tract. The life cycle is indirect, with crustaceans serving as the intermediate host. They can cause damage by puncturing the intestinal tract or blocking the passage of food through the gastrointestinal tract. Acanthocephalans rarely cause problems with captive fish, although wild-caught fish may harbor them when first obtained. Because of the complex life cycle involving intermediate hosts, re-infestation is not likely to occur in the fish tank. After some months, these parasites are likely to die off and disappear. As long as the worms do not perforate the intestines, the damage is self-limiting.

Some crustaceans are parasitic to fish. The fish louse, *Argulus*, and the gill maggot, *Ergasilus*, are both crustaceans that parasitize fish. The fish louse attaches itself to the skin and fins by means of twin suckers, and it feeds on blood of the host. The intense irritation may cause the fish to scratch against rocks and heavily infested fish may even jump out of the water. Red lesions occur where the lice have attached, and this opens the skin up to secondary bacterial and fungal infections.

Roundworms may also be found in fishes; also called nematodes, they have complex indirect life cycles, and fish can serve as both intermediate and final hosts. The developmental cycle of nematodes comprises egg, four larval stages and adults. Nematodes can cause problems to fish, and also to humans who become infected by eating raw or poorly prepared fish. Roundworms are red-brown in color and vary in size. In severe infections, they may be seen protruding from the vent. Nematodes can be found in the digestive system, swim bladder and body cavity. The nematode infection causes anaemia, proliferation of cells in internal organs, muscles and skin or even the formation of the tumour (van Dujan, 1956). The study of nematode parasitic fauna of South Asia was done by Senadhura (1967), Bashirullah (1973), Natarajan and James (1977) and Ahamed (1981).

Roundworms called nematodes are the most common parasite found in marine fishes. Some people call these nematodes herring worms or cod worms. Actually, several different species exist and it is hard to distinguish between them. All are in the family Anisakidae and are anisakid nematodes. Wild fish normally carry variety of parasites, and always show negative effects when the infection is extremely high. Parasites form an integral part of an ecosystem, especially in the water bodies with poor environmental condition that can exert a strong regulatory effect on population. Environmental stress can depress host immunity. Fish become more susceptible to these infections, which can be more severe or even fatal. Parasites appear to be an important stressor affecting fish population and sometimes a major cause for the mass mortality of juvenile fish (Kuperman and Matey, 2001).Fishes serve as potential host for many different species of parasites. Small numbers of parasites are common and probably do little harm if the number exceeds beyond a particular level, can cause adverse effects to host fishes. All parasites have tremendous reproductive potential and under ideal conditions they can overwhelm the fish.The last few decades have witnessed extensive research on parasite and disease problems of fishes. In addition, infestation can cause heavy economic loses through mass mortalities, parasites and disease affect growth and reproduction of fish. The fish becomes vulnerable to predation and susceptible to infections by pathogens (Hoffman, 1974).

Mass mortalities of fish both in culture and capture fisheries owing to parasitic invasions are now on the increase and being frequently reported from many south east Asian countries (Kabata, 1985; Arthur, 1987). Microbial pathogens and metazoan parasites often produce considerable damage to host animals and cause major diseases in fish (Sindermann, 1990). Severe infections by protozoans, digeneans and crustacean parasites are serious concern in tropical region because the development of parasites was temperature dependent (Sarig, 1976). Parasites in fish represent an ecological co-existence of two organisms of different types in close contact; on the other, which is the host. The living togetherness is beneficial to the parasite, but is detrimental to the fish and can leads to its mortality (Scharperclaus, 1991).

Monogenean Parasites

Monogenea (Gill and Skin Flukes)

Monogeneans (flukes) are a group of parasites best described as flatworms. Monogeneans are commonly found on the gills, skin or fins of fishes and lower aquatic invertebrates. A few may invade the rectal cavity, ureter, body cavity and even the blood vascular system. There are more than 100 families of monogeneans found on fishes of the world, in fresh and salt water, and at a variety of temperatures. Most monogeneans are browsers, moving about the body surface and feeding on dermal (skin) mucus and gill debris. Monogeneans have a series of hooks that enable them to attach while feeding. Most species are host- and site-specific, requiring only one host to complete an entire life cycle. In fact, some adult monogeneans will remain permanently attached to a single site on the host. Morbidity and mortality epidemics in cultured fish caused by excessive parasite loads are associated with crowding, inadequate sanitation and deterioration of water quality. Although monogeneans are commonly found on wild fish, they are rarely a direct cause of disease or death in free-ranging populations.

Many monogenetic trematodes have been transferred with their hosts (Hoffman 1970; Bauer and Hoffman 1976). They are easily transferred because no intermediate host is necessary. Some destroy their hosts, but most are specific for one species of fish or for two or more closely related species. Nevertheless, transfers to new locations are undesirable (Hoffman and Schubert 1984). Because of high host specificity, most monogenea do not spread to other fish species when they are introduced with their hosts into a new habitat. But they may become more dangerous to their hosts in the new surroundings (Hoffman and Schubert 1984). Monogenea of food fishes have also been transferred to many countries, mostly in the temperate zone–Asia, Europe, and North America–where fish culture has increased. Such transfers were reviewed by Hoffman (1970) and Bauer and Hoffman (1976). Monogenea of ornamental fishes have been transferred throughout the world. They usually do not attack other fishes, but transfer should be avoided because the parasites of many tropical fishes have not been thoroughly studied (Hoffman and Schubert 1984). Skin and gill flukes are common fish parasites and in small numbers probably cause little harm. In large numbers both skin and gill flukes represent a serious threat to fish health. Flukes can cause lesions and tissue damage as well as producing side affects such as hyperplasia of both skin and gill epithelium and creating entry sites for secondary infections. The connection between flukes and bacterial fish diseases such as ulcers is well established. Although *Gyrodactylus* and *Dactylogyrus* are classed as skin and gill flukes respectively, they are not exclusive to these areas and flukes of either persuasion may be found at either site.

Flukes are small worm-like parasites up to 2mm in length. Technically they are monogenean trematodes, which describe their biological classification (trematodes) and the fact that they only need one host to complete their life cycle (monogenean) whereas many parasitic trematodes need two or more hosts and are thus digeneans. Flukes have an array of hooks at one end of their body which they use to anchor themselves to the fish and to move around in a jerky, caterpillar-like fashion. These hooks can be seen clearly in the photos incorporated in later pages. The gyrodactylids differ from dactylogyrids in that *Gyrodactylus* has a v-shaped head, no eyespots and is viviparous (live-bearing), whereas *Dactylogyrus* has a scalloped head, distinct eyespots and is oviparous (egg-laying). It is generally agreed that heavy fluke infestations affecting more than one fish are usually caused by overcrowding, poor water quality or polluted water - that is water with a high organic content and stress. Under ideal circumstances, these parasites can rapidly multiply. With this head start - three generations in one - the doubling time for *Gyrodactylus* can be as little as 24-hours. It is much the same with dactylogyrids, which, in warm water and ideal conditions can reproduce very rapidly and quickly overwhelm any sick or stressed fish.

The fish response to the irritation caused by flukes is similar to any other irritant, whether it is parasite or chemical. Initially there will be rubbing and flashing. As the disease reaches a more advanced stage the fish will become lethargic - which is the point when some people think the problem has gone away. At a very advanced stage the fish will isolate itself and spend long periods laying on the bottom with its fins clamped to its body. Other signs may be skin cloudiness resulting from excess mucus production, skin hyperplasia, or focal reddening. A definite diagnosis can only be made via a skin scrape or gill biopsy. In large numbers, flukes will kill fish either directly, or indirectly through secondary infections. Right from the start let me say that in many cases flukes are hard to treat and complete eradication is virtually impossible. For individuals affected with flukes, consecutive salt baths over 2-3 days can be useful. Using a quaternary ammonium compound in conjunction with, but not at the same time as, salt baths can be useful, in both clearing the skin and gills of excess mucus and debris as well as soothing damaged tissues.

Malachite and formalin are often affective, although my experience is that the stronger dose needs to be used. It most cases, where the flukes are a nuisance rather than a real danger the old M&F will probably suffice. For all out war, the best treatments are organophosphates such as masoten. Another promising treatment is high dose, bath treatments with chloramine-T. A few trials I have carried out show that in most cases chloramine-T is quite effective, but it is early days to draw any firm conclusions. The draw-back of bath treatments is that it does mean handling the fish and having a large treatment tank. Due to the complications of using chloramine-T in a typical pond, it is unlikely to be so successful as a pond treatment. Bath treatments also allow the life phases which exist in the ponds (eggs and embryos) to continue to survive. Your freshly "cleaned" individual fish will be returned to a pond with emerging stages in it.

Hoffman (1970) listed nine species that are known to have been transferred: *Anacanthorus anacanthorus, Dactylogyrus anchoratus, D. vastator, D. wegeneri, Gyrodactylus bullatarudis, G. elegans, Urocleidoides reticulatus, Urocleidus crescentis* and *U. orthus.*

Anacanthorus anacanthorus, A. brazielensis and *A. neotropicalis* have been on the gills of the redbreasted piranha (*Serrasalmus nattereri*), which was introduced into the United States from South America (Mizelle and Price, 1965).

Anacanthorus brevis (Mizelle and Kritsky 1969a). This species was taken from the gills of *Brycon melanopterus.* It was transferred from Brazil to the United States (Hoffman and Schubert, 1984).

Archidiplectanum archidiplectanum (Mizelle and Kritsky, 1969b). This species was taken from the gills of *Gnathonemus petersi*. It was transferred from western Africa to the United States (Hoffman and Schubert, 1984).

Cichlidogyrus spp.This parasite was on the gills of Tilapia species and was transferred from Africa to United States in 1960 on *T. mossambica* (Hoffman, 1970).

Cleidodiscus amazonensis, C. piranhus and *C. serrasalmus.* This parasite was on the gills of the redbreasted piranha that were transferred from South America to United States (Mizele and Price, 1965).

Dactylogyrus anchoratus Dujardin 1845; Wegener 1857. This parasite was on goldfishes. It probably originated in Asia and was transferred to the United States (Price and Mizelle 1964). *Dactylogyrus extensus* Mueller and Van Cleave 1932. This parasite was on the gills of common carp. It has been transferred from Europe to the United States and Israel (Paperna 1964). *Dactylogyrus vastator* Nybelin 1924. This species was on the gills of goldfishes. It probably originated in Asia (Price and Mizelle 1964). *Dactylogyrus wegeneri* Kulwiec 1927. This parasite was on goldfishes and was probably transferred from Europe or Asia (Price and Mizelle 1964). *Dactylogyrus minutus.* This parasite of the common carp is well known in Europe and in Central Asia and has been found in the United States (Hoffman and Schubert 1984). *Dactylogyrus baueri* and *D. formosus.* Both parasites are on the gills of goldfishes and were transferred from Japan to the United States (Rogers 1967).

Gyrodactylus cyprini (Diarova 1964). This species is a parasite of the common carp and is well known in Europe and Central Asia. It occurs in the United States (Rogers 1968). *Gyrodactylus elegans.* This species is a parasite of the goldfish, probably originated in Asia, and then transferred to Europe and to the United States (Malmberg 1962).

Heteronocleidus gracilis was on the gills of *Colisa labiosa* (Mizelle and Kritsky 1969b) and probably transferred from India to California (Hoffman and Schubert 1984). There is no record of it in Florida.

Longihaptor longihaptor was on gills of *Cichla ocellaris* (Mizelle and Kritsky 1969a) and was transferred from Brazil to the United States (Hoffman and Schubert 1984).

Pseudacolpenteron pavlovskyi is a parasite of the common carp and was transferred to Israel and the United States (Hoffman and Schubert 1984).

Trianchoratus acleithrium Price and Berry 1966. This parasite was taken from the gills of *Helostoma rudolfi* and was imported from Malaysia into the United States (Mizelle and Kritsky 1969a).

Urocleidoides amazonensis (Mizelle and Kritsky 1969a) was on the gills of *Phractocephalus hemiolopterus* and was transferred from Brazil to the United States (Hoffman and Schubert 1984). *Urocleidoides catus* (Mizelle and Kritsky 1969a). This parasite was on the gills of *Phractocephalus hemiolopterus* and was transferred from Brazil to the United States (Hoffman and Schubert 1984). *Urocleidoides megorchis* (Mizelle and Kritsky 1969a) attacks the gills of *Sorubim lima* and was transferred from South America to the United States (Hoffman and Schubert 1984). *Urocleidoides reticulatus* Mizelle and Price 1964. This parasite was on a guppy (*Lebistes reticulatus*) and was transferred from Trinidad to California (Mizelle and Price 1965). *Urocleidoides robustus*. This parasite was on the gills of *Symphysodon discus* (Mizelle and Kritsky 1969a) and transferred from Brazil to the United States. *Urocleidus crescentis* and *U. orthus*. These parasites were on the gills of redbreasted piranha and were transferred from South America to the United States (Mizelle and Price 1965).

Trematoda, Digenea

Manter (1963) believed that the intestinal trematode, *Crepidostomum farionis* of salmonids was transferred from Europe to North America in trout and became established (Hoffman and Schubert 1984). Because this pathogen is restricted to salmonids, it is not established in Florida.

Bolbophorus confusus. This Eurasian strigeid trematode probably came to the United States in a stray pelican, its natural final host (Hoffman 1970). Hoffman and Schubert (1984) found it in fathead minnows (*Pimephales promelas*) that were shipped from South Dakota to Arkansas. This species has not been seen in Arkansas since its reported introduction and may not have become established. No record of this species in Florida could be found, but fathead minnows are frequently shipped into Florida from Arkansas as bait, and the trematode could have been introduced into the state. *Cryptocotyle lingua*. This species has been in coastal marine fishes in the United States. It was probably carried by European snails on ships from Europe to the East Coast of the United States about 100 years ago (Sindermann and Farrin 1962).

Acathocephala

Adult members of the Acanthocephala are highly specialized heterosexual, intestinal parasites that take up nutrition parenterally since they have no intestine. Vertebrates are used as final (definitive) hosts, arthropods as intermediate hosts. The body consists of two major parts, the praesoma and the metasoma. The praesoma comprises the proboscis, armed with a set of specific hooks (Figure 11.1), a more or less pronounced neck, the proboscis receptacle, and the two lemnisci (Figure 11.4, Figure 11.6), which are cylindrical appendages of the praesomal tegument. The tube-shaped metasoma (= trunk) is bounded by a solid body wall, enclosing the pseudocoel, which is mainly filled with male or female sexual organs.

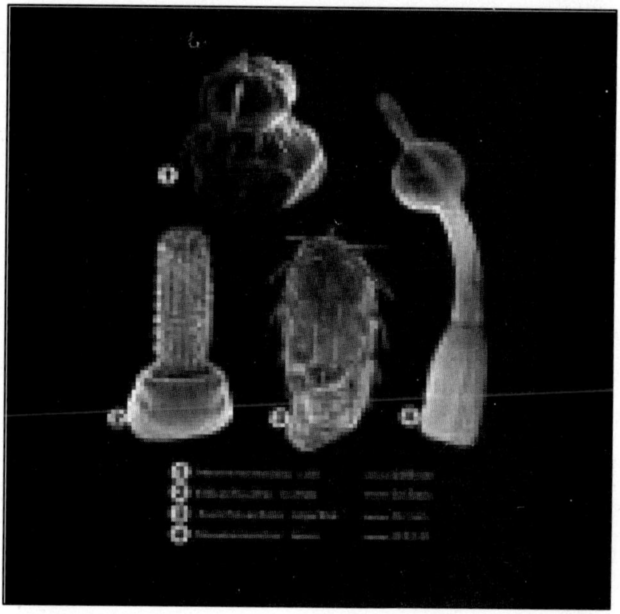

Figure 11.1: Scanning Electron Micrographs of *Acanthocephalan praesomae*

Additional morphological features as well as biological characteristics determine the affiliation to one of the three classes: Archiacanthocephala, Palaeacanthocephala, and Eoacanthocephala.

System

Class 1: Archiacanthocephala Meyer 1931: species have terrestrial life cycles; mammals (birds) are final hosts, and insects (millipedes) intermediate hosts; in addition, paratenic hosts are often involved; main longitudinal vessels of the lacunar system run dorsally and ventrally; usually eight uninucleate cement glands; few tegumental nuclei; ligament sacs inside the pseudocoel, also in adult worms (Figure 11.16). The important orders are:

Order: Apororhynchida

Order: Gigantorhynchida

Order: Moniliformida

Family: Moniliformidae

Genus: Moniliformis

Order: Oligacanthorhynchida

Family: Oligacanthorhynchidae

Genus: Macracanthorhynchus

Genus: Prosthenorchis

Class 2: Palaeacanthocephala Meyer 1931: in general, aquatic life cycles; fish (and also waterbirds, seals) are final hosts, crustaceans intermediate hosts; main vessels of the lacunar system run laterally; two to eight multinucleate cement glands; numerous tegumental nuclei; ligament sacs ruptured in adult worms.

Order: Echinorhynchida

Family: Echinorhynchidae

Genus: Acanthocephalus

Genus: Echinorhynchus

Family: Pomphorhynchidae

Genus: Pomphorhynchus

Order: Polymorphida

Family: Centrorhynchidae

Family: Plagiorhynchidae

Family: Polymorphidae

Genus: Corynosoma

Genus: Filicollis

Genus: Polymorphus

Class 3: Eoacanthocephala Van Cleave, 1936: aquatic life cycles; fish (also reptiles, amphibians) are final hosts, and small crustaceans (mostly Ostracoda) intermediate hosts; main vessels of the lacunar system run dorsally and ventrally, only a single giant, uninucleate cement gland; tegument with giant nuclei; ligament sacs generally persistent in adults.

Order: Gyracanthocephala

Order: Neoechinorhynchida

Family: Neoechinorhynchidae

Genus: Neoechinorhynchus

Family: Tenuisentidae

Genus: Paratenuisentis

Life Cycle

See Figures 11.2 and 11.3.

Attachment

Concerning the attachment to the host's intestinal wall, two groups can be distinguished: perforating and non-perforating acanthocephalans.

Non-Perforating Acanthocephalans

Generally, acanthocephalans that have a short neck do not penetrate deeply into the host's intestinal wall with their praesoma, but display some mode of shallow

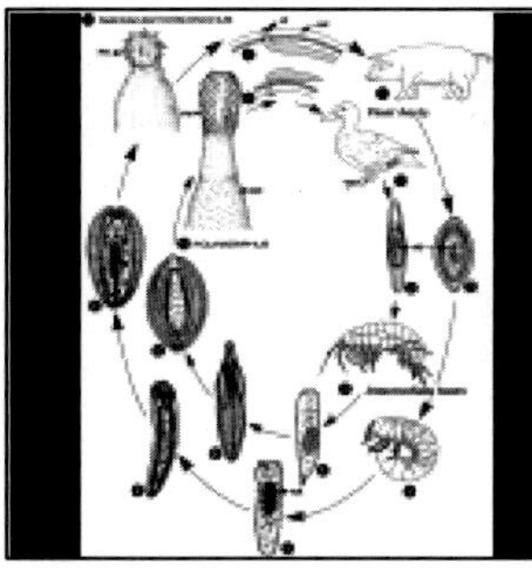

Figure 11. 2. Life Cycle of Common Acanthocephalan Species (Table 11.1)

A *Macracanthorhynchus hirudinaceus; B Polymorphus minutus.* 1: The adults live in the intestine of their final hosts, being attached by their hooked proboscis. The penetration of the intestinal wall leads to inflamed protrusions (IP) appearing along the outer side. 2: After copulation the adult females excrete eggs for several months (patent period). These eggs are passed fully embryonated (*i.e.* they contain the hooked acanthor larva) with the feces of the host. 3-6 Intermediate hosts (*Gammarus* spp. or beetle larvae) become infected by ingesting infective eggs. Inside the intestine the acanthor is released from the egg (4; RA), enters the body cavity and is transformed into an acanthella larva (5). The latter matures within 60-95 days (in *M. hirudinaceus*) and is described as an infective larva (cystacanth). Infection of the final hosts occurs when they swallow infected intermediate hosts. The young worms reach sexual maturity within 60-90 days in *M. hirudinaceus* (after 20 days in Polymorphus minutus) and start egg production (= end of prepatent period). AC, acanthor; BH, body hooks; IP, inflamed protrusion of IW; IW, intestinal wall; PH, proboscis hooks; RA, released acanthor

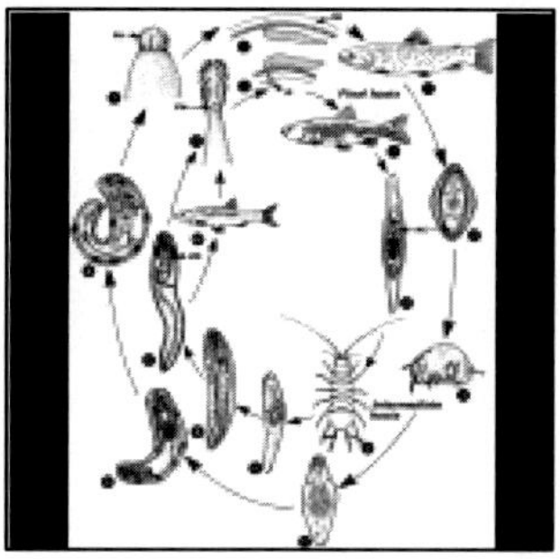

Figure 11.3: Life Cycle of Two Common Acanthocephalan Species Parasitizing Fish

A *Neoechinorhynchus rutili*; B *Acanthocephalus anguillae.* 1: Adults are attached to the intestinal wall of their final hosts, trouts (A) or chubs and other fish (B). 2: Fully embryonated eggs are passed with host's feces. 3-6 Intermediate hosts (A Asellus aquaticus, B ostracod crustaceans) are infected by uptake of eggs. Inside their intestine the acanthor larva (4) is released from its eggshell, enters the body cavity and becomes transformed into the acanthella larva (5). This stage differentiates to the infective larva without encystation in about 30-60 days (6) depending on outer conditions. Final hosts are infected by swallowing intermediate hosts. In *A. anguillae* a paratenic host may also become involved. When bleaks and some other fish ingest intermediate hosts (Asellus aquaticus), the infective larva enters the fish viscera, but there is no further development, but quick depeneration. Infected paratenic hosts may be a second source of infection for the final host. Neoechinorhynchus rutili and A. anguillae reach sexual maturity in about 20-30 or 40-60 days respectively (prepatent period). Adults live only for about 2-3 months (patent period). AC, acanthor; IP, inflamed protrusion of IW; IW, intestinal wall; LM, lemniscus; PH, proboscis hooks

Table 11.1: Some Important Species of the Acanthocephala

Class/Species	*Size (Adults, mm; Eggs, E, µm)*	*Final Host*	*Intermediate Host*	*Paratenic Host*	*Geographic Distribution*
Archiacanthocephala					
Moniliformis moniliformis	m 30–45 f 140–270 E 90–125 × 50–62	*Rattus* spp., other rodents, occasionally humans, monkeys etc.	Cockroaches	–	Worlwide
Macracanthorhynchum hirudinaceus	m 50–90 f 200–650 E 90–100 × 50–56	Pigs, occasionally humans, dogs, etc.	Beetles (larvae)	–	Worlwide
M. ingens	m 130–150 f 180–300 E 96–106 × 51–54	Raccoons, other mammals	Beetles	Amphibia, reptilia	North America
Prosthenorchis etegans	m 20–40 f 30–55 E 60–65 × 41–43 (78–81 × 49–53)	Monkeys, other mammals	Cockroaches, beetles	–	South America, domestic cycle worldwide
Palaeacanthocephala					
Acanhocephalus anguiliae	m 5–7 f 10–35 E 100–125 × 12–14	Chub, barbel	*Asellus aquaticus*	Small cyprinid fish*	Europe
A. ranae	m 5–12 f 20–60 E 100–130 × 13–16	Amphibia	*Asellus aquaticus*	–	Holarctic
Echinorhynchus truttae	m 8–11 f 15–20 E 100–110 × 23–26	Salmonid fish	*Gammarus* spp.	–	Europe
Corynosoma semerme	m 3–5 f 3–5 E 79–100 × 16–29	Seals, birds, occasionally dogs, etc.	*Pomoporeia affinis* (Amphipoda)	Varoius fish	Holarctic

Contd...

Table 11.1–*Contd...*

Class/Species	*Size (Adults, mm; Eggs, E, µm)*	*Final Host*	*Intermediate Host*	*Paratenic Host*	*Geographic Distribution*
Pomphorhynchus laevis	m 6–16 f 10–30 E 110–121 × 12–19	Chub, barbel, trout	Gammaridae	Small cyprinidae and other fish*	Palaearctic
Filicollis anaris	m 6–8 f 10–25 E 75–84 × 27–31	Ducks, other water birds	*Asellus aquaticus*	–	Palaearctic
Polymorphus minutus	m 2–3 f 6–10 E 100 × 11–12	Ducks, other water birds	*Gammarus* spp.	–	Holarctic
Eoacanthocephala					
Neoechinorhynchus cylindracus	m 4.5–8.5 f 10–15 E 51–61 × 17–28	Predatory fish (bass, etc.)	*Cypria globula* (Ostracoda)	Small fish	North America (bluegills, etc.)*
N. emydis	m 6–15 f 10–22 E 20–25 × 20–22	Turtles	*Cypria globula* (Ostracoda)	Water snails	North America
N. rutili	m 2–6 f 5–10 E 26–27 × 14–17	Salmonid fish and other fish	Ostracoda	–	Holarctic
Paratenuisentis ambiguus	m 2.5–8 f 8–14 E 62–72 × 26–31 (fixed in formalin; 48–58 × 20–28)	Eels	*Gammarus tigrinus*	–	North America, Europe (recently introduced)

m: Male: f: Female; E: Egg

* This host category is not yet sufficiently investigated. Thus, it remains doubtful whether these species are true paratenic hosts.

attachment, *i.e.* they do not create lesions reaching as deep as into the muscular layers of the intestinal wall (Figures 11.4, 11.5 and 11.6), Acanthocephalan Infections/ Figure 11. 8). Accordingly, often even the posterior half of the proboscis does not become surrounded by host tissue (Figure 11.6). Layers of connective tissue within the hosts' intestinal wall often appear to function as penetration obstacles. This might be the stratum compactum in salmonids retaining *Echinorhynchus truttae* in superficial positions or a collagen layer interiadly lining the intestinal mucosa of perch (*Perca fluviatilis*) affecting the mode of attachment of *Acanthocephalus lucii* (Figure 11.5). On the other hand, the tipped proboscis hooks seem to use the collagen layers as suitable substrates of anchorage (Figure 11.6). In Figure 11.6 a necrotic tissue with a slight infiltration of granulocytes and haemorrhagic involvement, typical of the attachment site of *Acanthocephalus lucii* and other non-perforating species, is shown (Figure 11.4). When non-perforating species were experimentally inoculated into small specimens of fish not comprising penetration obstacles in their gut wall, three non-perforating species did not try or succeed in perforating either. And accordingly such species usually cannot be found in extraintestinal locations like perforating species. *Paratenuisentis ambiguus* and *P. lucii*, both non-perforators, do not possess colagenolytic proteinases useful in chemical support of penetration activity. So paratenic hosts do not occur in the life cycles of non-perforating acanthocephalans, but postcyclic transmission of intraintestinal worms like *Neoechinorhynchus rutili* in sticklebacks to predatory brown trout seems to be very common.

Such species either continuously or occasionally change their point of attachment, exposing them to the posteriadly directed intestinal peristalsis. In infrapopulations of *Echinorhynchus truttae* in brown trout all specimens have arrived at the posterior end of the small intestine by the time the worms have matured. As has been shown for *Neoechinorhynchus cylindratus*, a species that is potentially perforating, infrapopulations with high worm densities lead to enhancement of change of the point of attachment and consequently to greater posterior shift. An interesting feature can be observed in other neoechinorhynchids. Although they occupy superficial positions, they do not seem to migrate or become shifted after an initial period of establishment, due to a firm capsule of collagen fibres enclosing their small, roundish proboscis. A negative point of the proboscis remains in the intestinal wall after deattaching a worm using forceps (Acanthocephalan Infections/Figure 11.5). Not unlikely, this massive collagen formation is provoked by the excretion of proline (amino acids) or other substances by the praesoma.A typical non-perforating species is the archiacanthocephalan *Moniliformis moniliformis* displaying a deep proboscis cavity and shallow attachment (Acanthocephalan Infections/Figure 11.8).

Perforating Acanthocephalans

Many Acanthocephalans possess a long neck which may comprise a bulbus such as the *Pomphorhynchus* species (Palaeacanthocephala) with an inflated neck region (praesoma) (Figure 11. 1). In the eoacanthocephalan *Eocollis arcanus* it is the anterior part of the metasoma which forms a bulb. In both cases the bulbus functions as a dowel enabling the worm to occupy a permanent point of attachment at a specific site. The deep and quick perforation of the intestinal wall may be supported by a

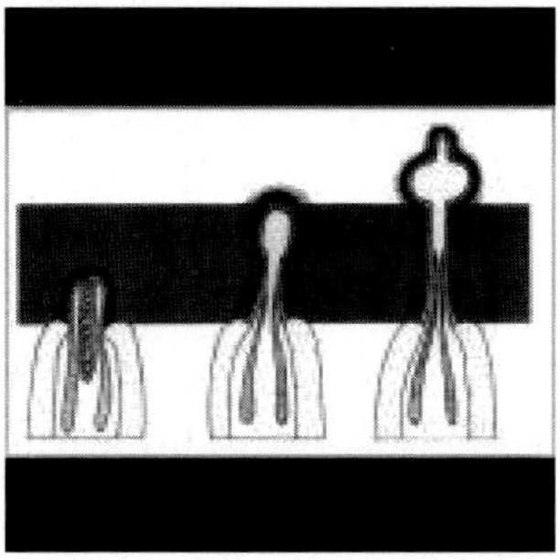

Figure 11.4 A–C: Schematic Drawings of the Acanthocephalan Praesoma and the Corresponding Mode of Attachment

Black area: tissue necrosis, hatched area: tissue neoplasia. A Acanthocephalan with a short neck, and lemniscis branching away from the posterior end of the praesomal tegument. A deep proboscis cavity is formed. Necrotic host tissue (black area) can be found all around the parasite's proboscis. B Perforating acanthocephalan with a long neck, and lemnisci branching away from the praesomal tegument at the mid neck. In the chronic mature stage of infection the proboscis becomes deeply embedded in the intestinal wall and is usually kept fully evaginated. Necrotic host tissue in mainly confined to the proximity of the proboscis. Tissue neoplasia is pronounced also proliferating into the peritoneal cavity. C Perforating acanthocephalan with a long neck, a bulbus and lemnisci branching away from the praesomal tegument at the mid neck. In the chronic (mature) stage of infection the parasite is "doweled" in the intestinal wall with the bulbus. The proboscis is usually kept fully evaginated. Necrosis is mainly confined to the tissue close to those parts of the worm that project into or border the peritoneal cavity. Tissue neoplasia especially in the peritoneal cavity is conspicuous.

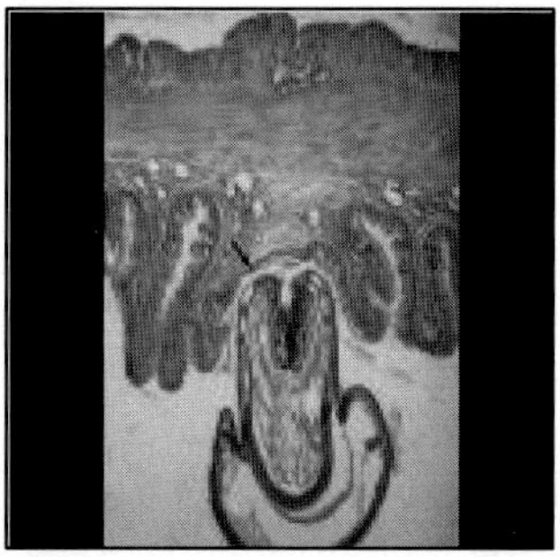

Figure 11.5: Light microscopical micrograph of a longitudinally sectioned (paraffin section) anterior body of *Acanthocephalus lucii* attached to the intestinal wall of a river perch (Perca fluviatilis). Only the intestinal mucosa has been ruptured. A thin layer of collagen fibres (arrow) interiadly lining this epithelium was not perforated. Also note the proboscis cavity at the anterior tip of the worm showing the normal condition of the proboscis of this non-perforating species. × 20.

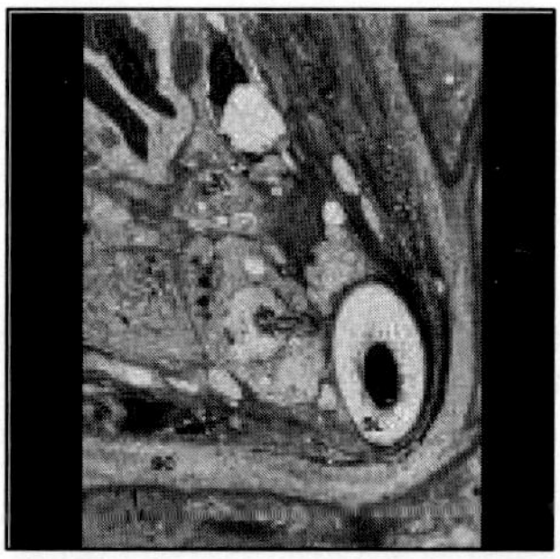

Figure 11. 6. Transmission electron micrograph of a transversally sectioned proboscis hook of *Acanthocephalus lucii* surrounded by necrotic and inflamed intestinal tissue of Perca fluviatilis. The hook: note the perforations in its striped layer (SL); the connective tissue of the hook does not reach into the tip sectioned here. It has punctually perforated the layer of subepithelial connective tissue (SC) that can be seen in Figure 11.5 at a lower magnification. But the proboscis in toto did not perforate this layer (LA: lamina of fine amorphous material lining the mucosal epithelium). The proboscis (not seen) is either in front or behind the plane of the section. × 2,000

proteolytic enzyme as shown for *Pomphorhynchus laevis* which excretes a trypsin-like proteinase into the culture medium. It has a collagenolytic activity and the molecular mass differs slightly among infectious larvae and adult worms removed from fish. The long-necked species *Acanthocephalus anguillae* does not display such abilities in lysing collagen and accordingly the collagenic stratum compactum of salmonid fishes retains most worms in rather superficial connections with the intestinal wall. In experimental infections in adult rainbow trout the worms take about 60 days to perforate the stratum compactum, in juveniles of the same salmonid host it occurs around 20–30 d. p. i. And in the long run, only those worms which succeed in penetrating seem to survive for several months in this host, while the others probably do not have the potential to withstand the intestinal drift. As shown for *Pomphorhynchus laevis*, typical perforating species do not change their site of attachment and thus do not become backwards shifted. In natural populations of fish hosts, species like *Pomphorhynchus laevis, Eocollis arcanus* or *Acanthocephalus anguillae* are not only found in positions with a praesoma deeply inbedded inside the intestinal wall, but also partly lying in toto inside the peritoneal cavity or viscera especially in small specimens or species. Obviously, in such hosts the intestinal wall or the collagen layers within it are not strong enough to withstand the worms' penetrating activity. In juveniles of goldfish experimentally infected with *A. anguillae*, the first worms started projecting into the peritoneal cavity with parts of their praesoma up from about 10 d. p. i., worms of about 20–30 d. p. i. were mostly found in various positions like lying with parts of their bodies inside one intestinal loop and projecting into another with the proboscis or anterior body (Acanthocephalan Infections/Figure 11.5). In contrast, at 50 d. p. i. all worms recovered had taken intraperitoneal positions and most of them were already degenerating. Due to this quick death of the worms in extraintestinal positions, one may conclude that they did not leave the intestine "by purpose" but slipped into the peritoneal cavity in toto by lack of penetration obstacles or other features. Thus, the small fishes that became infected in these experiments should not be called paratenic hosts. However, true paratenic hosts exist in the life cycles of certain perforating acanthocephalans – *Oncicola pomafostomi*, for instance, is parasitic in the intestine of felidae and canidae in Southeast Asia and Australia while it has been found under the skin of 19 species of passerine birds where it probably has a certain longevity making the birds true paratenic hosts.

Often *Macracanthorhynchus hirudinaceus* occupies extraintestinal positions in humans. The migration of this perforating acanthocephalan through the gut wall is very painful. Such infected humans with an intraperitoneally located worm might be named accidental hosts since they do not play a role in the transmission of the acanthocephalan.

Among perforating species a proboscis cavity is formed mainly during the early phase in the final host when the worm has not yet penetrated, later on the cavity's depth and frequency of invagination are progressively reduced (Acanthocephalan Infections/Figure 11. 5).

Food Uptake

In non-perforating acanthocephalans the proboscis itself is usually kept in a more or less invaginated condition creating a deep proboscis cavity (Figure 11.4) which obviously functions as a funnel collecting remnants of cells and nutrients leaking into the lesion that has been created by the worm. In eo- and palaeacanthocephalans, especially lipid substances such as triacylglycerols are highly abundant as storage lipids in the intestinal wall of fish, ducks or seals serving as final host. As shown in Figure 11.10 lipid matters as well as, for instance, peptides deriving from the granules of eosinophilic granulocytes contribute to the efflux from the necrotic tissue. However, autoradiographic studies by Taraschewski and Mackenstedt with two species of eoacanthocephalans and four palaeacanthocephalans (two non-perforating and two perforating species) show that predominantly lipid substances are absorbed at the worms' praesoma (Figures 11.7 and 11.8). The "apical organ" of eoacanthocephalans, a structure not yet well understood at the tip of the proboscis, *i.e.* at the bottom of the proboscis cavity (Figure 11.7), and the tegument of the anterior half of the proboscis (Figure 11. 8) play the most active role in lipid uptake. Interestingly, the proboscis hooks, too, can be considered organs well adapted to the task of lipid uptake (Figure 11.7 and Figure 11.8). In accordance with their tapered and tipped construction (Figure 11.7B) the hooks are in reach of lipid deposits which are not (yet) in contact with the surface of the praesoma. Behind the septum between prae- and metasomal tegument the uptake of a triacylglycerol as well as of vitamin A was very low in *in vitro* trials. However, if "shoulders" of the metasoma were in contract with the praesomal surface during the exposition of a worm to the labelled nutrient, the shoulders too revealed a markable label (Figure 11.8), suggesting that enzymes localised at the praesomal surface were involved. Uptake of amino acids as well as monosaccharides also occurs at the surface of the praesoma (Figure

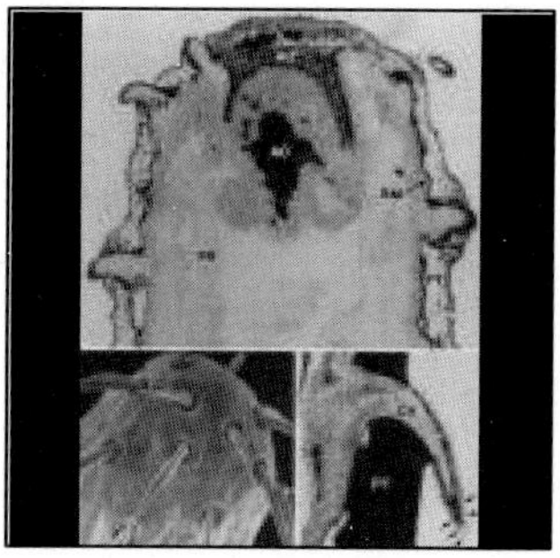

Figure 11. 7 A–C: Micrographs Showing the Proboscis Tip and Hooks of Various Acanthocephalans

A Longitudinal semithin section of Paratenuisentis ambiguus that had been *in vitro*-exposed to [3H]-glyceroltrioleate for 5 minutes. The proboscis is fully everted, which is not the normal condition in vivo. Note the intense label of the "apical organ" (AO) (cf. tegument). The lipid that has been taken up by the hooks and the surrounding tegument (PT) seems to be transported along the outer membrane and the basal membrane (BM) of the tegument; CH: connective tissue of a hook, PR: proboscis retractor muscle. × 3,500. B Scanning electron micrograph of the tipped hooks at the anterior proboscis of Acanthocephalus anguillae. × 1,600. C Longitudinal semithin section of a proboscis hook of *A. anguillae* that has been exposed to [3H]. Note the labelled tegument of the hook underneath its striped layer. The proboscis tegument (PT) has already absorbed huge quantities of the lipid. × 3,000. CH, connective tissue of the hook

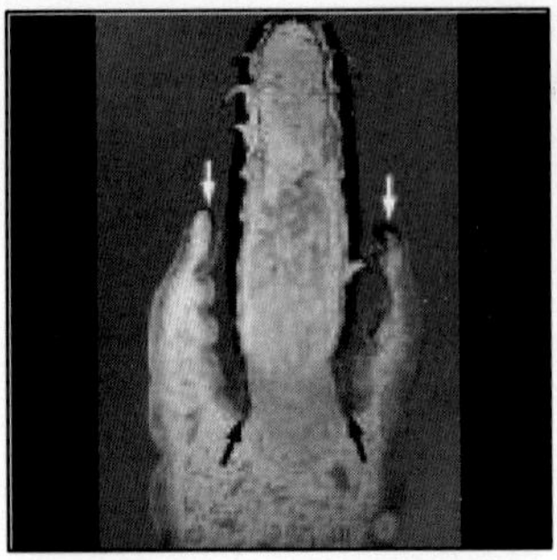

Figure 11.8: Longitudinally cut semithin section of the anterior body of an *Acanthocephalus lucii* that had been in-vitro exposed to [3H]-vitamin A for 15 minutes. Note the intense label of the anterior half of the proboscis tegument which lines the proboscis cavity normally formed. Behind the septum between the praesomal and the metasomal tegument (black arrows) almost no absorbance of the substance offered has taken place. However, it appears that the accidentally formed "shoulder" of the metasoma that obviously was in contact with the praesoma during the exposure has attained some label in its tegument (white arrows), suggesting that this part of the metasoma took advantage of enzymatic activity prevailing at the praesomal surface. × 60

11. 9), but the metasomal tegument seems to be the major absorptive surface for these substances.

Concerning the uptake of nutrients by adult perforating species, the mechanisms do not basically deviate from those described for non-perforating species. Since the whole praesoma is deeply embedded in the gut wall, a funnel for substances leaking into intestinal lumen is not very large. Intraintestinal attached worms that have a bulbus can be easily recognized at the gut's exterior side (Acanthocephalan Infections/

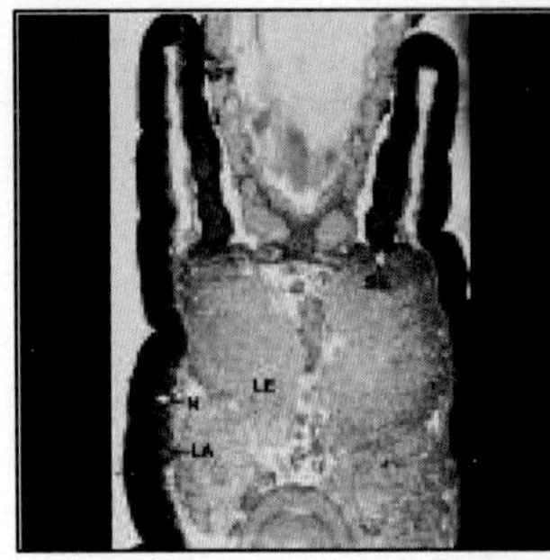

Figure 11.9. Autoradiographically treated longitudinal section of the anterior body of a female *Echinorhynchus truttae* that was exposed to [3H]-lysine for 8 minutes. Note the less intense label of the presomal tegument compared to the metasomal tegument. × 100. SE, septum; LA, tegumental lacunar system; LE, lemniscus; N, nucleus

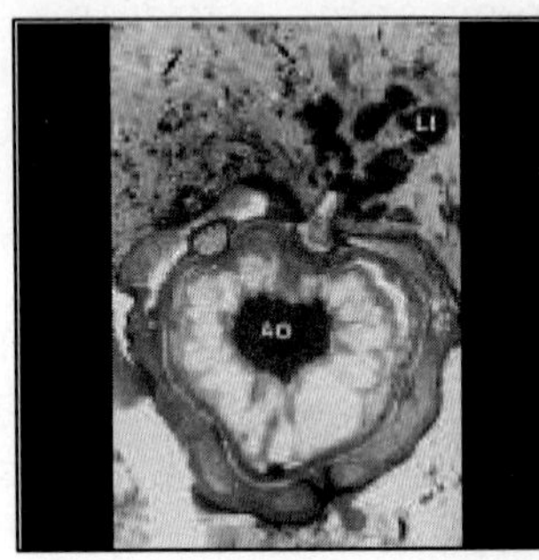

Figure 11.10: Semithin section of the proboscis of a specimen of *Neoechinorhynchus rutili* attached to the gut wall of a naturally infected juvenile rainbow trout. Note the conspicuous quantities of lipid (LI) accumulating at the worm's proboscis. They seem to derive from the surrounding necrotic tissue. Also, fused granules from degranulated eosinophilic granulocytes (arrow) are abundant, fusing with the lipid drops. In such a methylene-blue-stained section the lipid attains a greenish-golden colour while the fused granules have a deep blue. Thus both substances can be distinguished well. × 100. (AO, apical organ).

Figure 11.6) but also in species without a bulbus, like *Macracanthorhynchus hirudinaceus*, a fibrous whitish nodule with reddened annulation around it can be seen.

Integument

The following descriptions of the acanthocephalan tegument are based on reviews by Starling, Miller and Dunagan, and Taraschewski. The tegument of acanthocephalans is a syncytium of up to 2 mm in thickness (*Macracanthorhynchus hirudinaceus*). It either contains numerous small nuclei (Figure 11.13C) or specific numbers of giant nuclei in eoacanthocephalans (Table: 11.1). The nuclei of the tegument of the metasoma (trunk) are not immersed below the tegument (Figure 11.13C). In the praesoma (proboscis and neck), however, the nuclei are harboured by the lemnisci, sack-shaped outgrowths of the praesomal tegument projecting into the body cavity (Figure 11.1). The tegument is supported by underlying fibres of connective tissue, partly identified as collagen, of equal thickness in all parts of the body, and by cords of circular (only in the metasoma) muscles and longitudinal muscles (in both parts of the body; Figure 11.13C). These components together build up the body wall (Figure 11.13C). The tegument shows a typical stratification and a differentiation related to the praesoma-metasoma organisation of the acanthocephalan body.

Metasoma

The syncytial tegument of the Acanthocephala is delimited by a plasma membrane carrying a filamentous surface coat (Figure 11.12A) which has a similar appearance in all systematic groups of these worms regarding the surface of the metasoma (Figure 11.12 and Table 11.1). This glycocalyx also covers the pores (openings of the outer membrane's crypts) of the tegument. It may reach a thickness of up to 1 µm or more and obviously proteoglycanes are present in it. Infective larvae inside the intermediate host's hemocoel carry the most conspicuous surface coat. The plasma membrane itself forms densely set crypts projecting into the outer part of the tegument. Their greater density in the metasoma compared to the praesoma might have to do with the heavy competition pressure between the trunk surface being located inside the gut lumen, and the host's intestinal mucosa for the absorbance of nutrients. The crypts have been calculated to increase the worm's "outer" surface 20- to 80-fold. The crypts have slender necks with electron-dense annulations underneath their outer openings (seen as pores by SEM) and they form branches directly underneath the pores or further inside the tegument. The crypts are considered extracytoplasmic digestive organelles which under the influence of surface hydrolytic enzymes maximise the opportunities of food absorption by these gutless worms. So it might be a point of debate whether the membrane limiting the lumen of the crypts really is an "outer" surface (Figure 11.12).

Interiadly, the outer membrane is supported by an electron-dense layer of about 5 µm in thickness. Due to perforations of this "cuticle" by the crypts, this layer obviously having stabilising functions is seen as a stripped layer in TEM micrographs (Figure 11.12). Usually the longitudinal extension of the membrane-crypts considerably exceeds the diameter of the stripped layer, forming a spongy belt rich in pinocytotic activity (Figure 11.12). This layer is considered a "vesicular layer" by a few authors.

The feltwork layer adjoining underneath again seems to contribute to the skeletal task of the tegument but its diameter is about five times larger than that of the stripped layer (cf. Acanthella/Figure 11.1, showing this feature in the praesoma). It is characterised by fibres displaying no particular order, and normally it contains large amounts of glycogen. Metasomal spines, present in many palae- and eoacanthocephalans, have been described as outgrowths of the feltwork layer and are thus invested by a thin cover of a somewhat condensed stripped layer. These trunk spines, often ornamenting a wider part of the ventral surface than of the dorsal side, are thought to act as additional holdfast organs.

The more proximal radial layer (Figure 11.13C) occupies more than 70 per cent of the tegumental diameter. This layer is considered to be the main metabolic centre of the acanthocephalan body. During the (allometric) growth of the worms it is mainly the increase in diameter of the radial layer which leads to a thicker tegument. This layer is characterised by radially arranged fibres connected to the basal membrane. Electron dense matter accumulates around the spokes (Figure 11.13C). This layer harbours the tegumental nuclei (Figure 11.13C) and the major canals of the lacunar system (Figure 11.14C). The lumen of this system is poor in organelles and electron dense matter (Figure 11.14C) and in autoradiographic trials it takes up and/or retains less nutrients than the surrounding cytoplasm. Also its glycogen content is low in contrast to the true radial layer, which usually stores plenty of this carbohydrate. Apparently contractions of the subtegumental musculature (Figure 11.13C) act as the motive force for fluid flow inside the lacunas. However, it remains unclear whether these caverns fulfil functions of a circulatory system. The basal membrane, interiadly bounding the tegumental syncytium, shows a typical labyrinthine structure. Along its distal surface it is supported by amorphic matter, whereas a thin lamina of a fine matrix lines the proximal side of the membrane (Figure 11.13C).

Praesoma

The tegument of the praesoma is separated from that of the trunk by a septum composed of fibres and adherent amorphic matter (Figure 11.12B). So even the lacunar cavities are part of two different systems, which might make sense considering the assumed involvement of hydrostatic pressures in the protrusion, invagination and retraction of the proboscis or the entire praesoma. Within the praesoma mainly the neck possesses lacunar cavities. The praesomal tegument reveals major differences compared to that of the metasoma and these features become more prominent towards the anterior part of the proboscis. Generally, the praesomal tegument contains more amorphous, electron dense matter, more mitochondria and rough and smooth endoplasmic reticulum (Figure 11.13A) as well as lipid (especially in eo- and palaeacanthocephalans, Figures 11.13A and 11.15) than the metasomal tegument. Interestingly, a submersion of the tegumental nuclei only occurs in the praesoma. The lemnisci harbouring the nuclei (Figure 11.16) do not show a specific stratification like the tegument they branch away from. They too contain lacunar spaces, and are very rich in lipid.

The surface coat of the praesomal tegument reveals systematics-related specificities (Figure 11.15) and shows interesting links with the host-parasite

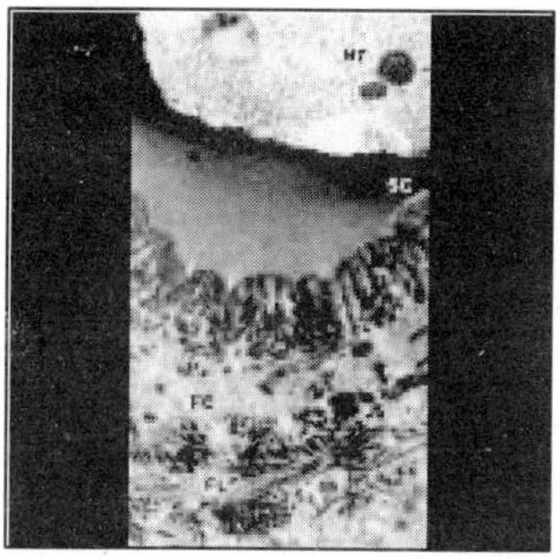

Figure 11.11: Transmission electron micrograph of a section through the distal part of the praesomal tegument of Acanthocephalus anguillae (in a rainbow trout, 90 dpi). Note the thick lipoid surface coat (SC) between the tegument and the necrotic tissue (NT) at the point of attachment, also the fused crypts of the outer membrane (FC) with supporting microfibres in it and the underlying feltwork layer (FL). A section with little osmiophilic content of the fused crypts was chosen in order to show the microfibres in it. × 25,000.

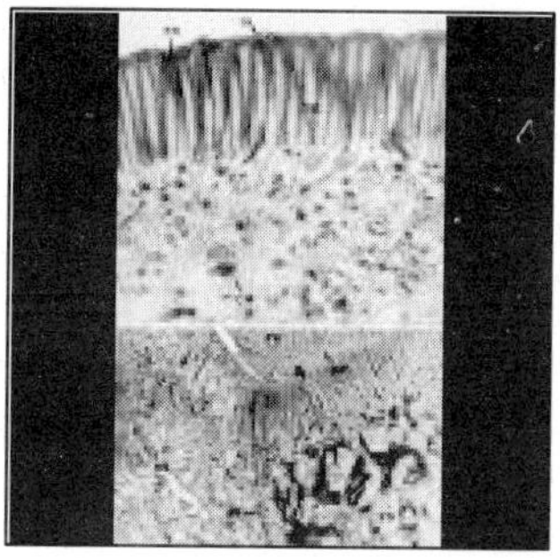

Figure 11.12A and B. Transmission electron micrographs of the distal part of the adult acanthocephalans' tegument. A Section through the metasomal tegument of *Acanthocephalus lucii*. Note the stripped layer (SL) functioning as a cuticle. It is perforated by densely set crypts of the outer membrane (CM). PO: pore of a crypt; GL: glycocalyx (which is thin on the specimen shown here). B Section through the area around the septum (SE) separating the praesomal tegument (PS) (which is folded in this worm shown *in vivo*) from the metasomal tegument (ME) of Acanthocephalus anguillae. Note the abundance of lipid (LI) in the praesoma and the osmiophilic film (OF) obviously shed from the praesomal surface into the thick, apparently liquid surface coat of the praesoma. × 28,000.

interactions (Acanthocephalan Infections). The fine structure and obviously also the chemical composition of the surface coat vary among the classes. Regarding archicacanthocephalans the optical impression of the praesomal glycocalyx resembles that of the metasoma, although it is more coarsely structured and more osmiophilic than the latter. Shedding of the surface coat frequently or often occurs and seems to follow a complexation of host's antiparasitic enzymes or antibodies with the surface coat. Eoacanthocephalans and palaeacanthocephalans reveal a lipoid, non fuzzy surface coat which may reach a thickness of several microns (Figure 11.11) and shows a matrix which suggests a liquid or semiliquid condition. In addition to lipid, mucus-like carbohydrates are also present in it. Often osmiophilic films, perhaps representing a "glycocalyx", can be seen in it, and it is rather likely that these films are shed into the voluminous coat once the outer membrane has become loaded with antiparasitic peptides of the host's defense system (Figure 11.12B). Unfortunately, the chemical properties of the acanthocephalan surface coat have not been extensively studied to date.

The (pores of the) praesomal crypts are less densely set and the striped layer measures half or less in diameter than the trunk surface. In Palaeacanthocephala the

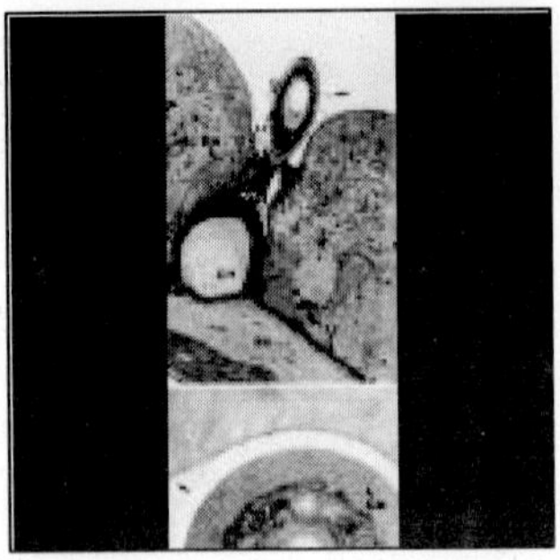

Figure 11. 13A–C: Transmission electron micrographs of sections through the proboscis tegument and hooks of the eoacanthocephalan *Paratenuisentis ambiguus.* The curved hook which is retracted in this micrograph is sectioned twice, at its connection with the subtegumental connective tissue (CH: connective tissue of the hook) and at its tipped outer portion. Note the lipoid substance (LI) being excreted through the pores in the hook, which (or a similar substance) is also abundant in the surrounding tegument; ER: rough endoplasmic reticulum. × 10,000. B Cross section through a retracted hook and neighbouring proboscis tegument. The section has been treated according to the electron microscopical PAS-staining method by Thiéry. Note the mucus-like carbohydrates inside (CM: crypt of the hook's outer membrane) and outside the hook (arrow). × 56,000.

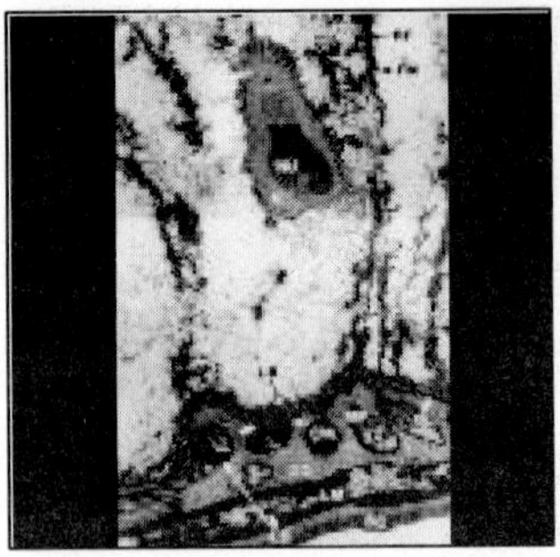

Figure 11.13 C: Transmission electron micrograph of the basal part (radial layer) of the metasomal tegument of an adult *Acanthocephalus anguillae.* Note the radially arranged fibres (RF) with fibrous, electron-dense material (FM) attached to them, the labyrinthine basal membrane (LB) and the cords of subtegumental musculature (CM: circular muscle, LM: longitudinal muscle) consisting of an outer myogenic portion (MP) and an inner non-contractile portion (CP); BL: basal lamina, LD: lipid drop, N: nucleus, NU: nucleolus, SC: subtegumental connective tissue. × 50,000.

single crypts are fused underneath the striped layer, forming large caverns with stabilising fibres in them (Figure 11.15B). The other systematic groups have retained their individual crypts (Figures 11.12A and C). Generally, the strata of the tegument as described from the trunk cannot be well distinguished: due to the abundance of fibres they often all together appear like a feltwork layer. The metasomal labyrinthine structure of the basal membrane is considerably reduced and instead its coating with amorphous material is more pronounced.

Proboscis Hooks

Irrespective of the systematic affiliation of the worms the hooks (Figures 11.13A and B, Figure 11.15) possess a central cone of connective tissue which partly has been demonstrated to contain collagen and/or chitin. This major part of the hook arises

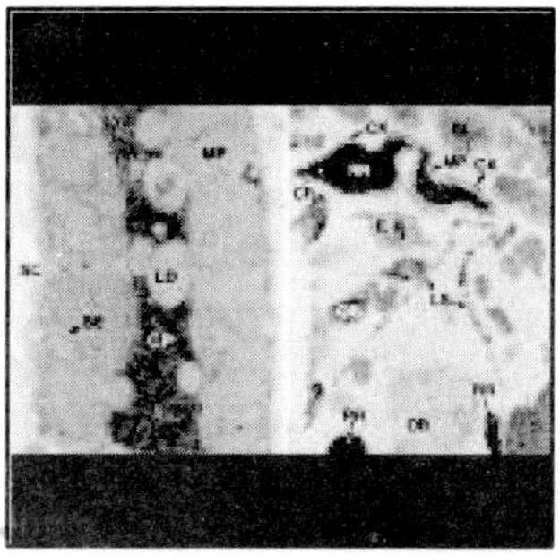

Figures 11.14A and B. Micrograph of acanthocephalan muscles of *Paratenuisentis ambiguus* showing their bi- component construction comprising an outer myogenic belt (MP) and an enclosed cytoplasmic portion (CP). A Longitudinally ultrathin-sectioned subtegumental muscle of an infectious larva. The section has been treated according to the electron microscopical PAS method of Thiéry in a mode to visualize glycogen. Note the intense Thiéry label in the core of the muscle; LD: lipid drop, SC: subtegumental connective tissue, SE: septum. × 260. B Transversally semithin sectioned body cavity of an adult worm that was exposed to 3H-glucose and then autoradiographically treated. The intense label in the cytoplasmic portion of the receptacle retractor muscles (RR) seems to be due to glucose-metabolites (probably mainly glycogen) incorporated in these muscles; CK: knobs on the muscle's surface also showing the bi-portion structure, E: egg, LS: ligament strand, OB: ovarian ball, SL: subtegumental longitudinal muscle. × 300.

from the subtegumental connective tissue. In its proximal part it encircles a finger-like projection of the subtegumental longitudinal musculature (Figure 11.15). But this musculature tie is not present in all hooks of all species, implying that not all hooks can be individually retracted.

In eo- and palaeacanthocephalans the fibrous core of the hooks carries a condensed tegumental cover making these holdfast organs pointed (Figure 11.13, Figures 11.15A and B). The stripped layer does not markably differ between eo- and palaeacanthocephalans but in eoacanthocephalans the crypts are not fused but entangle with fingerform protrusions of the tegument's basal membrane inside the hooks (Acanthocephalan Infections/Figure 11.7). In both of these subclasses the tipped hooks are capable of discharging lipid substances through their pores (Figure 11. 13A and B). Mucus-like carbohydrates also contribute to the excreted matter (Figure 11.13B) and, rather likely, also enzymes are contained in it which thus far can only be hypothesised. The grease-like surface coat of the hooks may be very voluminous (Figure 11.15). Amazingly, however, the hooks are also capable of absorbing nutrients from the host tissue surrounding them (see Food Uptake).

In archiacanthocephalans the hooks do not bear a tegumental vestment and the naked cone of the connective tissue, thus piercing the tegument, is less pointed than the hooks of the two other systematic groups (Figure 11.15C). Obviously, the hooks attain their slippery surface cover by dipping into a pit encircling them – this annular cleft filled with a highly osmiophilic lipoid paste deriving from the surrounding tegument (Figure 11.15C).

Uniformly in all systematic groups of the Acanthocephala, the close tegumetal surrounding of the hooks is rich in lipid droplets, mitochondria and rough (Figure 11. 13A and B) as well as smooth endoplasmic reticulum, indicating elevated metabolic activity.

Proboscis Cavity

Contrary to the way how acanthocephalans usually are shown in drawings made from dead worms (Figure 11.16), *in vivo* and *in situ* the acanthocephalan proboscis normally is kept in a semi-invaginated position, especially among species with superficial attachment (Figure 11.5). Thus a proboscis exhibiting a more or less deep anterior cavity resembling a mouth opening should be part of our idea of these gutless worms. Among eo- and palaeacanthocephalans inside the proboscis cavity the tegumental surface including the hooks appears as a labyrinth with remnants of host cells and tissue between its curves and with grease occupying all external niches of the labyrinth at its bottom plane. In Archiacanthocephala the proboscis cavity is not filled with grease in its inner part and the labyrinth is lined by the fuzzy glycocalyx described under praesoma.

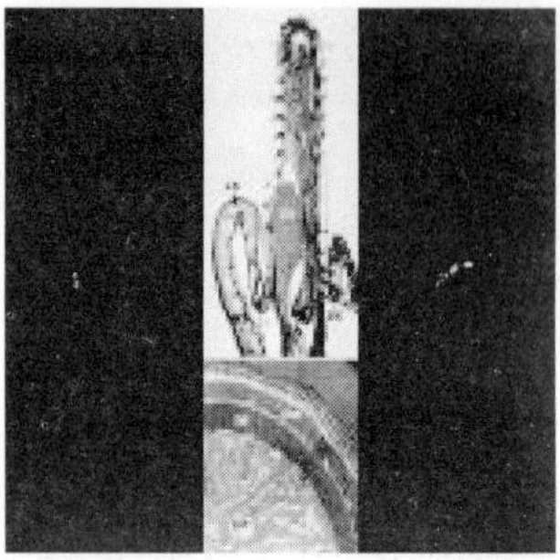

Figure 11.14C-D: C, D Semithin sections of the praesoma (and partly the metasoma: A) of adult acanthocephalans. A Longitudinal section showing the cytoplasmic "finger" (CP) of the (inner) receptacle wall projecting into the proboscis. Also note the lacunar system (LS) inside the metasomal tegument (MT); SE: septum between praesomal and metasomal tegument. × 30. D Transverse section through the receptacle wall musculature (IW: inner wall, OW: outer wall) of Acanthocephalus lucii. Note the tubular structure of the proboscis retractor musculature with its cytoplasmic cores (CP) of low density. × 150.

Musculature

Relatively few investigations have dealt with the fine structure of the acanthocephalan musculature. Some muscles, such as the receptacle retractor muscles, appear obliquely striated, *e.g.* fibres are connected to Z-line-like structures. The basic feature of the acanthocephalan musculature is their two-component structure composed of an outer myogenic, contractile belt and a cytoplasmic core enclosed by it (Figure 11.13C and Figure 11.14). The interior part seems to have a function in energy storage.

Usually glycogen is very abundant in it (Figure 11.14A) and in autoradiographic experiments with labelled glucose, the glucose, or more likely metabolites of it like glycogen, accumulates in the cytoplasmic core (Figure 11.14B). The cords of subtegumental musculature follow this bi-component composition (Figure 11.13C), as do the retractor muscles (proboscis retractor-, Figure 11.14B, receptacle-, Figure 11.14B and neck retractor musculature), for instance. In addition the receptacle retractor muscles also carry small knobs on their surface which have non-contractile cores (Figure 11.14B). In all these muscles the central non-contractile portion may contain plenty of organelles, mainly mitochondria, or may be rather electron-lucent, suggesting a higher fluidity than the latter cytoplasm. Inside the proboscis retractor musculature a low viscosity core should enable a quick directional shift of the enclosed

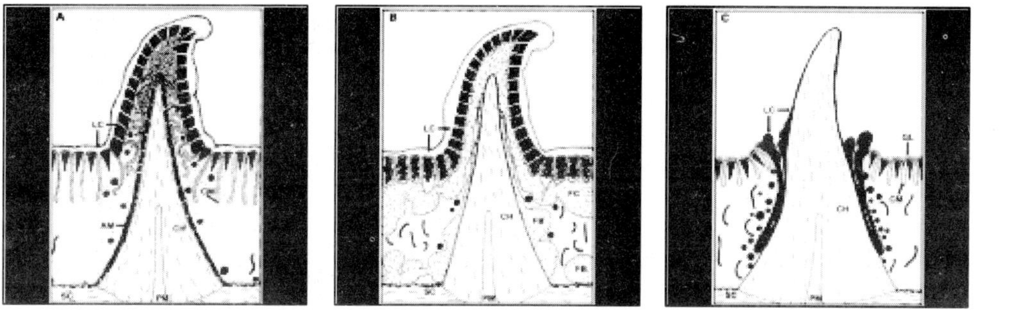

Figure 11.15A–C

A Schematic drawing of an eoacanthocephalan proboscis hook and surrounding tegument. Note the tegumental cover of the hook's connective tissue (CH), and the crypts of the outer membrane (CM) entangling with "crypts" of the tegument's basal membrane externally lining the connective tissue of the hook. These crypts perforate a layer of amorphic matter (AM) covering the connective tissue of the hook's tegument. LC: lipoid coat on the hooks and on the tegument; PM: finger-shaped protuberance of the subtegumental musculature; SC: subtegumental connective tissue. B Schematic drawing of a palaeacanthocephalan proboscis hook and surrounding tegument. Note the tegumental cover of the hook`s connective tissue (CH), the lipoid coat (LC) on the tegument and the hook, the fused crypts of the outer membrane (FC) supported by fibres, and the fused crypts of the basal labyrinth (FB) which possibly are continuous with the latter fused crypts. PM: finger-shaped protuberance of the subtegumental musculature; SC: subtegumental connective tissue. C Schematic drawing of an archiacanthocephalan proboscis hook and surrounding tegument. Note that the connective tissue of the hook (CH) has no tegumental cover and is invested only by a lipoid coat (LC) which is discharged by the tegument into the pouch surrounding the hook. The proboscis tegument carries a fuzzy glycocalyx (GL), and the crypts of the outer membrane (CM) are not fused. PM: finger-shaped protuberance of the tegumental musculature; SC: subtegumental connective tissue.

cytoplasm when the proboscis cavity is formed or discontinued. An interesting differentiation is shown by the proboscis receptacle musculature enclosing and thus forming the hollow into which the proboscis can be retracted. In palaeacanthocephalans it consists of a double wall which has almost no non-contractile portion (Figure 11.14D). In eoacanthocephalans it is considered single-walled but the "receptacle protrusor musculature" exteriadly surrounding the receptacle without being firmly connected to it probably represents the outer wall of the receptacle. It reveals the described two-portion structure. The inner wall basically consists of a firm, contractile wall but on its dorsal inner side a conspicuous sack-shaped cytoplasmic outgrowth with a very narrow contractile outer cover projects into the posterior half of the proboscis (Figure 11.14C). The cytoplasmic finger seems to function as the major glycogen deposit of the praesoma. In archiacanthocephalans the inner wall consists of plane myogenic tissue whereas the outer wall is formed by spirally arranged single muscle cords with non-contractile cores. Due to this spiral

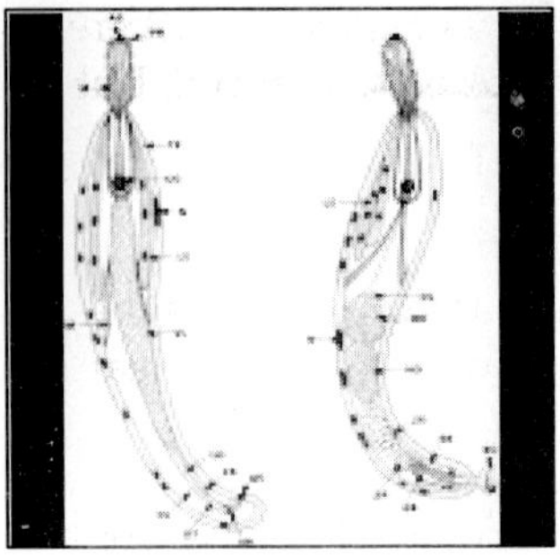

Figure 11. 16A

Diagrammatic representation of a female acanthocephalan (*Paratenuisentis ambiguus,* Eoacanthocephala) with emphasis on the sexual organs (most muscles omitted). It has been reduced in size (length) compared with the male worm. The ventral ligament sac leading into the uterine bell is specific to eoacanthocephalans. Note the lack of genital ganglia in the female worm. For the inscriptions of the nonsexual organs see B. Eggs and floating ovaries inside the ligament sacs are not shown (Figure 11.18A). B Diagrammatic representation of a male acanthocephalan (*Paratenuisentis ambiguus,* Eoacanthocephala) with emphasis on the sexual organs. Most muscles are omitted. The single polynucleate cement gland and the presence of a cement reservoir and a seminal vesicle are specific to eoacanthocephalans. AS, apical sensory organ; BC, bursa copulatrix (evaginated); CG, cerebral ganglion; DL, "dorsal" ligament sac; ES, egg-sorting apparatus; GÖ, genital opening; GS, Saefftigens pouch; HO, testes; LE, lemnisk; M, muscle; N, giant nucleus; PR, proboscis with hooks (evaginated); SB, seminal vesicle; SP, sphincter; TE, tegument; UG, uterine bell; UT, uterus; VL, "ventral" ligament sac; ZD, cement gland; ZR, cement reservoir.

arrangement the retraction and protrusion of the praesoma (not only the proboscis can be invaginated) are performed in a torsion-like, screwing fashion.

Excretory System

Excretory products of most acanthocephalans seem to be released exclusively through the body wall, but it is not known whether this takes place through the whole tegument or through special regions. In addition, oligacanthorhynchids and probably other archiacanthocephalans have protonephridia. Their efferent canals either lead into the vas deferens (male) or into the uterine bell (female). Two types of protonephridia are known:

Dendric Type

Numerous flame cells drain into branched canals which lead into a central canal.

Saccular Type

The flame cells drain into an encapsulated bowl and a subsequent central canal.

Excretory products of acanthocephalans seem to be similar to other helminths, containing lactate, succinate, etc. Ethanol, however, has been described as the main excretory product of *Moniliformis moniliformis*. There is still controversy over whether acanthocephalans are osmoconformers or not, but most species seem to have little osmoregulatory ability.

Reproduction

Acanthocephalan reproduction as well as the fine structure and genesis of the oocytes and spermatocytes show some unique features.

Reproductive Organs

Acanthocephalans are dioecious. Male worms are usually smaller than females, and in addition sexual dimorphism may affect other features such as trunk spination. Only males have a pair of genital ganglia (and a bursal ganglion, so far only described for *Moniliformis moniliformis*), whereas both sexes have a cerebral ganglion. Sensory papillae of the genital region are confined to males. And indeed only males seem to be active finding in a sexual mate and copulation. In female worms the sexual organs lie within two ligament sacs (Figure 11.16) which rupture in palaeacanthocephalans and some eoacanthocephalans. The male sexual organs are located within only one ligament sac (Figure 11.16B).

The male gonads and accessory organs are enclosed by the dorsal ligament sac (the ventral sac does not persist in males), and further posteriorly, by the muscular genital sheath (Figure 11.16B). The organs are attached to the ligament strand which keeps them in position. Males normally have two testes, but monorchidism is rather frequent. A seminal vesicle may be present (Eoacanthocephala). The vasa efferentia fuse to form a vas deferens, which fuses with one or several ducts of the cement gland(s) to form a genital canal. Cement glands are significant accessory organs (1–8 in number), and eoacanthocephalans have a separate cement reservoir (Acanthocephalan Infections/Figure 11.6). The cement locks the female vagina after copulation until the first embryonated eggs are released, and forms typical copulatory caps on the posterior tips in inseminated females. If protonephridia are present, the genital canal is joined by the (ciliated) excretory canal. The genital (or urogenital) canal leads into the bursa copulatrix (Figure 11.16B). The muscular terminal part of the genital canal inside the bursa is considered a penis. Additional accessory organs are the Saefftigen's pouch and a few glandular structures associated with the bursa that is not yet well known. The fluid-filled muscular Saefftigen's pouch is connected with the lacunar system of the bursa tegument. By its contraction it regulates the hydrostatic pressure of the bursa and thus its protrusion or invagination (Figure 11.16B).

The Female Reproductive System Consists of Two Major Tubes

The ligament sacs (or the pseudocoel if the sacs are ruptured) that contain the floating ovaries (ovarian balls). An efferent duct system including a complex egg-sorting apparatus which is unique among helminths. The two ligament sacs are interconnected at their anterior end. Posteriad, one sac leads into the uterine bell while the other is connected to a lateral opening of the subsequent apparatus (Figure 11.16A).

The muscular efferent duct consists of the uterine bell, the egg-sorting apparatus, the uterus, and the vagina which is enclosed by one or two genital sphincters (Figure 16A). Eggs from the dorsal (Archiacanthocephala) or ventral (Eoacanthocephala) ligament sac are "sucked" into the funnel-shaped bell which leads into a narrow duct. The subsequent egg-sorting apparatus of *Moniliformis moniliformis* consists of two lateral pockets, two dorsal median cells, two anterior ventral median cells, two posterior ventral median cells, and two lappet cells. It ensures that normally only embryonated eggs are found in the host's feces. By a complex interaction between the muscular activity of the bell wall and the cells and pockets of the apparatus, only embryonated eggs are allowed to enter the uterus, while immature ones are forced back into the ventral (Archiacanthocephala) or dorsal (Eoacanthocephala) ligament sac. The egg-sorting mechanism is not fully understood. The uterus is surrounded by layers of muscles and fibrous material. The vagina is a narrow duct which connects the uterus with the gonopore (Figure 11.16A) and was found to carry glandular appendages in some species. The gonopore of a few species is surrounded by genital spines, and after insemination is generally blocked by a copulatory cap imposed on it by the male until eggs are released.

Gametogenesis

Acanthocephalan reproduction as well as the fine structure and genesis of the oocytes and spermatocytes show some unique features. Acanthocephalan spermatozoa are filiform (Figure 11.17) and consist of a nucleocytoplasmic spermatozoan body rich in glycogen, and a flagellum (Figure 11.17). They measure 20–80 ìm in length depending on the species, and obviously do not posses mitochondria or acrosomes. They contain a longitudinal chromatin strand (which is not membrane bound), two lateral rows of "dense inclusions" of unknown function and a centriole which gives rise to the flagellum. The axoneme of the free flagellum consists of microtubules which in most species are arranged in a $(9\times2)+2$ pattern (Figure 11.17), but also either one or three central tubuli have been found even within one species. The microtubules may show typical dynein arms, but the pattern is not consistent (Figure 11.17). Among the phases of spermatogenesis the spermiogenesis is best described. It is characterized by several events:

The centriole of the flagellum migrates from the posterior to the anterior region of the spermatid while the spermatids are still connected in clusters (Figure 11.17B) by cytophoral stalks. The flagellum then slightly extends posteriorly (while the nucleus becomes elongated) and finally it extends greatly anteriorly (Figure 11.17C). Thus, as a result of this extension the spermatozoan body becomes reversed in relation to the free flagellum.

The nuclear membranes disintegrate to form the nucleocytoplasmic body (Figure 11.17A and C). Only a remnant of the nuclear envelope remains.

The mitochondria disappear from the spermatozoan body. The spermatozoan body detaches from the spermatid's residual body containing the mitochondria. Several species of acanthocephalans have been found to be precocious, *e.g.* mature spermatozoa have been found in male larvae.

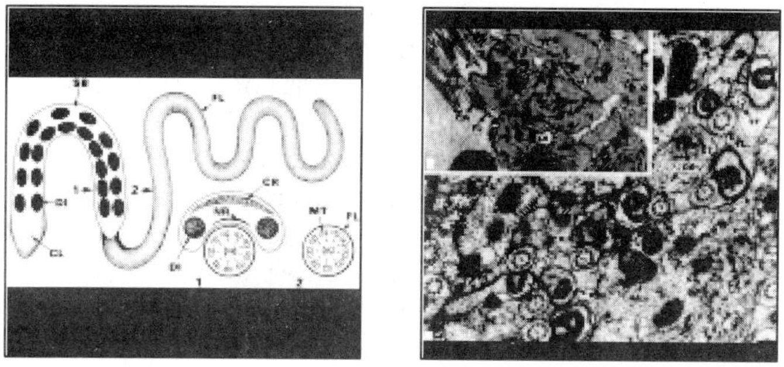

Figure 11.17 A–C: Acanthocephalan Spermatozoan and Spermatid Morphology

A Diagrammatic presentation of an acanthocephalan spermatozoan (after Whitfield 1971, Marchand 1981, cited by Crompton and Nickol 1985, and our own observations). 1, 2 Transverse sections through the spermatozoan body (× 2) and through the flagellum (× 3). B Transmission electron micrograph of a cluster of spermatids of *Echinorhynchus truttae* in the process of nuclear and flagellar elongation, *i.e.*, spermiogenesis. Note the numerous mitochondria (MI) inside the spermatids. × 6.840. Transmission electron micrograph of a transverse section through spermatids (ST) and early spermatozoans (SB) of E. truttae. Note the apparent lack of mitochondria in the spermatozoan body and the rupturing nuclear envelopes (RN) in some spermatozoans. × 39.900. CL, centriole; CR, chromatin; DI, electron-dense inclusions; FE, flagellum extending from the anterior part of a spermatid posteriad; FL, flagellum; GB, Golgi body; MI, mitochondrion; MT, microtubules with dynein arms; N, nucleus; NE, nuclear envelope; NR, remnant of nuclear envelope; RN, rupturing nuclear envelope; SR, spermatozoan body; ST, spermatid; TF, terminal flagellum

Mature oocytes are spherical cells that lie below the surface of the free-floating ovaries (ovarian balls) and show typical electron-dense inclusions (Figure 11.18A and C). The floating ovaries derive from the ovarian primordium of some larval stage. Immature floating ovaries have a thick surface coat and lack microvilli-like structures of their outer membrane. Mature ones consist of two syncytia, *i.e.*, the central oogonial syncytium and the peripheral supporting syncytium. Furthermore they contain developing oocytes which seem to derive from the oogonial syncytium. The superficial supporting syncytium reveals microvilli-like outgrowths of its surface which absorb nutrients from the body cavity, as can be demonstrated by autoradiographic experiments. Fertilised ovaries (and unfertilised mature ovaries of a few species) apparently lose their microvilli (Figure 11. 18C). The actual process of oogenesis from oogonia to mature oocytes is not yet well known.

Fertilization

Acanthocephalan females may become inseminated subsequently several times. But little is known about how the worms attract each other – if they do so – prior to copulation and insemination. According to observations by Richardson *et al.*, in the palaeacanthocephalan Leptorhynchoides thecatus, parasitising in green sunfish,

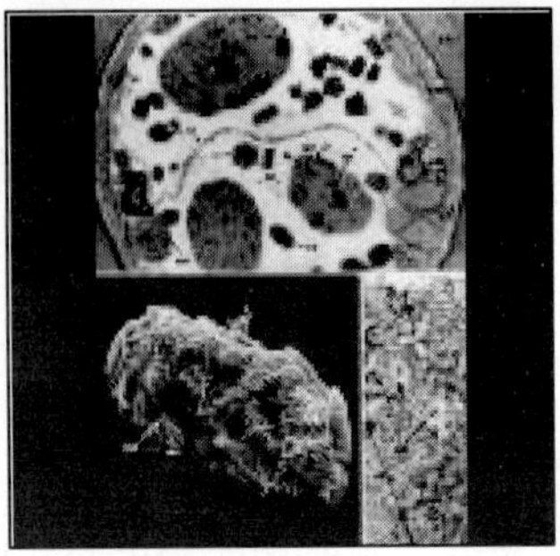

Figure 11.18 A–C: Micrographs of Acanthocephalan Floating Ovaries and Fertilization of the Enclosed Oocytes

A Light micrograph. Floating ovaries (FO) are contained within the two ligament sacs which are separated by the fibrous ligament strand (LS) (Paratenuisentis ambiguus, Eoacanthocephala). Note the oocytes (O) lying underneath the surface syncytium (SS) of the ovaries, adhering sperms (SP), the detached zygotes (DZ), and the shell-coated developing eggs (E). × 260. B Scanning electron micrograph. A floating ovary from the body cavity of Acanthocephalus anguiliae (Palaeacanthocephala) shows numerous sperms at its surface. The flagellum is visible as a slender prolongation of the spermatozoan body. Note the detaching zygotes (DZ) which already resemble the spindle shape of the mature eggs. × 4.560. Transmission electron micrograph. Sperms have penetrated the surface syncytium and the underlying oocyte (zygote?) of a floating ovary of Neoechinorhynchus rutili. The accumulation of "inclusions" (IN) at the oocyte's margin seems to follow fertilization and possibly initiates the formation of the egg shell. × 16.450. CM, circular musculature; CP, cytoplasmatic part of muscle; CT, connective tissue; DZ (black), detached zygote; DZ (white), detaching zygote; E, shell-coated developing egg; FO, floating ovaries; IN, inclusion; LM, longitudinal musculature; LS, ligament strand; MI, mitochondria; MP, myogenic part of muscle; N, nucleus of oocyte; NU, nucleolus of oocyte; O, oocyte; RR, receptacle retractor muscle; SP, spermatozoa; SS, surface syncytium of floating ovary; TE, worm's tegument.

mate finding follows a very simple pattern. Individuals of both sexes usually are positioned inside the pyloric ceca in a mode such that their posterior ends extend into the intestinal lumen within the small area from which the ceca orginate. So emigration to find a mate is unnecessary (see also Behaviour). There is still a lack of evidence about the function of the copulatory cap which locks the vagina of inseminated females during the prepatent period. Some philosophical debate has been held about the applicability of the "selfish gene theory" preventing males with inferior genes from reproduction. There are many open questions on acanthocephalan copulation and fertilisation.

The following steps of fertilisation have been documented: oocysts become fertilised whilst lying underneath the surface (syncytium) of an ovary. The flagellum of the sperm attaches to the surface of the ovary (Figure 11.18B), which leads to an inflation of the flagellar apex. The subsequent penetration of the spermatozoan through the supporting syncytium into the oocyte (Figure 11.18C) apparently initiates meiosis and the formation of polar bodies. The electron-dense inclusions of the mature oocyte move to the periphery and initiate the formation of a fertilisation membrane around the zygote. The zygote now becomes ovoid and gives rise to a fertilisation gap between its surface and the supporting syncytium of the ovary.

Postzygotic Development

The first step of postzygotic development is the formation of the first larva, the acanthor.

Females of all acanthocephalans release fully embryonated eggs. Prepatent periods of worms from homoiothermic hosts last between 22 days (*Polymorphus minutus*) and 70 days (*Macracanthorhynchus hirudinaceus*); in poikilothermic hosts development depends on the temperature. These same two species may remain patent for a maximum of only 25 days (*P. minutus*) or for up to 10 months (*M. hirudinaceus*).

After being taken up by the intermediate host the acanthor changes its morphology and becomes an Acanthella – the stage between the acanthor and the larva that is infective to the final (or paratenic) host.

Leeches

Synonym

Hirudinea

Classification

Subclass of Annelida

General Information

In general, members of the Hirudinea, which are endowed with a complete intestinal tract, look like plathelminths and are dorsoventrally flattened monoecious animals (hermaphrodites). Mostly they develop 32 inner segments (except for 29 in Acanthobdella), which each correspond to 2–14 outer annuli (some authors count 34 segments). Except for Acanthobdella the Hirudinea, parasitic or not, are characterized by an oral sucker and a posterior disk-like acetabulum, which act as strong holdfast organs supporting movements on the surface of animal hosts or plants, etc. (Figure 11.1, *Hirudo medicinalis*/Figure 11.1, *Hirudo medicinalis*/Figure 11. 2). Bloodsucking leeches produce enzymes (*e.g.*, hirudin) preventing coagulation of blood during feeding. Typically, closed circulatory vessels and a definitive respiratory system are lacking in most species and the coelom cavities are restricted to a few blood-transporting lacunae. The ontogenesis of the Hirudinea proceeds directly, without involving any larval stages (such as the trochophora larva which is characteristic for other annelids).

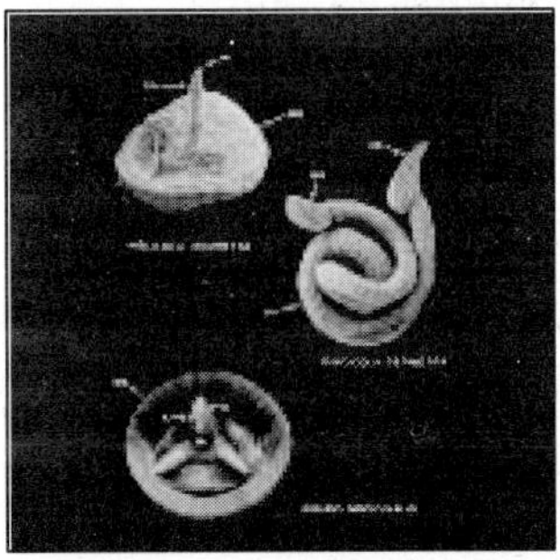

Figure 11. 18D: SEM Micrographs on Leeches

a) Anterior pole of *P. geometra* with the protruded pharynx (× 100). b) Total lateral view of *P. geometra* (× 10). c) Mouth part of the medical leech, *Hirudo medicinalis* (× 20). AS, anal sucker; K, jaw; M, mouth; MS, oral sucker; PH, pharynx; Z, row of teeth.

Thus, juvenile worms hatch immediately from eggs which are joined in packs of 1–200 into a cocoon; the latter is secreted 2 days to months after copulation of two individuals by their clitellum region (segments 9–11).

System

The classification of the Hirudinea is still a matter of debate; however, the following groups are widely accepted:

Subclass: Hirudinea (some parasitic species)

Order: Acanthobdellida

Unlike other orders, they show only the posterior acetabulum, are endowed with some surface bristles and are composed of 29 inner segments (each with four outer annuli)

Order: Rhynchobdellida

The anterior part of the intestine forms a large, protrusible and retractile proboscis without teeth; specimens suck blood and thus may transmit several endoparasites; they are eyeless, their blood colourless; host location occurs by body contact

Order: Gnathobdellida

Anterior part of intestine is noneversible, but is armed by three pairs of jaws with varyingly strong cuticular teeth, the members of this group that are up to 25 cm long are characterized by five pairs of eyes, five annuli per segment and red blood in a lacunar system; host finding occurs by means of olfactory systems

Order: Pharyngobdellida

Specimens have no teeth and a nonprotrusible pharynx; they suck blood by means of their strong muscular pumping pharynx (armed with one or two stylets but no teeth), are provided with three or four pairs of eyes and contain red blood in lacunae

Integument

As in other members of the phylum Annelida, the leeches are covered by a unilayered, cellular epidermis which excretes to its surface a thin cuticle consisting of a proteinaceous ground substance including collagen fibers. Chitin is only present in the rigid bristles (chaeta) which are, however, absent in most hirudineans except Acanthobdella. This cuticle remains very elastic and is molted several times (not at definite intervals) in Hirudinea, but not in other annelids.

Intestine and Food Uptake

The intestine of parasitic leeches consists of a frequently armed mouth, a pharynx, an esophagus, a stomach, an intestine with 6–7 large paired diverticles, a hindgut and the rectum ending with the anus (*Hirudo medicinalis*/Figure 11. 1). In the medicinal leech *Hirudo medicinalis* the teeth of the 3 jaws are pored by the canaliculi of unicellular glands, which produce besides anesthetic and vasodilatory compounds, an anticoagulant (hirudin) which acts as an antithrombokinase. Due to this fact a large leech may ingest about 15 g blood (*i.e.* tenfold his own body weight) within a few minutes. This phenomenon was used for centuries in human medicine as a general remedy. The blood cells are stored in the intestinal diverticles, while the fluid compounds are rather quickly excreted. This amount of blood allows the leeches to survive starving periods of up to two years. The blood cells inside the intestine are retained due to the activity of the hirudin in full shape for a long time and become digested bit by bit due to the activity of particular bacteria (*Pseudomonas* and *Aeromonas hirudinis*) instead of its own digestive enzymes (which, however, are not completely absent). The bacteria are packed by the adult worm on its eggs and are thus transmitted to the next generation. Investigations of our group prove the long (1/2 year) survival of many agents of diseases (viruses, bacteria, protozoan parasites) inside the stored blood. Furthermore, the antibodies of HIV- and Hepatitis B viruses were found in African leeches, so that leeches pose a considerable danger for their hosts.

Excretory System

The excretory system of leeches consists of 10–17 pairs of metanephridia (*Hirudo medicinalis*/Figure 11.1) which each open with a porus at the ventral body surface. They start with a ciliated funnel that opens into the body cavity or into a spherical nephridial capsule (depending on the species) and collect materials to become excreted.

Nervous System

The nervous system of leeches (*Hirudo medicinalis*/Figure 11.1) corresponds in principle to that of other annelids, while being composed of a ventral chord system with a typical broad fused ganglion in each segment. From there 3–4 ring-like running pairs of nerves extend to the periphery. The ganglia of the segments of the posterior sucker are fused to form a large terminal ganglion. Anteriorly, a supraesophageal ganglion is developed innervating sensillae of various types. This epi- = supraesophageal ganglion is connected to the subesophageal ganglion which innervates the mouth and the anterior intestine.

Copepod Parasites

General Information

The crustaceans were termed with respect to their often very strong chitinous cuticle, which is strengthened in comparison to that of insects by the inclusion of calcium salts; thus growth can only proceed after repeated molts. The heteronomously segmented dioecious crustacea appear in a great variety, although lower (former: Entomostraca) and higher (Malacostraca) groups can be clearly differentiated according to their larvae (nauplius or zoea). Common to all crustaceans are the following features:

They have two pairs of antennae (possibly of different size).

The uptake of O_2 occurs by means of inner or outer gills.

A pair of mandibles and two pairs of maxillae act as mouth parts. They are located ventrally at the head, which is often closely connected to the thorax (cephalothorax) and covered dorsally by a carapax in some groups.

Each of the heteronomous body segments is ventrally provided with a pair of extremities, which appear as maxillipodia, pereiopodia or pleopodia (depending on their location along the body).

These extremities show a typical biramous organization and are composed of an inner endopodite and an outer exopodite.

Eyes of the compound type are situated, if present, dorsally on stalks; eyes of the median type (*e.g.*, nauplius eye) are composed of three or four small pigment-cup ocelli of the inverse form.

The circulatory system is open involving a dorsally located heart (in thorax or abdomen).

Antennal or maxillary nephridial glands work as an excretory system supported by different cells (nephrocytes) opening at the body's surface.

The development occurs in general as a metamorphosis involving different larval stages (nauplius or zoea) and growth by repeated molts.

System

The following taxa include some parasitic forms, although most members of the crustacea are free-living:

Class: Crustacea

Subclass: Ostracoda (some species are parasitic on gills of fish)

Subclass: Copepoda (some are ectoparasites of fish, some are intermediate hosts for cestodes and nematodes)

Subclass: Branchiura (some species are blood suckers on fish, thus named fish lice, Argulus).

Subclass: Cirripedia (some species are endoparasites in marine crabs)

Subclass: Malacostraca (higher Crustacea)

Order: Amphipoda (free-living, some are intermediate hosts for endoparasites)

Order: Isopoda (some species are ectoparasites of fish)

Order: Decapoda (many species are intermediate hosts for endoparasites)

Life Cycle

Lepeophtheirus salmonis.

Integument

The cuticle of the crustaceans is constructed according to the same plan as in ticks, mites and insects. However, in crustaceans minerals such as $CaCO_3$ and $Ca_3(PO_4)_2$ may occur additionally, thus leading to a further hardening of the cuticular exoskeleton (needed *e.g.* to overwhelm high water pressures). The molt (ecdysis) in crustaceans occurs due to activity of hormones such as in other arthropods. Here the so-called Y-organs are steered by neurohormones. Ecdysis starts in crustacea when the excretion of the molt-inhibiting hormone (MIH) is stopped and the Y-organs produce crustecdyson (ecdysteroids).

Nervous System

The nervous system shows trends towards concentrations at the anterior pole to form a large brain and in the thoracal region, where a large thoracic ganglion is formed by fusion of segmental ganglia during evolution. In the highly aberrant body shapes of the parasitic crustaceans many variations of the nervous system occur.

Argulus

A major threat in a pond or tank

Argulus, or fish lice, represent a major threat to fish health; both as a result of direct tissue damage and secondary infections. Fish lice are one of the biggest parasites (5-10 mm) and visible with the naked eye.

Argulus feed by first inserting a pre-oral sting which injects digestive enzymes into the body. They then suck out the liquidised body fluids with their proboscis-like mouth. (Take a look at the movie to see the mouth in close-up.) Feeding can take place on the skin or in the gills

A close up shot showing its eyes at the top of the picture (Figure 11.19). The two prominent circular structures near the top are the suckers used to hold onto the fish while feeding. Just below and in-between the suckers is the proboscis-like mouth that it uses for feeding

This feeding activity causes intense irritation. Fish are damaged by the constant piercing of the skin by the stylet and there is often localised inflammation. The other danger is that opportunistic bacteria such as *Aeromonas* or *Pseudomonas* sometimes

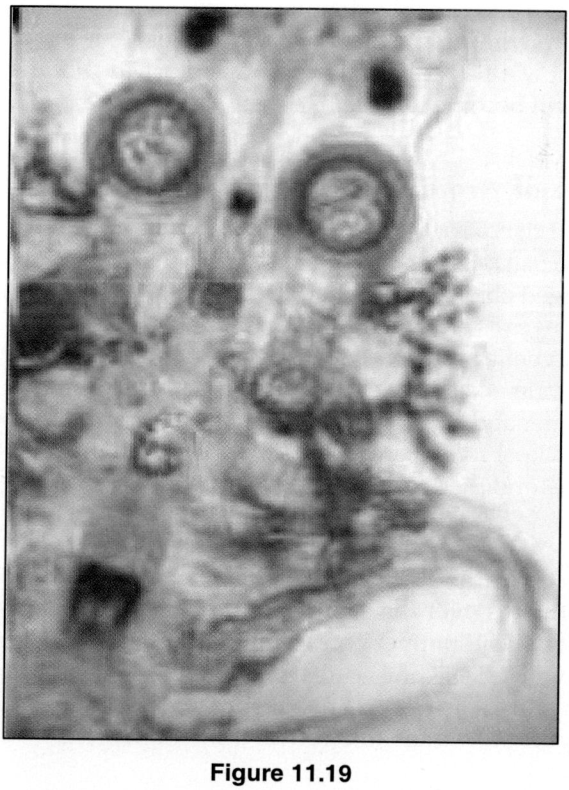

Figure 11.19

infect these damaged areas leading to skin ulcers and gill disease. It is also believed that the stylus may occasionally 'inject' viruses and bacteria into the fish. The various spines, suckers and hooks that lice use for attachment may also cause additional tissue damage. So all-in-all a thoroughly nasty parasite!

In addition to physical damage, affected fish are subjected to severe stress, which often leads to secondary parasite infestations such as white-spot and Costia. This type of combined attack on stressed and often weakened fish can result in high numbers of fatality. So quite clearly, even finding one louse warrants immediate treatment and a follow up examination to check for secondary health problems

Identification

Biologically, *Argulus* spp.,are crustacean parasites in the subphylum crustacea which means they are grouped along with shrimps, prawns and water fleas etc. Animals in this group have a rigid or semi-rigid chitin exoskeleton, which has to be moulted as they grow larger. They are in the class Branchiura, a group of crustaceans with very similar features; all branchiurians are fish parasites.Although it is easy to spot lice when you know they are there, they are easy to miss in the rush to take skin scrapes. To the naked eye they appear as very small dark spots that are easy to overlook unless they move. They are often found in relatively sheltered areas behind the fins or around the head. They are usually easier to spot on fins rather than the body, as they tend to show up more against a plain transparent background. Lice are oval-shaped and flat and capable of moving very quickly. In an aquarium, they can sometimes be seen swimming as they move from host to host.Fish with a heavy lice infestation will show a classic irritation response such as rubbing and flashing. At a later stage they will become lethargic. Affected fish may have focal red lesions on their body.

The Life Cycle of *Argulus*

As with most fish parasites, they have a high reproductive potential. Mating takes place on the fish, after which the female swims away and lays eggs on plants and other submerged objects. When the eggs hatch the juvenile passes through several metamorphic changes as it develops into an adult. Around 4 days after hatching, the newly-hatched juvenile actively seeks a host and continues its development on the fish. The whole cycle takes between 30 – 100 days depending on temperature. The eggs can over-winter and hatch in spring as water temperatures increase. Adults can survive without a host for several days. Any treatment plan has to take account of emerging juveniles and therefore prevailing temperatures.

Treatments

The most successful and effective treatments against lice are organophosphates. Using three treatments over the estimated life cycle of the parasite almost always eradicates lice. At typical summer pond temperatures of 20°C or higher, treatments at 10-day intervals will kill existing adults and juveniles as well as emerging juveniles. The down-side is that in the UK organophosphates are banned for use as fish disease treatments. They are still obtainable - but at a sky-high price.

There are no other treatments currently available that are likely to be totally effective. There is some suggestion that using a chitin inhibitor such as dimilin will stop the juveniles developing as they moult their exoskeleton but there has been no real testing done on this proposal. (dimilin)

More environmentally friendly alternatives are currently undergoing licensing evaluation tests for use in the food-fish industry. However, the draw back is again liable to be costs. Initial reports suggest that these alternatives may be better at controlling rather than eradicating lice.

On a final point, in one incident last year examined a small aquarium that was overrun with lice. There had been no new additions of fish or plants. The only possible source was live *Daphnia* put in the tank from time to time.

Parasites and fungus

Common problems facing koi keepers

Koi parasites and fungus

These are two problems that regularly affect koi health; both can be dealt with fairly simply when they are at an early stage. Fungus, if allowed to spread, will quickly overwhelm fish. Large numbers of parasites can cause physical damage to both skin and gills and are often the precursor to bacterial disease. Diagnosis and treatment is often a chicken and egg situation.

Disease Diagnosis

Diagnosis is really 'detective work'. The clues are usually there for the careful observer and if you look carefully enough. If you are methodical and try to eliminate as many possibilities as you can, without jumping to conclusions, then you may well be led to the correct conclusion. At the very least, you will have gathered important observations that will assist someone else, more expert than you, in reaching a diagnosis or coming to a conclusion. There are two main problems. The first is that there are many clinical signs that can have more than one cause. For example, 'flashing' and leaping out of the water could be caused by irritation from parasites, but it may just as well indicate a water quality problem. For this reason it is important to carry out regular water testing and to keep a record of readings in order to assist in arriving at the correct diagnosis.

The other difficulty is deciding whether the apparent cause of poor health is the primary or secondary factor. For example, a fish that is unwell is more prone to parasites, which may begin to multiply and become a health problem in their own right. But this would be a secondary infection brought about by the original problem. This is what meant when compared making a diagnosis with being a detective. It is often very easy to jump to conclusions and thereby overlook the initial cause.

However, bearing in mind these potential complications, we can, for the sake of simplicity, divide health problems into three general categories, namely:

- External or skin problems: fungus, parasites, bacteria
- Gill problems
- Other problems

Fungus

The most common fungus to affect fish is *Saprolegnia*. Often it is first noticed as white or grey cotton wool-like tufts sticking out of the skin or fins of fish. It is sometimes tinged green with trapped algae. Fungus is nearly always a secondary problem, meaning it has occurred because of some other initial damage. The primary cause may be bacterial, parasitic, water quality or the fish may have physically damaged it's skin or fins in some way. It can be a serious problem because the fungus will spread and invade healthy tissue. Treatment with a proprietary anti-fungus compound will normally clear the problem, if applied early enough. But it is just as important to treat the underlying primary cause to prevent the fungus reappearing.

Parasites

Fish attract a whole range of microscopic parasites, some of which live on the outside of the body on the skin, fins and gills, while others invade the internal body tissues and organs. Those living on the outside are known collectively as ectoparasites, while internal parasites are called endoparasites. Although some endoparasites can be quite serious, it is fair to say that most parasitic problems in ponds involve ectoparasites. The latter vary in size and structure, from minute single- celled protozoa, such as *Costia* and *Trichodina*, to worm-like skin and gill flukes which may be up to 1 mm long (see the main health section for more details).

In small numbers they are usually no problem but, given their enormous reproductive capabilities and the right conditions, they can rapidly multiply and overwhelm the fish. In severe infestations fish can become literally smothered. Large numbers can endanger the health of fish because of the damage they do to delicate skin and gill surfaces. For instance. flukes and leeches have attachment devices - normally suckers and/or hooks - which enable them to stay attached to the fish and move freely over the body. The damaged tissues then provide ideal entry sites for secondary fungal or bacterial infections.

As you can probably imagine, the presence of large numbers of parasites is extremely irritating and one of the main signs of a ectoparasite infestation is the sight of fish rubbing themselves against a solid object, often bottom drains, in an attempt to ease this irritation. This rubbing action, during which the fish often turn on their sides, is called flashing. Other signs of irritation are fish leaping out of the water and swimming at speed around the pond, apparently in great irritation.

The other response is an increase in mucus production, often leading to 'slime disease', whereby the mucus forms a grey/white layer over parts of the body. If fish lie motionless in the pond or gasp at the water surface, or hang around water returns, this may also indicate a parasitic problem, but it is also indicative of several other serious problems, and the beginner would be well advised to seek more expert help in this situation.

Although the above signs of irritation are useful indicators of an ectoparasite disease, it should be remembered that poor water quality, especially high ammonia or nitrite levels, will cause the same response or 'clinical signs'. In reality, the only certain way to make a diagnosis by carrying out a microscopic examination of a skin scrape, which has the advantage of determining the actual species involved and the severity of the infection.

Other specific problems worthy of mention are 'white spot', which is fairly easily cured but can be fatal if not rectified quickly. This is easy to identify, as the fish looks as though it has been sprinkled with fine grains of salt. Small, white cysts up to 1 mm in diameter are clearly visible over the body and fins of the fish.

When the trophonts inside these cysts mature, the cysts rupture, releasing the mature trophonts into the water. The damage done as the cyst ruptures can provide a site for secondary infection. Meanwhile, the mature white spot trophont reproduces, releasing hundreds of 'swarmers' or tomites into the water to re-infect the fish. Larger ectoparasites, such as *Argulus* (fish louse) and *Lernaea* (anchor worm), are visible without magnification.

These are two real parasites which should be treated as soon as possible. *Argulus* is up to 10 mm across and is often overlooked because at a glance it appears to be just a small black mark. However, if your fish are 'examined' regularly these lice should be easily spotted and if watched for a few moments these black' spots will be seen to move. If *Argulus* is suspected, a close examination in a shallow bowl will quickly settle any doubts. The first sign of *Lernaea* is what appears to be fleshy pieces of cotton, up to 20 mm long, attached to the skin or fins of the fish. Closer examination shows the body of the anchor worm, with its head buried into the flesh of the fish, with a trailing body and two egg sacs at its tail end.

With the exception of a few species, such as white spot, *Argulus* and *Lernaea,* most ectoparasites are not harmful in small numbers, being similar to fleas on a cat or dog. The major danger comes when, for some reason, the number gets out of control. Without the aid of a microscope it is difficult to determine just how severe the infestation is. However, it would be a mistake to treat the pond with an anti-parasite treatment every time a fish rubbed against the side or jumped out of the water. Equally, it would be wrong to ignore things if the fish were obviously suffering from great irritation and were constantly leaping around or tearing around the pond, rubbing along the bottom and sides.

Chapter 12
Microbial Diseases

Introduction

Isolation and identification of etiological agent(s) of any disease is essential for diagnosing and treating disease(s). For every disease there must be a cause or a combination of causative factors (mixed etiology) and detection of these factors is imperative for a rapid diagnosis and treatment of the disease(s) and also to take suitable timely prophylactic measures.

Microbial diseases are the predominant group of diseases among finfish and shellfishes. Obviously, identify of the disease causing microbe is absolutely essential to provide means for treating diseases and for preventing disease outbreak. Occurrence of infection and diseases are noticed not only in the culture system, but also in the natural environment. However, severity of the disease is left to the virulence of the pathogen. Mild cases of infection often escape attention and are detected when large scale mortality occurs among the animals and it is too late for any remedial measures to be effectively taken. Whether finfish or shellfish, early detection of infection may be difficult partly due to variation in the refractive index between the water and air which results in changes in the visual properties of the objects. For example, disease spots in the body of these animals become unnoticed when they are removed from their habitat. Moreover, diseased fishes become incapacitated and easily fall prey to the predators. This may minimize or completely wipe out the infected stock even before the fish farmers realize it.

So, in large scale culture systems, it is essential that continuous monitoring be carried out to detect finfish and shellfish diseases at the very outset.

In the present investigation, diseased finfishes and shellfishes were mainly obtained from the culture systems at the Central Marine Fisheries Research

Institute(CMFRI) Prawn Research Centre, Narakkal (Cochin); the CMFRI Mussel farm, Vizhinjam, the CMFRI Oyster farm, Tuticorin and the CMFRI Kovalam Research Centre, Madras. A few cases were also detected from the natural populations of finfish from the inshore waters off Cochin and from the Vellayani Lake, near Trivandrum.

The diseased finfish and shellfishes under moribund conditions (except the case of tumour in *Arius jella*) were aseptically collected in suitable containers which were clean and sterile. When needed, long sterile forceps were used for handling the materials to minimize damage and contamination. The collected samples were taken to the laboratory immediately.

To clearly pinpoint and confirm the microbial causes of the diseases, the Koch's postulates to be satisfied are:

1. The suspected organism should be associated regularly with all cases of the suspected disease and in logical relationship to the disease, its symptoms and lesions.
2. The suspected organism should be isolated in pure culture from the suspected case.
3. When such pure culture is inoculated into suitable animals, the disease should remanifest itself.
4. The same organism should be reisolated in pure culture from the inoculated animals.

The different species of diseased finfishes and shellfishes, analyzed for investigating the causes of diseases, are listed in Table 12.1.

Pre-Sterilization of Surgery Cell

During the present investigation, a large surgery cell in which the candidate could enter and carry out the work was used. The cell was first rendered dust free and clean with a neat towel. The surgery cell was then disinfected with dettol followed by absolute alcohol by the swab method. The cell was then sealed and sterilized by switching on ultraviolet tube lights for a period of 60 minutes.

Surgery

The diseased specimens were placed, in a clean and decontaminated dissection tray, from the container. The specimens found alive were killed by a hard blow on the head or by decapitation. The pathological samples from the external parts such as skin, fins, tail, eyes, gills and tumours were removed as follows:

External Parts

Skin Scrapings

Skin scrapings in the case of finfish and shell scrapings from the shellfish (prawns) from the infected area were taken aseptically using sterile surgical blades with necessary help of the sterile scalpel, forceps scissors.

Table 12.1: List of Diseased Finfishes and Shellfishes

Sl.No.	Scientific Name	Popular name	Habitat
1.	*Anquilla bicolor bicolor*	Level finned eel	F
2.	*Arius jella*	Small eye cat fish	M
3.	*Clarias batrachus*	Teysmann's spotted cat fish	F/E
4.	*Chanos chanos*	Milk fish	F/E/M
5.	*Etroplus suratensis*	Pearl spot	F/E
6.	*Tilapia mossambica*	African mouth breeder	F/E
7.	*Sillago sihama*	Lady fish	E/M
8.	*Chirocentrus dorab*	Wolf herring	M
9.	*Pampus argenteus*	Silver pomfret	M
10.	*Mugil cephalus*	Grey mullet	F/E/M
11.	*Lates calcarifer*	Bekti	E/M
12.	*Lactarius lactarius*	White fish	M
13.	*Megalops cyprinoids*	Indian tarpon	E/M
14.	*Wallago attu*	Wallago	F/E
15.	*Penaeus indicus*	Indian white prawn	M
16.	*Penaeus monodon*	Giant tiger prawn	M
17.	*Metapenaeus monoceros*	Speckled shrimp	M
18.	*Panulirus homarus*	Spiny lobster	M

F: Freshwater; E: Estuarine; M: Marine.

Fins/Tail/Uropods/Telson

Infected fins/tail/uropods/telson was clipped off aseptically with the help of sterile surgical instruments.

Eyes

The infected eye tissue was removed aseptically from the affected areas.

Gills

The operculum was lifted first by using sterile forceps and the infected gill filaments and tissues were cut off and transferred aseptically.

Tumours

The tumours were excised, with the help of sterile scissors, surgical blades and forceps and transferred aseptically.

Internal Parts

Using sterile instruments, the internal parts were removed as follows:

Kidney

The ventral area of the specimen was sterilized first by carefully applying absolute alcohol and burning off the excess, after descaling the region. The body was then slit open mid ventrally from operculum to vent exposing the visceral organs. The kidney was aseptically removed without any contamination from other internal parts, with the help of surgical instruments. During the operation, the colour of the various internal organs and other changes if any were also noted.

Blood

Blood was removed directly from the open heart by using a sterile one ml tuberculin syringe with a one inch needle of gauge size no.22 or by cardiac puncture from the anaesthetized fish using a similar syringe. Sterile platinum wire was also used to transfer the blood.

Hepatopancreas

By dissecting open the carapace, along the post rostral carina, the hepatopancreas was removed aseptically.

Mussel and Oyster Tissue

By aseptically dissecting open the bivalves, the infected tissue was transferred.

All the pathological samples transferred were immediately placed in sterile Petri dishes. In the case of blood, the sample was removed into sterile glass tubes having sterile anticoagulant (EDTA) to avoid clotting.

For the sterility check, inside the surgery cell, open nutrient agar plates were also kept exposed inside the cell during surgery. The surgery nature of the surgery cell was confirmed by absence of microbial growth in the plates.

All the surgical instruments were either autoclaved or dipped in absolute alcohol and the excess burned off.

Chapter 13
Microbial Spoilage in Fish and Shellfish

Introduction

Seafood is an excellent choice of food, for protein and other essential nutrients. In many regions of the world, seafood and other aquatic products form a major portion of the diet, while in other parts of the world it is viewed as a delicacy. Today more and more people are turning to fish as a healthy alternative to red meat. The low fat content of many fish species and the effects against coronary heart diseases by the n-3 polyunsaturated fatty acids found in fatty pelagic fish species are extremely important aspects for health conscious people, particularly in affluent countries where cardiovascular disease mortality is high (Huss, 1994). Fish and fishery products are in the forefront of food safety and quality improvement, because they are among the most internationally traded food commodities. Besides supporting the health lives of millions of people, seafood also happens to support the growth of pathogenic bacteria, acting as a vehicle for the transmission of pathogenic organisms that can cause infection or intoxication. So it is responsible for a significant percentage of food borne diseases. The principal biological agents that cause seafood borne disease are bacteria, viruses and parasites. The products of microorganisms such as bacterial toxins, algal toxins or products of bacterial metabolism such as histamine can cause food poisoning when fish or shellfish harboring these products are consumed. The incidence of disease causing agents in seafood has become a reason for the rejection or detention of these products in the international markets.

Seafood differs from other foods in a number of ways. The type of microorganism associated with seafood may vary depending upon whether fresh or processed.

Seafood is processed in to a wide range of products and is consumed in many forms like smoked, canned, salted, dried, frozen and raw. Most seafood is harvested from a wild population and the fishermen are hunters with no influence on handling of their prey before it is caught. The seafood processor is limited in his choice of raw material to what is available in respect of size, condition and fish species landed by the fisherman. The microbial flora present in the intestine, surface and gills of fresh fish is an indication of the bacterial flora present in the aquatic environment from where it is caught (Huss, 1994).

The major determinant both in the quality of seafood and safety is the presence of various types of bacteria present in the raw material. Depending on the route of entry in to the fish body microorganisms can be grouped into three classes as bacteria present in the aquatic environment, bacteria derived by way of pollution of aquatic environment and bacteria acquired by seafood from terrestrial environment through contamination (Thampuran, 2003).

The bacterial species that cause human illness encountered in seafood can be classified into three groups (Feldhusen, 2000). These include the members of the family Vibrionaceae, such as *Vibrio cholerae, Vibrio parahaemolyticus* and *Vibrio vulnificus,* virulent strains of *Aeromonas hydrophila* and *Clostridium botulinum,* which are indigenous to the environment. Second category comprises of bacteria introduced via faecal contamination, which are mainly members of the family Enterobacteriaceae, such as *Escherichia coli, Salmonella, Shigella* and pathogenic serotypes of *Yersenia enterocolitica*. The third group consists of bacteria introduced while processing and storage such as *Listeria monocytogens, Bacillus cereus, Clostridium perfringens* and *Staphylococcus aureus.*

Thorough cooking of seafood products would virually eliminate all microbial and parasitic pathogens. It will not destroy some microbial toxic metabolites. Some of these toxic metabolites are produced by Gram-positive bacterial genera like *Bacillus, Clostridium* and *Staphylococcus*. These metabolites show various degrees of heat resistance. Among these toxins Staphylococcal enterotoxin is the most heat resistant one, which is produced by *S.aureus*.

A number of studies have been carried out on the incidence and characterization of *Staphylococcus* isolated from different foods, clinical, bovine and other sources. The present work is an attempt to know the incidence of *Staphylococcus* particularly *Staphylococcus aureus* in various fish and fishery products both meant for internal consumption and export market. The main objectives of the study are:

Investigation of the microbial quality of various fish/shellfish and fishery products.

1. Study the incidence of *S.aureus* in various fish and fishery products.
2. Study the *toxigenic* potential of *S.aureus* and *Staphylococcus* species from various fish and fishery products.
3. Biochemical characterization of *S.aureus* and other *Staphylococcus* species.
4. Assess the pathogenic potential of *S.aureus* and other *Staphylococcus* species.

5. Assess their antibiotic sensitivity.
6. Study the effect of environmental conditions on growth and toxin production by *S.aureus* in *in-vitro* conditions.
7. Assess the temperature on growth in *in-vivo* conditions.
8. Comparison of molecular method and immunological method for the detection of toxigenicity of *Staphylococcus* strains.

Chapter 14
Bacteriology of Fresh Fish

Introduction

Quantitative Aspects

The flesh and internal organs of freshly caught healthy fish are normally sterile but bacteria may be found on the skin, gills and in the intestine (Shewan, 1962). The skin of fish usually carries a bacterial load in the order of 10^2-10^7cfu/cm^2. The gills and intestine also carry a bacterial load of the order 10^3-10^9/g and 10^3-10^9/g respectively.

Most of the studies reported a total bacterial count in the range of 10^4-10^6cfu/g in fresh fish and shellfish. The microbial quality of retail fish fillets in the Netherlands had bacterial counts of 10^6cfu/g (Vanden Broek *et al.*, 1984). Similar results were reported by Chang *et al.* (1984) in *Sardinops melanostica* from Korea Torres-Vitela *et al.* (1997) from Mexico.

Qualitative Aspects

Several investigators have concluded that the microorganisms associated with most fishery products reflect the microbial population in their aquatic environment (Liston, 1980; Colby *et al.*, 1993; Ashie *et al.*, 1996; Gram and Huss, 1996). Variations in marine environment affect the type of bacteria present on the skin and gills and surface of newly caught fish. Thus, in the cooler seas of the Western hemisphere, bacterial flora is dominated by psychotropic Gram- negative rods.

In the warmer waters of India, east coast of South Africa, Australia and in the Adriatic sea, flora was dominated by mesophiles, among which *Micrococcus* was the most important species (Wood, 1953; Gillspie and Macrae 1975). The qualitative composition of fish from temperate waters mainly comprised of the genera

Pseudomonas, Moraxella, Acinetobactor, Flavobacterium, Cytophaga and *Vibrio*, which constituted about 80 per cent of the population (Liston 1980).

The flora on tropical fish often carries a slightly higher load of Gram-positive bacteria compared with fish from colder waters (Liston, 1980). Studies by Karthiyani and Iyer (1975); Surendran(1980); Surendaran and Gopakumar (1983); Surendaran *et al.* (1989); Abraham *et al.* (1992) and Shetty and Shetty (1990) indicated the predominance of Gram negative asporogenous rods like *Vibrio, Pseudomonas, Acinetobactor, Flavobacterium, Cytophaga, Aeromonas* etc. in tropical fishes like oil sardine, Indian mackeral and Jew fish.

Bacteriology of Fresh Crustacean Shellfish

Qualitative Aspects

The main groups of bacteria comprising the flora of crustacean shellfish are *Micrococcus, Corynebacterium, Achromobactor* and *Pseudomonas* together with fewer numbers of *Flavobacterium/Cytophaga* and *Bacillus* (Cann, 1977). Their proportion in composition varies with the temperature of waters in which the animals live, cold-water species having largely *Pseudomonas* and *Achromobacter* (Liston, 1980), while *Micrococcus* and *Corynebacterium* dominate in crustaceans from warm waters (Sreenivasan, 1959). Gulf shrimp contaminated largely of *Achromobactor, Micrococcus, Pseudomonas* and *Bacillus* (Williams and Ree, 1952; Williams *et al.*, 1952). *Micrococcus* dominated in the flora of Lousiana crab (Alford *et al.*, 1941) while *Achromobactor* dominated in Pacific crab (Lee and Pfeifer, 1975). Gram negative flora dominated in shrimps from Cochin (Karthiyani and Iyer, 1975; Surendaran *et al.*, 1985) while the dominance of Gram positive bacteria was noticed in shrimps collected directly at sea Off Mangalore (Krishnamurthy and Karunasagar, 1986).

Bacteriology of Molluscan Shellfish

Molluscan shellfish, except a few true marine species grow and reproduce in and are harvested from estuarine areas. The quality of shellfish depends on the quality of overlying estuarine waters. The bacterial counts in molluscan shellfish were in the range of 10^3 to $10^8/cm^2$ (Cann, 1977). The average bacterial counts of mussel (*Perna viridis*) from Calicut were in the range of 10^3 to $10^6/g$ (Surendaran *et al.*, 1986). In conclusion the bacterial load of molluscan shellfish is a reflection of the aquatic environment from where it is harvested. The shellfish samples collected from polluted waters especially from faecal contaminated waters generally harbour high bacterial load and pathogens in their meat due to its filter feeding nature.

Qualitative Aspects

The main groups of bacteria constituting the flora of clams are Gram-negatives (78 per cent) composed of *Vibrio, Aeromonas, Pseudomonas, Shewanella, Moraxella, Acinetobactor*, members of Enterobacteriaceae and a smaller portion by Gram positives like *Bacillus, Micrococcus, Arthrobactor, Chromobacterium* and *Staphylococcus* (Lalitha and Surendaran, 2005). *E.coli* and faecal coliforms were detected from Chesapeake Bay soft shell clam (*Myaarenaria*) (Chai *et al.*, 1990) and West African clam (*Egeria radiata*) (Ekanem and Adigoke, 1995).

Bacteriology of Freshwater Fish

Quantitative Aspects

Fresh water fish tend to have lower counts than marine species *viz.*, 102 to 105 cm^{-2} for skin and 104 to 107/g for gut content of temperate fish and 103-105 cm^{-2} of skin surface and 104-106/g guts of tropical species (Shewan, 1977).

Qualitative Aspects

Surendaran and Gopakumar, (1991) carried out the microbiological examination of the fresh water fishes rohu (*Labeo rohitha*), calbasu (*Labeo calbasu*), mrigal (*Cirrihinus mrigala*), and Catla (*Catla catla*). They reported the presence of *Pseudomonas, Moraxella, Acinetobactor, Vibrio, Micrococcus, Arthrobactor, Lactobacillus, Bacillus* and members of Enterobacteriaceae in these species. Lilabati and Viswanath, (1998) could isolate *Acinetobacter, Aeromonas, Flavobacterium, Micrococcus Moraxella, Pseudomonas* and *Vibrio* from the gills, intestine, skin and muscle of the iced fresh water fish (*Notopterus chitala*) sold in the retail markets of Imphal (Manipur).

Bacteriology of Frozen Fish and Shellfish

Quantitative Aspect

The microbiological quality of frozen fish depended upon various factors such as the nature of the raw materials, its pre- and post-harvest treatments and the sanitary condition of the processing factories, the rate and nature of freezing, the temperature and length of storage, the original numbers, growth during storage, thawing process and physical protection offered by the food (Luyet and Gebenio, 1940; Larkin *et al.*, 1955; Keroluk and Gunderson, 1959). The bacterial load of frozen fish, shellfish and fish products found to vary between 10^2 to 10^6cfu/g. A bacterial load of 10^2 to 10^6cfu/g in frozen fish (Thampuran *et al.*, 1981; Iyer and Shrivastava, 1989 b; Nambiar and Iyer, 1990; Lakshmanan *et al.*, 1991; Bandekar *et al.*, 2003; Gnanambal and Patterson, 2005).

Qualitative Aspects

The effect of freezing on the microbiological population is dependent on several factors. There is an initial mortality attendant on the freezing process itself. This varies with the type of organism. The initial reduction in number of bacteria immediately after freezing can range from only one or two per cent up to 90 per cent (Shewan, 1961; Simmonds and Lamprecht, 1980). Gram-negative organisms are generally more sensitive to freezing than Gram-positive organisms while spores and food poisoning toxins are unaffected by freezing (Shewan, 1961). The response to freezing varies greatly from strain to strain. *Pseudomonas* species are susceptible. *Salmonella* and members of Enterobacteriaceae are among the more sensitive types (Raj and Liston, 1961). The cocci including marine micrococci and enterococci are more resistant to freezing. The coliform count especially that of *E.coli* on frozen prawns have been found to drop by 95 per cent or more during freezing and frozen storage (Lekshmy, 1964) while faecal streptococci were relatively unaffected. For this reason, streptococci had been preferred as indicators of factory hygiene. Indicator bacteria like *E. coli* and *S. aureus* were isolated from many fish and fish products (Raj and Liston, 1961; Surkiewicz *et al.*, 1968; Baer *et al.*, 1976; Swartzentruber *et al.*, 1980). The presence of pathogens like *Salmonella* and *Vibrio cholerae* non-O1 were reported from different products (Varma *et al.*, 1989; Iyer and Shrivastava, 1989a; Lakshmanan *et al.*, 1993; Aravindan and Sheeja, 2000; Gnananmbal and Peterson, 2005).

Bacteriology of Dried Fish and Shellfish

Quantitative Aspects

Bacterial counts on fully cured seafood are generally low, unless there has been extensive surface contamination. Only halophilic bacteria that have no public health significance can grow on such foods. The Gram-negative bacteria, which are predominant in fresh fish, are relatively sensitive to high salt concentrations and so bacterial numbers decline. Most of the studies reported a bacterial count in the range of 10^3 to 10^5 in different cured dried fish and shellfish (Sreenivasan and Venkataraman, 1958; Solanki and Sankar, 1986, Basu *et al.*, 1989).

Qualitative Aspects

The presence of indicator bacteria like *E.coli*, faecal streptococci and *S.aureus* were reported in few cases (Valsan *et al.*, 1985; Joseph *et al.*, 1986). None of the studies reported the presence of any other bacteria of public health significance from these products.

Bacteria of Public Health Significance from Fish and Shellfish

Fish and shellfish are occasionally found to be the source of food poisoning caused by pathogenic bacteria. The pathogens are generally derived from three sources *viz.*, (a) agents naturally present in the environment, (b) derived from pollution and (c) acquired by the fish by way of handling, processing and marketing (Thampuran, 2003). The pathogens in seafood that are naturally present are the members of the family Vibrionaceae, Aeromonadaceae and the species *Clostridium botulinum*, particularly type E. The water temperature naturally has a selective effect on the pathogens present. Thus the psychrophilic organisms like *Clostridium botulinum* and *Listeria monocytogens* are common in Arctic and other colder climates, while the mesophilic types *Vibrio cholerae* and *Vibrio parahaemolyticus* are part of natural flora on fish from coastal and estuarine environments of temperate and warm tropical zones (Huss, 1994).

Seafood Borne Pathogenic Bacteria

Non-Indigenous Bacteria

Escherichia coli

The natural habitat of *Escherichia coli* is the intestine of human and vertebrate animals. Generally *E.coli* strains that colonize the gastrointestinal tract are harmless commensals (Huss, 1994). But some of the *E.coli* strains particularly those belonging to the classes of enteropathogenic *E.coli* (EPHC), enteroinvasive *E.coli* (EIEC), enterotoxigenic *E.coli* (ETEC) and enterohemorrhagic *E.coli* (EHEC) are causing diseases (Doyle, 1989).

E.coli may be isolated from environments, polluted by faecal material or sewage. The organism can survive and even multiply in warm tropical waters (Rhodes and Kator, 1988; Jimnez *et al.*, 1989). In temperate waters, this organism is absent from fish and crustaceans at the time of capture. The organism is particularly useful as indicator of contamination (small numbers) or mishandling such as temperature abuse in product handling (large numbers). Contamination of food with *E.coli* implies a risk that one or more enteric pathogens may have gained access to the food. However, failure to detect *E.coli* does not ensure the absence of enteric pathogens (Silliker and Gabis, 1976). It is also a recognized indicator of possible contamination with enteric pathogens (Bonnel, 1994; Liston, 1980; D'Aoust, 1989). The resistance of *E.coli* to adverse physical and chemical conditions is low. There is no good correlation between *E.coli* and enteric pathogens because it dies off faster than the latter (Temple *et al.*, 1980; Burton *et al.*, 1987). This makes *E.coli* less useful as indicator organism in examination of water and frozen or otherwise preserved fish products. The incidence of *E.coli* in fresh finfish has been reported by many workers (Lakshmanan *et al.*, 1984; Anand *et al.*, 2002; Rao and Gupta, 1978). Anand *et al.*, (2002) reported an *E.coli* count in the range of 0-100cfu/g in fresh finfishes landed in Tuticorin fishing harbour.

E.coli is commonly found in different seafood's like scallops, mussels and clams (Vijayan *et al.*, 1982; Gorczyca *et al.*, 1985). Their presence in fresh and frozen shrimps from Asian countries has been reported (Varma *et al.*, 1985; Lekshmy and Pillai, 1964a,b; Sumner *et al.*, 1982; Jayaweera and Subashinghe, 1990; Jeyasekaran *et al.*, 1990; Anand *et al.*, 2002). The prawn from deeper waters showed the absence of *E.coli* (Fonseka and Rana, 1990; Cann, 1977). Contrary to this result, Rao and Surendran, (2003) reported the presence of *E.coli* in deep-sea fish (*Psenopsis cynea*) and shrimp (*Heterocarpus woodmasoni*). They also reported the high prevalence of *E.coli* in fish collected from the retail markets of Cochin. Since this organism is present in the intestine of human and other warm-blooded animals, it can be concluded that the presence of *E.coli* in different fish, shellfish and fishery products indicates faecal contamination.

Salmonella

Salmonella spp. are members of the family Enterobacteriaceae and occur in more than 2000 serovars. These mesophilic organisms are distributed geographically all over the world. The principal symptoms of Salmonellosis (non-typical infections) caused by *Salmonella* are non-bloody diarrhoea, abdominal pain, fever, nausea and vomiting which generally appear 12-36 hours after ingestion. However, symptoms may vary considerably from grave typhoid like illness to assymptomatic infection. The infective does in healthy people varies according to serovars, foods involved and susceptibility of the individual. There is evidence for a minimum infection does (M.I.D) of as little as 20 cells while other studies have consistently indicated a higher M.I.D (Varnam and Evans, 1991).

The natural habitat of *Salmonella* is the gastrointestinal tract of mammals, birds, and reptiles (Pelzer, 1989), but principally occurring in the gut of man and animals and in environments polluted with human or animal excreta. Open marine waters are free from *Salmonella*, but estuaries and contaminated coastal waters may harbour the pathogen. Also, poor personal hygiene may transmit this organism. *Salmonella* species reach aquatic environment through faecal contamination. This accounts for the occasional detection of *Salmonella* from fish and fishery products.

Indigenous Bacteria

Vibrio spp.

The genus *Vibrio* contains a number of species (Figures 14.1 and 14.2), which are pathogenic to man. The pathogenic species are mostly mesophilic, ubiquitous in

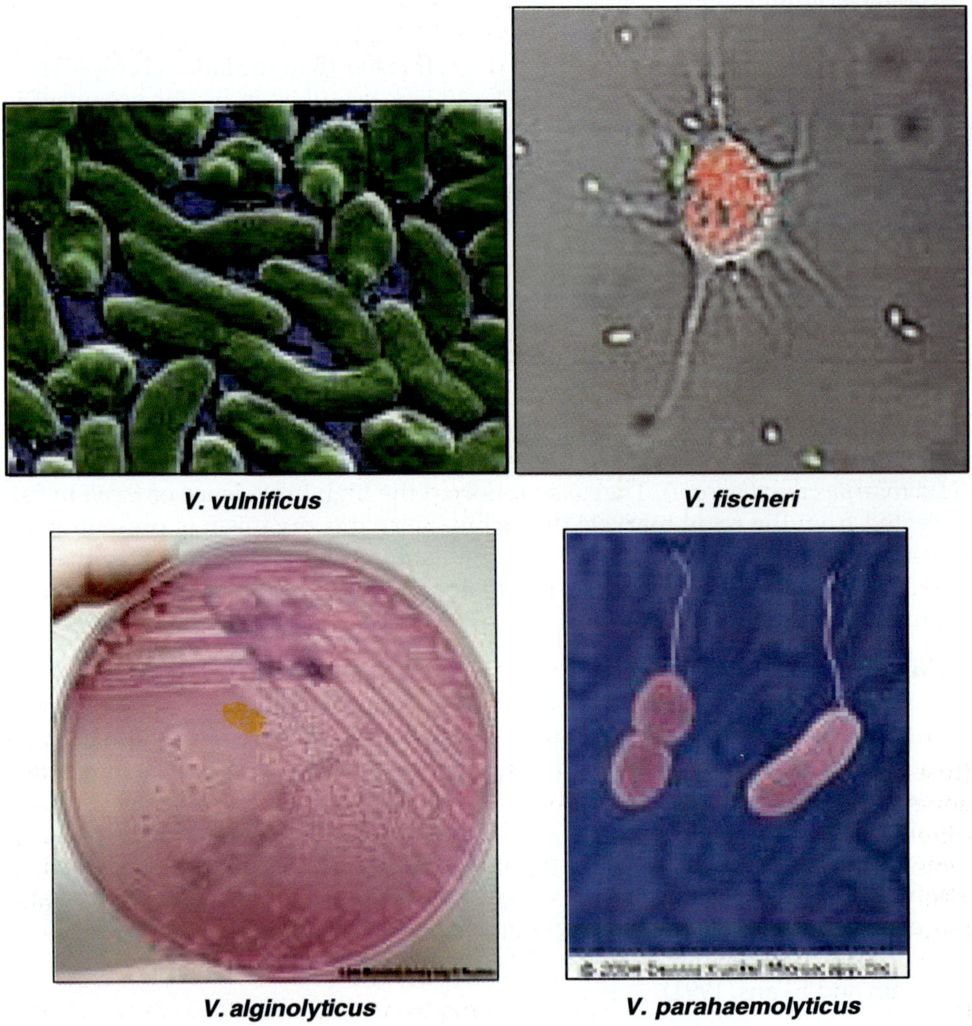

V. vulnificus ***V. fischeri***

V. alginolyticus ***V. parahaemolyticus***

Figure 14.1

tropical waters and in high numbers in temperate waters during late summer or early fall. The disease associated with Vibrios is characterized by gastro enteric symptoms, varying from mild diarrhea to the classical cholera with profuse watery diarrhea. One exception is infection with *Vibrio vulnificus*, which are primarily characterized by septicemia. Most Vibrios produce powerful enterotoxins. As little as 5 micro gram cholera toxin (CT) administered orally caused diarrhoea in human volunteers (Varnam and Evans, 1991). A number of other toxins also are produced by Vibrios. Though named pathogenic, *Vibrio* species are not always pathogenic. The majority of environmental strains lack the necessary colonization factors for adherence and penetration, appropriate toxins or other virulence determinants necessary to cause disease. Vibrios are frequently isolated from seafood and in particular from shellfish.

Vibrio cholerae

It is the etiological agent of cholera, a devastating disease that causes severe dehydrating diarrhea and death in healthy people. *Vibrio cholerae* occurs in two serotypes, the O1 and non-O1. The O1 serotype occurs in two biovars, the Classical and E1-tor. The Classical biovar serovar-O1 is restricted to parts of Asia (Bangladesh). The E1-tor biovar causes most of the cholera outbreaks. The organism may be transmitted from the environment to humans by a number of routes including consumption of under cooked seafood or shellfish.

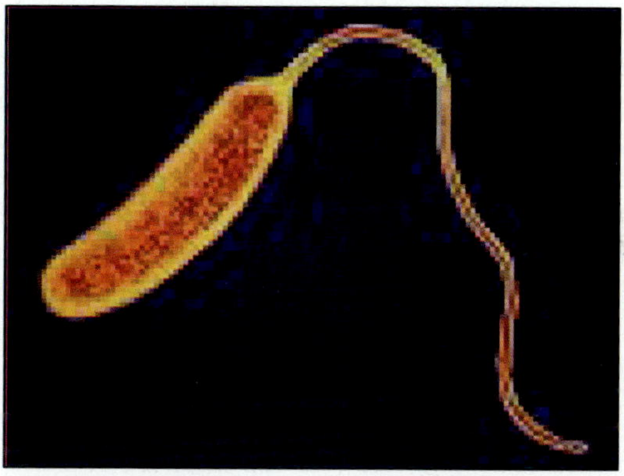

Figure 14.2: *V. cholerae*

Fish and other seafood have been implicated as a source of cholera outbreaks (Joseph *et al.*, 1965). This organism gain access to these products, either from the aquatic environments or through different stages of processing. Their prevalence in different aquatic environments is well documented. It has been isolated from brackish water (Lee *et al.*, 1982; Blake *et al.*, 1980), marine water (Kenyon *et al.*, 1984; Mathew 1986), fresh water (Blake *et al.*, 1980). Their presence in these water bodies indicates the autochthonous nature of these organisms.

V.cholerae non-O1 is a less virulent strain of *V.cholerae*, which lacks the genetic potential for causing epidemic diseases. It is more prevalent in seafoods than *V.cholerae*

O1. Different studies revealed the possibility of contamination of fishery products with *V.cholerae* non-O1 during different stages of handling and processing (Delacruz *et al.*, 1990; Iyer and Varma, 1990b; Iyer *et al.*, 1990b). Many studies reported the incidence of *V.cholerae* in fresh and frozen seafood (Berry *et al.*, 1994; Wong *et al.*, 1995; Baffone *et al.*, 2000; Elhadi *et al.*, 2004 and Varma *et al.*, 1986). Comparing with finfish, the incidence of *V.cholerae* in shellfish is high This can be attributed to their chitinase activity, which plays an important role in this association. The chitinase activity may increase the affinity of Vibrios to crustaceans. This explains why *V.cholerae*, a chitin digester, is found more frequently in crustacean shellfishes like shrimps and crabs than other seafood. The incidence of *V.cholerae* in molluscan shellfish like oysters and clams has been reported (Twedt *et al.*, 1981; Wallace *et al.*, 1999; Namdari *et al.*, 2000). Molluscan shellfishes are generally filter feeders. The presence of *Vibrio cholerae* in these products might be due to the uptake and accumulation in the gastro intestinal tract of these organisms. It can be concluded that the presence of *V.cholerae* in fish and fish products are inevitable. Only way out is the hygienic handling, processing and proper cooking of fish and fishery products.

Vibrio parahaemolyticus

Vibrio parahaemolyticus is a halophilic bacterium that occurs world wide in estuarine environment (Joseph *et al.*, 1983; Karunasagar *et al.*, 1986). Studies reveal that thermostable direct haemolysin (TDH) and the related haemolysin (TRH) encoded by the tdh and trh genes respectively are the major virulent factors of this organism (Nishibuchi and Kaper, 1995; Okuda *et al.*, 1997a). It can cause gastroenteritis in humans and the illness is most frequently associated with the consumption of raw or undercooked seafood and seafood contaminated with bacterium after cooking (Rippey 1994). It has been isolated from a high percentage of fresh and refrigerated seafood (Wong *et al.*, 1992) and frozen seafood (Wong *et al*).This organism has been recognized as an important cause of food borne illness from consuming raw or undercooked seafood in Asia and North America (Bag *et al.*, 1999; De Paola *et al.*, 2000; Matsumoto *et al.*, 2000; Okuda *et al.*, 1997a). It has been revealed that there is a high percentage of incidences of *V.parahaemolyticus* in seafood imported from Asian countries (Wong *et al.*, 1999)

The incidence of *V.parahaemolyticus* was reported from crustacean and molluscan shellfish (Colakoglu *et al.*, 2006; Ottaviani *et al.*, 2005; Brenton *et al.*, 2001; Chang *et al.*, 1996; Broek *et al.*, 1979). Even though this organism is considered as halophilic in nature, its seasonal distribution in fresh water environments and association with fresh water fishes were reported by Sarkar *et al.*, (1985) from Calcutta. Many studies from India reported the isolation of these organisms from different fresh and frozen fish and shellfish (Chatterjee, 1980; Bandekar *et al.*, 1982; Pradeep and Lakshmanaperumalsamy, 1986; Karunasagar *et al.*, 1990; Thampuran, 1996; Sanjeev, 1999; Sanjeev *et al.*, 2000 and Prasad and Rao, 1994).

V. parahaemolyticus is a halophilic pathogen, which becomes a health hazard only under certain conditions like raw consumption of seafood or recontamination of cooked fish and fish products.

Clostridium botulinum

Clostridium botulinum is widely distributed in soil, aquatic sediments and fish (Huss, 1994; Huss and Pedersen, 1979). It is also found in sewage, rivers, lakes, seawater, fresh meat and fish (Haagsma, 1991). Human botulism is a serious but relatively rare disease. The disease is an intoxication caused by the *C.botulinum* toxin pre-formed in food. Symptom may include nausea and vomiting followed by a number of neurological signs and symptoms, visual impairment (blurred vision), loss of normal mouth and throat functions, weakness of limbs or total paralysis and respiratory failure which is usually the cause of death (Huss, 1994). The majority of the food borne botulism associated with fish in the Northern and temperate regions are associated with *C.botulinum* type E, where as in tropical area *C.botulinum* type C and D predominate (Huss, 1981; Dodds 1993a, b). Sometimes fish and fish products have been a vehicle for type A and B botulism. The food poisoning caused by *C.botulinum* type F by the consumption of herring was reported from Czech Republic (Sramova and Benes, 1998). The severity of botulism makes *C.botulinum* an important food borne pathogen. World wide approximately 450 outbreaks with 930 incidents are recorded annually, 12 per cent of which are caused by psychrotrophic serotype of *C.botulinum* type E (Hatheway, 1995).

C.botulinum becomes a hazard when processing practices are insufficient to eliminate the *C.botulinum* spores from raw fish, particularly because of improper thermal processing (Chattopadhyay, 2000). The growth of *C.botulinum* and toxin production depend on appropriate conditions in food before eating, the temperature, oxygen, pH, water activity and the presence of preservatives and competing microflora (Johnson, 2000). Majority of the fish botulism are associated with fermented fish (Huss, 1981). Botulinum toxin formed in the raw material will be found again or even increase *in situ* in the final products such as heavily salted, marinated or fermented fish (Huss and Rey Petersen, 1980). The prevalence of *Clostridium* particularly non-proteolytic type E in different fish and shellfish samples are extensively studied and reported (Christiansen *et al.*, 1968; Hielm *et al.*, 1998; Hyytia *et al.*, 1999; Korkeala *et al.*, 1998). In India, *Clostridium botulinum* has been reported from fish and environments (Lalitha and Iyer, 1990; Lalitha and Surendran, 2002).

Listeria monocytogens

Listeria monocytogens is a Gram-positive motile bacterium that grows well at 37°C but which at the same time is psychrotolerant and halotolerant. Seven species of *Listeria* are known and of these, only *L.monocytogens* is pathogenic to humans (Farber and Peterkin, 2000). *L.monocytogens* is an organism indigenous to the general environment where it is typical of decaying plant material. In addition, it occurs in the gastrointestinal tract of 2-6 per cent of humans, who are healthy carriers. The organism has not been isolated either from free open waters or from fish caught or cultured in such waters. In contrast, water close to agricultural run-off harbour the organism and in principle, the bacterium must be assumed present albeit in low levels on raw fish (Gram, 2001, Huss *et al.*, 1995).

Many studies reported the incidence of *Listeria monocytogenes* in different fish, shellfish and fishery products (Lennon *et al.*, 1984; Weagent *et al.*, 1988; Boerlin *et al.*,

1997; Gram, 2001; Fuchs and Reilly, 1992; Fuchs and Surendran, 1989; Facinelli *et al.*, 1989; Frederiksen, 1991). Review on incidence of Listeria in seafood world wide (Embarek, 1994) found the prevalence of *L. monocytogenes* varied from 4-12 per cent in temperate areas and the low prevalence of (0-2 per cent) in tropical waters. The presence of this organism in processing environment also has been reported (Autio *et al.*, 1999; Vogel *et al.*, 2001).

Acute Infections

The initiation and progress of the disease are dependent on the resistance of the host and the ability of the organism to overcome these. *S.aureus* produces a very wide range of aggeessins and exotoxins that are chromosomal or plasmid mediated and include α, β, γ, δ toxins, leucocidins, staphylokinase, hyaluronidase, nuclease, protein A, coagulase, enterotoxin and various surface associated proteins promoting adhesion of cells to host epithelial cells and tissue components. The ability of *S. aureus* to cause infection is multifactorially mediated (Baird-Parker, 1990).

Acute Toxemia

The causative agents of acute staphylococcal toxemia are specific protein toxins. The specific toxins include toxic shock syndrome toxin 1 (exotoxin C) causing toxic shock syndrome, exfoliative toxin, causing the scaled skin syndrome and the enterotoxin, causing food poisoning (Baird-Parker, 1990; Sneath *et al.*, 1986).

Staphylococcal exotoxins

S.aureus produces a variety of exoproteins that contribute to its ability to colonize and cause diseases in mammalian hosts. Nearly all strains secrete a group of enzymes and cytotoxins which include four hemolysins (α, β, γ, and δ), nucleases, proteases, lipases, hyaluronidases and collagenases. Catalase production by *S.aureus* functions to inactivate the toxic hydrogen peroxide and free radicals formed by the myleoperoxidase system within host phagocytic cellsa after the ingestion of the microorganisms (Kloos and Banner, 1999). The free or bound coagulase sometimes called the clumping factor may act to coat the bacterial cell with fibrin, shielding them from opsonization and phagocytosis. Protein- A counters elimination of the pathogen by host polymorphonuclear cells. Thermo nuclease is a heat stable endo or exo nuclease that is capable of hydrolyzing DNA/RNA in host cells. Lipase initiates the spreading to cutaneous and subcutaneous, through phosphatidylinositolspecific phospholipase-C activity. Haemolysins produced by *S.aureus* influence a number of biological functions. α-haemolysin have a lytic effect on red blood cells from an array of animal species. β-haemolysins is a sphingomyelinase acting on lipocarbohydrate sphingomyelin complex (Sandel and McKillip, 2004). The main function of these proteins may be to convert local host tissue into nutrients required for bacterial growth. Some strains produce one or more additional exoproteins, which include toxic shock syndrome toxin (TSST-1), the staphylococcal enterotoxins, the exfoliative toxin and leukocidin (Dingis *et al.*, 2000).

Staphylococcal enterotoxins

Staphylococcal enterotoxins are a group of exotoxins included under the family of pyrogenic toxin super antigens (Dinges *et al.*, 2000). They are proteins produced by the Staphylococci under certain conditions in food and culture media. The ingestion of these substances by humans results in what is called Staphylococcal food poisoning. They are a family of majorm serological type of heat stable enterotoxins and function both as potent gastro intestinal toxins as well as super antigens that stimulate non-specific T-cell proliferation. Although these are two separate functions localized on separate domains of the proteins, there is a high correlation between these activities (Balaban and Rasooly, 2000).

Staphylococcal enterotoxins are a group of single chain low molecular weight proteins (26,900 - 29,600 daltons), produced by some species of *Staphylococcus*, primarily *S.aureus*. Thjey are similar in composition but are identified as separate proteins due to their difference in antigenicity (Casman *et al.*, 1963). To date, 14 different staphylococcal enterotoxin types have been identified which had been designated as A, B, C, D, etc., which share structure and sequence similarities (Le Loir *et al.*, 2003).

Staphylococcal Enterotoxin Types

Staphylococcal Enterotoxin A (SEA)

Staphylococcal enterotoxin A (SEA) is the most common toxin implicated in staphylococcal food poisoning (Holmberg and Blake, 1984). The gene for SEA (entA) is carried by a temperate bacteriophage (Betley and Mekalanos, 1985; Borst and Betley, 1994). Hybridization analysis of DNA from 'ent-A converting phage' suggests that this phage integrated into the bacterial chromosome by circularization and reciprocal crossover and that the entA gene is located near the phage attachment site. The 'sea gene' is composed of 771 base pairs and encodes an enterotoxin A precursor of 257 amino acid residue (Huang *et al.*, 1987).

Staphylococcal Enterotoxin B (SEB)

The coding region of the gene for SEB(ent B) contains- 900 nucleotides (Johns and Khan, 1988). The SEB precursor protein contains 267 amino acids (M 31,400 dalton) and includes an N-terminal signal peptide of 27 amino acids. The entB gene is chromosomal in clinical isolates of *S.aureus* from food poisoning cases (Shafer and Iandolo, 1978). However in other bacterial strains, the gene is carried by a 750kb plasmid (Shalita *et al.*, 1977).

Staphylococcal Enterotoxin Type C (SEC)

SECs are a group of highly conserved proteins with significant immunological cross reactivity (Bergdoll *et al.*, 1965). The three antigenically distinct SEC subtypes are SEC1 SEC2, and SEC3. The entC3 gene for SEC3 contains 801 bp and encodes a precursor protein of 267 aminoacids (Hovde *et al.*, 1990), containing a 27-residue signal peptide (Bohach and Schlievert, 1989; Hovde *et al.*, 1990). The entC3 gene is closely related to the gene for staphylococcal enterotoxin type C1 with 98 per cent nucleotide sequence identity (Couch and Betley, 1989).

Staphylococcal Enterotoxin D (SED)

Staphylococcal enterotoxin D (SED) is the second most common serotype associated with food poisoning (Chang and Bergdoll, 1979). The gene encoding SED is entD, which is located on a 27.6-kilobase penicillinase plasmid designated pIB485 (Bayles and Landolo, 1989). The entD gene contains 258 aminoacids including a 30-aminoacid signal peptide. The 228-aminoacid mature polypeptide shows sequence similarity to other staphylococcal enterotoxins (Balaban and Rasooly, 2000).

Staphylococcal Type E (SEE)

The gene for SEE (ent-E) encodes a 29kDa protein that is apparently processed to a mature extra cellular form with a molecular mass of 26kDa (Couch *et al.*, 1988). DNA sequence identity indicates that SEE, SED and SEA are closely related (Vanden Bussche *et al.*, 1993). SEA shares high sequence 81 per cent homology with SEA (Balaban and Rasooly, 2000).

Staphylococcal Enterotoxin Type G (SEG)

The entG gene encodes a 258-aminoacid precursor protein that is cleaved to form a toxin with 233 amino acids. The SEG is most similar to SEA SEB and SEC (Munson *et al.*, 1988).

Staphylococcal Enterotoxin Type H (SHE)

SEH is a recently discovered enterotoxin with a molecular mass of 27,300-dalton (Su and Wong, 1995). The amino-terminal of aminoacid sequence of the enterotoxin is unique and immunodiffusion assays do not detect cross- reactivity between SEH and previously identified enterotoxins.

Staphylococcal Enterotoxin Type I (SEI)

The '*ent I*'gene encodes a precursor protein of 242 amino acids (Munson *et al.*, 1998). The signal sequence of pre-SEI is cleaved to form a toxin containing 218 aminoacids. SEI has the lowest homology to other SEs.

Staphylococcal Enterotoxin Type J (SEJ)

Characterization of the enterotoxin D-encoding plasmid revealed the presence of an open reading frame (ORE) that encodes an enterotoxin, which has not been identified previously. It is designated staphylococcal enterotoxin J (SEJ) (Zhang *et al.*, 1998). The enterotoxin D and J open reading frames are transcribed in opposite directions and are separated by an 895 nucleotide intergeneric region which contains a perfect inverted repeat with each arm of the repeat, having a length of 21 nucleotides. The predicated 269-aminoacid SEJ protein has substantial sequence similarity to SEA, SEE and SED (64- 66 per cent). PCR amplification suggests that the 'ent j' determinant may be present on all SED- encoding plasmids (Balaban and Rasooly, 2000).

Resistance to Proteolytic Enzymes

Staphylococcus enterotoxins are highly stable. They resist most proteolytic enzymes such as pepsin or trypsin. This keeps their activity in the digestive tract unaffected after ingestion. They also chymotrypsin, papain and rennin (Le Loir *et al.*, 2003).

Heat Resistance

Staphylococcal enterotoxins are highly heat resistance (Le Loir *et al.*, 2003). Heat stability of the enterotoxins is one of the most important properties in terms of food safety (Denny *et al.*, 1971; Hernande *et al.*, 1993). They are thought to be more resistant to heat in foodstuffs, than in laboratory culture medium (Bergdoll, 1983). Heat stability seems to be dependent on the media the toxin is in, the pH, salt concentration and other environmental factors related to the level of toxin denaturation (Balban and Rassoly, 2000).

Immunological Assays

Double Gel Immunodiffusion

Double gel immunodiffusion, also called the Ouchterlony technique (Ouchterlony, 1949) was the first serological assay used for the detection of staphylococcal enterotoxins. Application of the Ouchterlony technique includes microslide, optimum sensitivity plate (OSP) and single radial immunodiffusion methods.

Radio Immuno Assay (RIA)

Radio immuno assay (Miller *et al.*, 1978) was the first sensitive method developed for the detection of staphylococcal enterotoxins, at levels of less than 1ng/ml. This assay is based on the competition for binding sites on specific antibody molecule between radio labeled standards and unlabelled antigens in samples. The major disadvantages of this method include requirement for purified staphylococcal enterotoxins and the use of radioactive materials and special detection equipment.

Enzyme Linked Immuno Sorbant Assay (ELISA)

The Enzyme Linked Immuno Sorbant assay was first introduced by Saunders and Burlett, (1977). It is currently the most sensitive methods used for the detection of staphylococcal enterotoxins. The major advantage of ELISA is its sensitivity, which is equivalent to the RIA, but without the health hazard associated with radioactivity. Various types of ELISA have been developed for staphylococcal enterotoxin detection, *i.e.* competitive ELISA (Kauffman, 1980; Stiffler-Rosenberg and Fey, 1978), non-competitive ELISA and Absorption-Inhibition ELISA (Morissette *et al.*, 1990).

Reverse Passive Latex Agglutination (RPLA) Assay

Reverse Passive Latex Agglutination (RPLA)assay was developed by Shingaki *et al.*, (1981). This is based on the visible agglutination of antibody-coated latex particles in presence of staphylococcal enterotoxins. The advantage of this method over ELISA is that it is simple and convenient to use. RPLA is more sensitive than the immunodiffusion assay and enable the detection of Staphylococcal enterotoxin in foods, at the lowest concentration levels implicated in outbreaks of food poisoning (Bankes and Rose, 1989).

Nucleic Acid Based Methods Used for the Detection of Enterotoxigenic *Staphylococcus*

The development of recombinant DNA technology provided a new tool for the detection of microbial pathogens in foods. The DNA based detection of enterotoxigenic staphylococci in foods have been reported by several investigators (Hatakka *et al.*, 2000; Johnson *et al.*, 1991; Zschock *et al.*, 2000). Advancement in molecular methods led to the usage of techniques like Pulse Field Gel Electrophoresis (PFGE), binary typing etc in the characterization of *S.aureus* from different samples (Zadoks *et al.*, 2000), but due to disadvantages like requirement of expensive equipments, time consumption, labor intensity, etc., simple techniques like Polymerase Chain Reaction (PCR) based ribotyping, Random Amplification of Polymorphic DNA (RAPD), etc. are the current methods of choice.

Incidence of Staphylococcal Food Poisoning

Staphylococcal food poisoning resulting from the consumption of food contaminated with staphylococcal enterotoxins is the major cause of food borne diseases world wide (Wieneke *et al.*, 1993; Knabel, 1995, Su and Wong, 1997; Dinges *et al.*, 2000). The foods that are most often involved in staphylococcal food poisoning differ widely from one country to another. Reports from UK showed that 53 per cent of the staphylococcal food poisoning cases were due to meat products, meat based dishes and especially ham, 22 per cent of the cases were due to poultry and poultry based meals, 8 per cent were due to milk products, 7 per cent to fish and shellfish and 3.5 per cent to eggs (Wieneke *et al.*, 1993). Haeghebaert *et al.*, (2002) reported the incidence of staphylococcal food poisoning in France resulted from the consumption of meat, 32 per cent, sausages, 22 per cent, pies, 15 per cent, fish and seafood, 11 per cent, egg and egg products, 11 per cent and poultry, 9.5 per cent. The reports from United States showed that 36 per cent of food poisoning was due to red meat, 12.3 per cent to salads, 11.3 per cent to poultry, 5.1 per cent to pastries, and only 1.4 per cent due to milk products and sea foods. In 17 per cent cases, the food involved was unknown (Genigeorgis, 1989). Staphylococcal food poisoning accounts for an estimated 14 per cent of the total out breaks of food borne illness within United States (Holmberg and Blake, 1984; Mead *et al.*, 1999). Staphylococcal food poisoning accounts for an estimated 40 per cent of food borne gastroenteritis in Hungary and 20-25 per cent in Japan (Bergdoll, 1979). Outbreaks resulting from improper manufacturing of canned corned beef have been reported in England, Brazil, Argentina, Malta, Northern Europe and Australia. In Great Britain 75 per cent of the 359 cases of food poisoning reported between 1969-1990 were associated with meat poultry or their products (Wieneke *et al.*, 1993).

Incidence and Enterotoxigencity of *S. aureus* in Fish and Fishery Products

Staphylococcal and especially coagulase positive staphylococci do not constitute the normal flora of fresh marine fish, but only get contaminated either from handlers or from the surface with which they come in contact (Bryan, 1973; Liston, 1980; Summer *et al.*, 1982; Sanjeev *et al.*, 1985a). Various authors reported the incidence of staphylococci especially *S.aureus* in fresh fish and shellfish and various fishery products (Ayulo *et al.*, 1994; Antony *et al.*, 2004; Jeyasekaran and Ayyappan, 2002; Lakshmanan *et al.*, 1984; Iyer *et al.*, 1986). Their presence in frozen fish and shellfish (Cann, 1977; Silverman *et al.*, 1961; Sanjeev *et al.*, 1986; Iyer and Shrivastava, 1988) and dried fish and shellfish (Sanjeev *et al.*, 1985; Prasad *et al.*, 1994) are well documented. Summer *et al.*, (1982) noted the presence of *S.aureus* above 100/g in headless (32 per cent), peeled (35 per cent) and deveined (67 per cent)cooked shrimps processed in Sri Lanka. Vanden Broek *et al.* (1984) showed *S.aureus* count in the range of 10^2/g in fish fillets from the retail markets of The Netherlands.

Chapter 15
Fish Vaccination

Introduction

Vaccines

Vaccines are preparations of antigens derived from pathogenic organisms, rendered non-pathogenic by various means, which will stimulate the immune system is such a way as to increase the resistance to disease from subsequent infection by a pathogen. Inducing a protective immune response to a pathogenic organism before the individual becomes naturally exposed to it (prophylaxis) seems at first sight an eminently sensible way of preventing an infectious disease. Vaccine must be safe (no side effects) and potent (induce a high level of protection). Vaccines are generally of two types: dead vaccines, which are composed of inactivated pathogens or extracts, and live vaccines, which are attenuated pathogens with no or low virulence.

The specific or adaptive immune response has two arms: humoral immunity (antibody production) and cell mediated immunity (CMI). Exposure to an antigen results in the stimulation of small number of virgin lymphocytes which are capable of recognizing the antigen through specific antigen receptors. These specific lymphocytes constitute a clone. Microorganisms have many different antigens on their surface and each antigen is capable of being recognized by different clone of lymphocytes. On stimulation a clone undergoes proliferation with the differentiation of daughter cells which have particular functions depending upon the population to which the clone of lymphocytes belongs.

There are two main populations of lymphocytes the 'T' lymphocytes which originate from the thymus and the 'B' lymphocytes which originate from the bursa of Fabricius in birds or the bone marrow of mammals. The source of the cell type in teleost fish is not definitely known though it is likely to be the kidney. A vaccine must simulate the natural infection in producing immunity but it must not produce clinical illness. Killed or inactivated vaccines are usually safer than live ones but even these can produce unwanted side effects. Living vaccines require special caution. They are usually attenuated but they are not avirulent. Before they can be used safely it is essential to prove beyond reasonable doubt that an attenuated strain will not mutate to one of enhanced or full virulence *in vivo*. Modern techniques of genetic engineering have made the process of attention more precise and irreversible and progress is taking place in the development of more effective and safe live vaccines for use in human medicine.

Vaccination for Controlling Disease

In areas where a disease is not endemic a policy of exclusion and eradication is the best means of control. However, such policy requires exceptionally high levels of vigilance, certification and cooperation between farmers and responsible authorities. On occasions it may fail and the introduction of a disease into a water course may result in it becoming endemic. Many diseases of fish are already endemic in wild fish populations which act as an unavoidable reservoir of infection for farms on surface water supplies. In these areas the risk of farmed stock becoming infected may be high and a slaughter/disinfection policy becomes extremely expensive and probably ineffective. For many of the major diseases of fish survivors of an outbreak become carriers which restricts the marketing potential of live stock and every new year class of fish is at risk. For bacterial diseases, outbreaks can be controlled by chemotherapy but this may fail for reasons outlined above. In such cases, when a disease is endemic, vaccination becomes useful, especially with killed vaccines. One precious problem with vaccination is the risk that vaccinated stock while being resistant to disease, may become carriers of a pathogen after exposure.

Thus vaccination has an important role to play in disease control but its use should depend upon various epidemiological considerations of each disease and the nature of vaccine. Presently only two vaccines have been proven effective on commercial scale; vibriosis and enteric redmouth vaccines. Many others are at an experimental stage.

History of Fish Vaccination

The first attempts to vaccinate fish against disease were published by Duff who attempted to orally vaccinate cut-throat trout against furunculosis with a diet containing chloroform-killed *Aeromonas salmonicida* cells.

The commercial product licensed for a vaccine for fish was granted in the USA in 1976 for use against enteric redmouth (ERM). Since then product licenses have been granted for *Vibrio* vaccines in many parts of the world and these two vaccines have proved of great commercial value.

Various techniques of mass vaccination of fish have been developed and these include immersion, hyperosmotic immersion, bath, spray and oral.

Vaccination Methods

Injection Vaccination

The method suffers from certain disadvantages, principally by requiring anaesthetization and handling, both of which can be quite stressful for the fish. It is also very labour intensive and cannot be used with fish much below 15g. However, by using repeater syringes and a production line system it is quite feasible for each operator to inject 1000fish/h. Basically the system consists of 'vaccinator(s)' seated at a vaccination table, a method of returning fish to the holding facility (a chute, conveyor belt) and an 'anaesthetist' controlling the supply of anaesthetized fish to the vaccination table for immediate injection thus preventing too great an exposure to the anaesthetic.

Oral Vaccination

The protocol of oral vaccination is very attractive since it is suitable for mass administration to fish of all sizes, imposes no stress on the fish because handling is not required and therefore does not interfere with routine husbandry practices. Furthermore, oral vaccination is the only method suitable for extensive pond rearing of fish where catching the fish prior to harvesting for injection or immersion vaccination is impractical. However, oral vaccination has some intrinsic limitations. In the case of killed vaccines large amounts are required thus increasing the cost and there is the uncertainty of individual dosage since fish consume different amounts of food.

A major problem seems to be the destruction of antigens in the stomach and foregut before they reach the immune sensitive areas of the lower gut. Future developments will need to overcome this problem, possibly by encapsulating the vaccine so that it survives passage through the stomach. There are few reports of oral vaccine trials in fish which lack a stomach, like the cyprinids, and the difficulties may be less or different for such species.

Immersion Vaccination

The several variations of immersion vaccination were developed mainly by Amend and colleagues in the USA for use of the *Vibrio* vaccines. The method began with hyperosmotic immersion (HI)but it was quickly realized that the hyperosmotic step (which was quite stressful) was unnecessary and direct immersion (DI), in the vaccine was equally successful. The method is simple and rapid, only a few seconds exposure to the vaccine are necessary. This method can now be automated by use of immersion vaccination machines. A further variation is the spray method, again developed for use of *Vibrio* vaccines, but has no advantages over DI and is more stressful to the fish.

While immersion (DI) vaccination has proved very efficacious for immunizing fish against vibriosis and ERM it has not been so successful for other diseases. The possible reason for this are many and diverse and are the subject of present research.

The uptake of antigens by fish following immersion has been demonstrated to be quite effective. The mechanism of uptake is not precisely understood but the gills appear to be the main route of entry, with the skin and lateral line system possibly playing a role. The failure to immunize fish against many diseases by the immersion method is, therefore, probably not mainly associated with inadequate uptake of the antigens but with the nature of the ensuing immune response (which is probably mainly integumentary and about which very little is known) or the inappropriate nature of the experimental challenge.

Chapter 16
Methods of Assessing Vaccines

Introduction

There are two basic factors relating to vaccine assessment: the challenge route and the method of calculating effectiveness.

The Challenge Route

The method of challenging fish may be important in assessing the efficacy of a vaccine. While immersion vaccination against vibriosis provides good protection to experimental water borne challenge and has proved effective against natural challenge in the field, it is much less effective when virulent-bacteria are injected into the fish. However, injection vaccination is effective against injected challenge. Injection vaccination results in high levels of circulating antibodies but immersion vaccination does not. This suggests that immersion vaccination stimulates integument immunity which is quite sufficient to protect against natural exposure.

The Method of Calculating Effectiveness

There are two accepted methods of calculating the effectiveness of a vaccine:

Relative Per cent Survival (RPS)

This method is suitable for experimental or field trials and takes the form of challenging populations of vaccinated and non-vaccinated fish and monitoring the mortalities over a set period. The method has certain baseline criteria: at least 25 fish should be used in duplicate for each test population: the challenge should cause at least 60 per cent morbidity or mortality in control fish on a time scale similar to that of the natural disease; the cause of mortality should be determined and non-specific infections must not exceed 10 per cent in any group and the mortality rate of vaccinated fish must be below 24 per cent if the test is deemed to be positive. The calculation is made using the following formula;

$$\text{RPS} = 1 = \frac{\text{Per cent vaccinate mortality}}{\text{Per cent control mortality}} \times 100$$

As can be seen from the above criteria it is extremely difficult to perform these tests successfully in the field because the challenge may not be the same for all groups, the minimum mortality rate may not be achieved in controls, and intercurrent diseases may account for more than 10 per cent of the mortalities.

Even under laboratory conditions it is difficult to achieve the required level of challenge and many statistical considerations must frequently be considered in interpreting results.

Increase in Lethal Dose 50 per cent (LD_{50})

This test can only be conducted in the laboratory and is based on the determination of the number of virulent organisms required to kill 50 per cent of vaccinated and non-vaccinated fish. The test gives a more quantitative assessment of the vaccine potency than the RPS but is more complicated to perform. Groups of vaccinated and non-vaccinated fish comprising at least six individuals are challenged with graded doses of a virulent organism. For bacteria *e.g.* A.*salmonicida*, injection challenge may use doses ranging from 10cells/fish to 10^6 with 10-fold stepwise increases. Immersion challenge may use the same number of bacteria/ml water. Following a period relating to the natural course of the disease (for furunculosis this would be 14 days) the mortalities for each group of fish are recorded. The cause of death is determined by re-isolation of the pathogen which must be specific. The number of organisms required to kill 50 per cent (the LD_{50}) of vaccinated and non-vaccinated fish is then calculated by the formula of Reed and Muench.

At least a 100-fold increase in the LD_{50} for vaccinated fish compared with non-vaccinated fish would indicate good potency of the vaccine.

A comparison of the two methods of assessing vaccine potency can sometimes reveal that an increase in LD_{50} of 10-fold in vaccinated fish compared with non-vaccinates would produce an RPS value of over 90 per cent. The latter value would be regarded as highly satisfactory but the former would cast doubt on the vaccine having a useful potency. Thus RPS values may appear high with only marginal increase in resistance to disease.

Vaccines in the Future

Finally, mention should be made of the recent advances in immunology and biotechnology which have revolutionized the development of vaccines for use in humans. Many of the techniques used are highly sophisticated and in view of the cost of research and the relatively limited market for vaccines in fish farming their applicability in this field will be slight. Nevertheless, they should be borne in mind and a short review of some of these recent developments follows:

Genetic Recombination

Genetic recombination is used to mass produce protective antigens by transferring the DNA coding the protective antigen into a suitable microorganism which is easily cultured and which then synthesizes the antigen.

Genetic Attenuation

Besides inserting pieces of DNA into the genome of a cell it is also possible to delete pieces. Specific deletion of a gene(s) which codes for a virulence factor in a pathogen, while preserving other protective antigens, allows the production of attenuated forms where reversion to the virulent wild type is impossible. Thus, the technique is valuable in production of live vaccines.

Protein Engineering

The aim here is to improve the immunogenicity of an antigen or to mass produce it cheaply. Protective antigens may be made more immunogenic by cleavage of epitopes which stimulate the T suppressor cells. The protective epitopes of an antigen may be composed of a short sequence of amino acids which can be artificially synthesized quite cheaply for use as vaccines (known as 'peptide vaccines') in combination with appropriate carrier antigens.

Anti-idiotype Vaccines

These are antibodies, the antigen-binding site of which mimics the structure of a protective antigen. They are produced by first raising antibodies to the appropriate antigen in a suitable animal. The specific antibodies produced (idiotypes) are then used to raise further antibodies which react with the antigen-binding site of the first (anti-idiotypic antibodies). The antigen-binding site of the anti-idiotype can be used as a vaccine since it resembles stoichiometrically the original antigen and will induce formation of antibodies which cross-react with it. Such vaccines have advantages when the protective antigen is difficult to obtain in large amounts or possesses undesirable side effects.

Chapter 17
Vaccination Against Vibriosis

Introduction

Geographical and Host Range

Vibriosis has a global distribution with epizootics on all continents and in a wide range of fish species. An excellent review of the distribution of vibriosis in fish is provided. The host range of vibriosis includes both cultured and feral fish. The disease causes significant losses in cultured Pacific salmon, Atlantic salmon and rainbow trout grown in marine cage farms and in cultured non-salmonids including eel, yellowtail, red sea bream, and sea bass.

There are several reports of vibriosis in a considerable range of feral fish. Reports of vibriosis have not been limited to marine environments. Epizootics of vibriosis in rainbow trout raised in fresh water possibly due to the feeding of contaminated marine trash fish. Hacking and Budd described an outbreak of vibriosis in tropical fishes maintained in a freshwater aquarium. In addition, problems of vibriosis caused by *Vibrio anguillarum* have occurred in freshwater farmed rainbow trout in Italy. Vibriosis is also a significant problem in farmed crustaceans *e.g.* prawns and shrimp.

Vibrio Species/Strain Variation

The most commonly encountered fish pathogenic *Vibrio* species is *V. anguillarum* and it is this species which is responsible for the majority of losses worldwide, and was probably the causative agent of the disease featured in the original reports of 'Red Pest' of eels in the 18th and 19th centuries.

V. anguillarum constitutes part of the normal microflora of the aquatic environment with maximum and minimum numbers in summer and winter respectively. *V.anguillarum* may constitute part of the natural microflora of marine fish and consequentially, fish may be continually threatened by the pathogen which may be able to cause disease under certain conditions.

The next most important member of the *Vibrio* species is *V. ordalii* (formely known as *V.anguillarum* Biotype II) which has been documented in Japan and the Pacific northwest, USA. Its occurrence is not as frequent as that of *V.anguillarum*. Other *Vibrio* species reported to be pathogenic to fish are as follows:

V. alginolyticus which has been implicated in vibriosis of sea bream (*Sparus aurata*) in Israel and of sea mullet. *V. carchariae* isolated from diseased sharks.

V.cholerae (NON-01) isolated from diseased ayu in Japan. *V. damsela* isolated from damselfish in southern California.

V. vulnificus Biogroup2 isolated from eels in Japan and the United Kingdom.

V. salmonicida recently characterized and named is the causal agent of 'Hitra' disease or coldwter vibriosis principally in Norwegian farmed salmon.

Pathogenicity

Symptoms of vibriosis are similar to those caused by *Aeromonas salmonicida*.The external pathology produced by *V.anguillarum* includes haemorrhaging at the base of the fins, around the vent and gills and inside the mouth. Petechiae, necrotic lesions and diffuse haemorrhages can appear on the body surfaces. Internally, the intestine is often inflamed with petechiae present on the viscera and musculature. The intestine may be distended and filled with clear viscous fluid. This histopathology produced by *V.anguillarum* and *V. ordalii* in salmonids is somewhat different.

Epidemiology

Vibriosis is a stress-related disease and is associated with a number of factors *e.g.*high water temperatures, rapid changes in water temperature or salinity, over-crowding, buffering by wave action and poor water quality including low oxygen levels and high suspended solids.

There also appears to be a seasonal effect on the disease, outbreaks occurring mainly in spring and autumn, though this may simply reflect rapid changes of temperature and salinity. Salmon smolts are particularly sensitive to vibriosis possibly due to some form of immuno-suppression caused by the rapid physiological changes taking place in these fish.

Chemotherapy

Prior to the introduction of immunization, the most common method employed to control vibriosis was treatment of diseased fish with antibiotics and antimicrobial chemicals.Antibiotics and antimicrobials are commonly administered as food additives or in bath treatments. Commonly used substances are tetracycline, sulphonamides, nitrofuran derivative, trimethoprim and, more recently the quinolines (*e.g.* oxolinic acid). However, antibiotic therapy for the control of vibriosis has met with variable success.

Nature of *Vibrio* Vaccines

Current commercial *Vibrio* vaccines in the Northern hemisphere contain mixtures of the most commonly encountered species which cause disease problems *i.e. V. anguillarum* and *V. ordalii*. The majority of these vaccines are simple inactivated cultures containing mixtures of whole cells and extracellular products. In a few cases the bacteria have been lysed either by chemical or physical methods.

The protective antigens of *V. anguillarum* appear to be heat-stable (to 100-121°C) lipopolysaccharides in the cell wall, which may also be released into the culture supernatant. These large molecular-weight compounds *i.e.* 100 kilodaltons (kD) are able to withstand severe extraction methods. Chart and Trust isolated from the outer membrane, two other minor proteins with molecular weights of 49-51kD which were potent antigens. A weakly antigenic protein with a molecular weight of approximately 40kD was also present. These antigens may be heat labile, which could account for the greater protection, achieved with formalin-inactivated vaccines compared to heat-killed products.

Chapter 18
Vaccine Development

Introduction

The evolution of commercial *V.anguillarum* vaccines commenced with the development and utilization of mass immunization methods, necessary to have practical application in modern aquaculture systems where large numbers of fish are involved. Several methods of administering commercial *Vibrio* vaccines are currently employed. These are: injection, oral, hyperosmotic immersion, immersion or dip, bath vaccination, spray or shower and automated immersion.

Injection

Intraperitoneal (i.p.)injection was used in the first reported effective immunization of fish against vibriosis. Handling is particularly important in the case of Atlantic salmon which are particularly sensitive to scale damage and infection at the smolting stage. Many of the disadvantages of injection vaccination can be overcome. For example, vaccination may be timed with grading, a time when the fish are normally being handled. The process can be speeded-up and made simpler for the operator by the use of multi-dose repeater syringes identical to those used in the poultry industry.

A further advantage of injection vaccination is that injectable vaccines permit the use of certain adjuvants that cannot be used with other delivery systems. Indeed an injectable *Vibrio* vaccine in combination with *Aeromonas salmonicida* is now commercially available and this vaccine contains an oily adjuvant to improve efficacy and duration of immunity.

Oral Vaccine

Oral vaccination was the first mass immunization method employed to vaccinate against vibriosis. This work demonstrated that both formalin-killed, lyophilized whole cells and wet packed whole cells of *V. anguillarum* provided some degree of protection. Maximal protection of Chinook salmon followed the feeding of 2mg of dried vaccine/ g of food for 15 days at temperatures even as low as 3.9°C. An important finding was the observation that longer feeding regimes did not result in enhanced protection. It is possible that the previously mentioned disadvantages of oral vaccines may preclude their use as a primary vaccination method. However, they may well have a place in fish health management when used as a booster on fish previously vaccinated by immersion. To expand this point further, many species of fish (*e.g.* salmon) are required to grow for long periods before reaching table size and this period often exceeds the 12 months protection afforded by immersion vaccination. There is, therefore, a requirement to give a booster vaccination at approximately 1 year after the primary vaccination. At this time it may not be practical or wise to handle fish for booster vaccination by either injection or direct immersion. Therefore, oral vaccination may be a useful method of booster vaccinating such fish.

Hyperosomotic–Immersion Vaccines

This method is purely of historical interest as it has been superceded by methods such as simple immersion or bath vaccination. The technique involves the immersion of fish in a hypertonic saline solution for a short time followed by a dip in an aqueous solution of the vaccine. The rationale behind the technique was that in the hypertonic salt solution fish lose water by osmosis, and when placed in the vaccine solution the fish replace water lost by osmosis at the same time taking up vaccine. While the hyperosmotic vaccination method gave good and reasonably long-lasting protection against disease, there were problems including hypertonic stress and gill damage which could reduce the growth rate of fish or make them susceptible to other infections. Because of these problems, research continued into modification of the hyperosmotic vaccination method and gave rise to the immersion or dip vaccination methods commonly in use today.

Immersion or Dip Vaccination

It subsequently became clear that the stressful dip in the hypertonic saline was unnecessary and equivalent protection rates could be obtained when the fish were dipped directly in a solution of vaccine. The technique simply involves dilution of the vaccine with hatchery water (usually 1:10 for most of the commercially–available vaccines), removal of the fish from the holding facility, immersion of the fish in the vaccine solution for up to 30s, draining of the excess vaccine from the fish and the return of the fish to the holding facility. The main portal of antigen uptake in dip vaccination is the gill tissue. Work with radio-labelled antigens suggests an active uptake mechanism by the gills possibly involving phagocytic cells on the gill surface. Immediately after vaccination, labelled antigens are concentrated in cells on the gill surface. A short while after vaccination these cells appear in the blood capillaries of the gill and some hours later in the anterior kidney and spleen, presumably where they are available for immune processing.

Bath Vaccination

Bath vaccination was developed because of the problems associated with net damage and handling stress when vaccinating salmon smolts by dip method. It is advisable to vaccinate salmon smolts against vibriosis while they are in fresh water, just prior to transfer to sea water. Unfortunately, smolts are particularly sensitive to scale damage from netting at this time and handling of this nature should be avoided if at all possible. Bath vaccination allows vaccination of fish within their holding tanks without removal. Basically, the technique requires reducing the level of the water in the holding to a known volume and adding vaccine to a known concentration (usually between 1:200 and 1: 1000 depending on the commercial product used). The fish are allowed to swim in this highly diluted vaccine for 1to 2h after which water flow is returned to the tank and the volume restored to normal.

Spray Vaccination

Spray vaccination was a development of a method used for marking fish with a dye and was developed simultaneously with immersion vaccination. This method involves the spraying of vaccine onto fish skin and can be semi-automated using a conveyor belt which passes under spray nozzles.

Chapter 19
Vaccine Trials

Introduction

Immunity

In the case of vibriosis a protective role of antibody has been clearly demonstrated, passively immunized rainbow trout with immune serum and observed protection upon challenge. The protection was specific as the effect could be abrogated by preincubation of the immune serum with *Vibrio* bacterin. Moreover, the decay of the injected immunoglobin correlated with the decrease in protection. Similar results using antibody were reported. In addition, transfer of lymphoid cells from the head kidney of immunized fish also conferred protection upon non-immunized donors although cells from the spleen and thymus did not. The above experiments show that the humoral response provides good protection against vibriosis. However, this correlation is not always clear. Immersion vaccination can be followed by serum antibody production but there are cases in which protection was observed in the absence of serum antibodies.

Effectiveness

As previously mentioned there are a number of vaccination methods available for use on a commercial scale. Some published trials have directly compared these methods and are useful for evaluating which method would be of most use on a commercial scale and for calculation of the economic benefits of vaccination.

Limitations of Present Vaccines

Adjuvants

Commercially available injectable vaccines contain various adjuvants depending upon the manufacturer of the vaccines. There is no doubt that the inclusion of the adjuvant increases the level and the duration of the efficacy.

Presently, adjuvant is available only for injectable vaccines. There are a number of water soluble adjuvant which could conceivably be added to immersion or oral vaccines to improve their protection and duration of protection and this is likely to be an important area of future research.

Vibrio vaccines are currently the most successful and show the most promise of any vaccine against the major diseases of fish. Commercial vaccines are well-established and well-researched and data from extensive commercial use are constantly being used to improve administration and administration methods. *Vibrio* vaccines have the double advantage of being simple to manufacture and simple to use in a variety of mass administration methods.

With the enormous expansion of the marine farming of trout and salmon in the last few years, vaccination has proven itself to be a successful and highly cost-effective tool in the management of the disease. The future for *Vibrio* vaccines also looks promising. Marine aquaculture is likely to expand rapidly and developments in technology are allowing more exposed areas of sea to be farmed or for sea water to be inexpensively pumped onto land-based marine farms. More and more marine species are being farmed worldwide and the constant presence of *Vibrio* bacteria in seawater makes the disease a constant threat. Moreover, the likelihood of legislation against the indiscriminate use of antibiotics in the aquatic environment will further encourage the use of vaccines. It is likely we will soon see multivalent *Vibrio* vaccines containing a mixture of *Vibrio* species and a further application of *Vibrio* vaccines is likely to be in the control of vibriosis in invertebrates with the vaccination of cultured shrimps and prawns.

Chapter 20
Viral Diseases of Fish

Introduction

Viral Haemorrhagic Septicemia (VHS)

VHS is a dangerous disease among cultivated rainbow trout. It is also known as Egtved disease, after the name of village Denmark, where the symptoms of the disease were first reported. Only rainbow trout contact VHS. The fish which are principally infected range from fingerlings. 5cm in length to large fish 200 to 300g weight. In fish older than one year, resistance generally increases with age.The disease usually occurs from winter through spring, when the water temperature is about 8°C or below. Out breaks usually do not occur when the water temperature exceeds 15°C, with exceptional cases. The mortality rate generally ranges from 20 per cent to 80 per cent and varies greatly with the environmental conditions.

Symptoms due to this viral attack include darkening of body colour, eye protrusion, severe anemia, haemorrhage in gills, eyes and at base of the pectoral fins and body surface and accumulation of abdominal fluid. In certain cases external signs of infected fish are little altered from uninfected fish, except for abnormal behaviour like swimming in circles or lazy swimming on their sides or darting through and out of the water and also their body becomes twisted.

Scattered haemorrhages in various skeletal muscles, perivisaral adipose tissue of the abdomen, air bladder, intestine, etc. also occur. The haemorrhaging is primarily an accumulation of erythrocytes, with few lencocytes.Liver exhibits a dark color or sometimes the colour fades. Haemolysis may occur. Kidney swelling may or may not be a symptom.

The size of the viral particle VHS or Eegated Virus is found to be 60nm in diameter and 180nm in length. It is a cylindrical or bullet shaped virus. VHS virus is sensitive to ether and stable against freeze-drying and can be preserved at -20°C. Its infectivity is lost in 50 per cent glycerine. Moreover, infectivity declines greatly if the temperature exceeds 21°C.For the isolation and identification of the virus, the kidney and spleen are considered to be the best organs. The source of infection is believed to be fish which have been infected but survived, retaining the virus in the body. Transmission occurs via water. Infection via eggs and the introduction of the virus accompanying transfer of eyed eggs are important.Lime disinfection is extensively used in infected ponds and on equipments believed to be contaminated. Treatment for 5 minutes with an active iodine concentration of 8ppm destroys VHS virus.

Infectious Haematapoietic Necrosis (IHN)

This disease mostly attacks rainbow trout and salmon fry. IHN develops primarily in fry and juvenile fish, where the mortality rate is high. It also develops in yearlings, but the mortality rate is not high. It is not reported to occur in two-year old fish.One characteristic of this disease is a sudden rise in mortality among fish. Diseased fish first acts lazily and float with the current. Then they listlessly swim with the current and behave as though suffering from cramps. Finally they float, turn over, swim about fiercely and then die.Diseased fish usually trait long, semi-transparent faucal casts from the anus. The abdomen swells and the eyes protrude. Anemia develops and the gills fade markedly. Hemorrhage often occurs at the base of the pectoral fins, posterior of the skull above the lateral line, body musculature near arms, walls of the oral cavity, in the intestines, in the yolk sac in sac fry which then swells with fluid. The principal pathological change in IHN is extensive necrosis of haematopoietic tissues in the kidneys and spleen. In advanced stages necrosis of granular cells in the lamina propria of the stomach and intestine and focal necrotic lesions in the liver occur.

IHN virus is an RNA virus belonging to the Rhabdovirus group. It is bullet shaped, with a mean length of 160nm and diameter of 90nm. This virus is sensitive to ether, glycerin and chloroform. It is inactivated in 15 minutes at 60°C and it loses infectivity at pH 3.0.The most important source of infection is believed to be fish carrying the virus which have been infected in their early lives and survived. Another source of infection is eggs or semen from viral carrying parent fish. Infection can also be caused via water and food and main entry of IHN virus is through gills and digestive tracts.

Treatment with sodium benzimidazole, an anti-viral agent is possible. Active iodine is also good for treatment of IHN. Temperature treatment for prevention of IHN is significant.

Spring Virenia of Carp (SVC)

SVC occurs in the spring when the water temperature is rising (13 to 20°C), but it occurs more readily when the water temperature is near 17°C. This disease occurs primarily in fish of one year of age and damage is greater in younger fish. Once the disease breaks out it spreads rapidly with rise in water temperature. There is no problem with SVC in fry and juveniles.Affected fish suffer darkening of their body and then tend to assemble at the water inlet. They exhibit sluggish breathing, and finally tilt to one side. Advanced symptoms are abdominal distension, protrusion of the eyes, reddening and swelling of the anus, anemia and haemorrhagic spots in the gills are characteristics. Serous fluid, mixed with pus or blood, collects in the abdominal cavity. The intestines undergo inflammation and dilate, and desquamation and disintegration materials from the mucosa, or pus-like materials are often found in the lumen. Haemorrhagic spots are found in almost all organs, especially on the walls of air bladder. Muscles exhibit reddening due to haemorrhaging. The spleen undergoes severe swelling. Signs of jaundice appear. Edematic swellings of various internal organs, necrosis of hepatic tissue, cardiac inflammation, pericarditis, splenic and renal interstitial infiltration are also seen.

SVC viral particles are bullet-shaped with a diameter of 70nm. They are sensitive to ether at pH3, chloroform and are heat labile. Virus bearing adult fish is known to be an important infection source for viral diseases in salmonids. Infection via eggs is negligible. Infection may occur via water, through the gills and the intestines and there is also a possibility that fish lice or leeches can be vectors.

Genomycin, Furagolidone, Chloranphenicol, Streptomycin and Oxytetracycline may be used for treatment of SVC.

Swim Bladder Inflammation of Carp (SBI)

Swim Bladder Inflammation of Carp (SBI) is an infectious disease of cultivated carp. SBI affects carp, regardless of age. In young fish the occurrence rate occasionally reach 100 per cent and in adult fish, this disease hardly occurs and the occurrence rate does not exceed 40 to 50 per cent.

This disease has two forms- acute and chronic. In the former, symptoms appear within one month of contraction at temperature near 20°C, but 2 to 3 months could be required. The swim bladder undergoes extensive inflammation, and the wall tissue of swim bladder frequently is destroyed, with complications of peritonitis occurring, as well as pathological changes reaching the kidney and other organs. The fish eventually die. The mortality rate reaches 100 per cent.The chronic form is seen throughout the year and the fish with this form heal naturally. In the initial stages, inflammation occurs in the wall tissue of the swim bladder and the swim bladder walls become turbid and hyperemic. Haemorrhagic spots gradually appear and the bladder walls thicken narrowing the internal space. Brown or black spots often appear due to the deposition of pigment derived from blood.

The inflammation gradually extends throughout the swim bladder walls. Serous fluid containing blood, purulent matter and mucoid, etc. accumulates in the swim bladder. Both the anterior and posterior chambers of swim bladder get affected and undergo enlargement or atrophy. Eventually, the tissue of the walls undergoes necrosis, which then gradually expand and the swim bladder is destroyed. The pathological changes extend to the peritoneum, kidney, etc. When extensive inflammation and necrosis of the swim bladder wall take place, fusion with surrounding organs also occur readily. The SBI virus is a rhabdovirus of 70nm diameter and 90 to 140nm length. It is sensitive to chloroform. It is inactivated at pH3 for 60 minutes; 50°C, 30 minutes.Fish, which have survived an epidemic, may be sources of infection as carriers of the virus. The spread of the disease does not occur through eggs or sperm.

Injection or oral administration of chloramphenicol is a treatment for this disease. Administration of food immersed in methylene blue solution is said to be effective.

Pike Fry Rhabdovirus Disease

Pike (*Esox lucius*), a sport fish contracts two kinds of diseases: Head Disease (Hydrocephalus) and Red Disease I, especially in fry. Head Disease is characterized by an accumulation of serous fluid in the mesencephalon, cranial swelling and protrusion of the eyes. Haemorrhaging occurs in the brain and spinal cord, and spleen and regressive degeneration of the uriniferous tubules occasionally occurs. Red disease is characterized by red swelling on the trunk and haemorrhaging of the spleen, spinal cord, pancreas, haematopoietic kidney tissue and skeleted muscles. Fading of gill colour is also extreme. Both Red Disease and Head Disease are different forms of the same viral disease caused by bullet-shaped rhabdovirus –like particle. The disease is commonly called Pike Fry Disease (PFD), and its pathogen is called Pike Fry Disease Virus (PFDvirus).

Epitheliocystis and Bluegill Virus

Large number of small cyst-like objects 15 to 85 μm in diameter appear on the skin, gills and fins. This condition was first found in young blue gill (*Lepomis macrochirus*). The cysts are actually enlarged epithelial cells or mucous cells in the epithelium and they are termed epitheliocyctis cells. In epitheliocysts cells large clumps of small, basophilic, egg-shaped or round granules occupy the cell, and the enlarged nucleus is flattened by pressure of such clumps. Epithelial tissue of the gills of stripped base (*Morone saxatilis*) and white perch (*Morone americana*) also develop epitheliocystis.

Cauliflower Disease of European Eels

Cauliflower disease of the European eel (*Anguilla anguilla*) is a disease characterized with the formation of large papilloma on the head, especially at the tips of the upper and lower jaws. This disease exhibits a conspicuous appearance and the rate of occurrence is high. The designation cauliflower disease came from the resemblance of the papilloma on the head to cauliflower. The disease is also known as stomato papilloma, epidermal papilloma or papillomatosis. Electron microscopic observation of papilloma tissue revealed the existence of "dark cells" containing layer of pentagonal or hexagonal particles. A polyhedral virus, 55nm in size has been isolated from the blood of diseased fish, but no virus has been isolated from papilloma tissue itself. Eels with cauliflower disease can be said to be in a condition of viremia. The virus may have some connection with the formation of papilloma, but there is still no proof for this. There is a suspicion that some contaminants are involved in the development of papilloma.

Infectious Pancreatic Necrosis (IPN)

IPN is a viral disease which attacks with high mortalities mainly fry of rainbow trout, brook trout and other salmonids. It was discovered in the United States. The most affected are the brook trout and the rainbow trout. One characteristic of IPN is that it induces sudden and intense mortality in populations of fry. It is common for the disease to subside after two weeks of intense mortality. Symptoms tend to appear first among fish with good growth in a population. The affected fish suffer inco-ordinated movement and tend to revolve perpendicularly, finally settling on the pond bottom and dying. The duration of time from the start of revolving movement until death is 1 to 2 hours at most. The disease symptoms usually seen are blackening of body colour, protrusion of the eyes and distension of the abdomen. Hemorrhaging is also seen in the abdominal walls and at the base of fins. Hemorrhaging in the pyloric caeca, as well as conspicuous fading in the liver and spleen are also noticed. The digestive tracts are usually free of food, and milky-white mucus accumulates in the stomach, and the anterior part of the intestines. Diseased fish often trial mucous stool like a thread from the anus. Necrosis of the pancreas and cells of islets of hangerhans are characteristic. Pathological changes also appear in hematopoietic tissue of the kidney at the terminal stage of the disease.

The causative agent, IPN Virus, is the smallest RNA virus among the known fish viruses. The virus particles have a size of 55 to 65 nm. Taxologically, they belong to the reovirus group. They have comparatively high resistance to high temperature. The most important infection source is adult fish carrying the virus. IPN virus particles deposited in water can retain infectivity for a long time. Transmission via water, as well as via air, may be possible. IPN virus can exist in eggs or on the surface of eggs, where it is difficult for chemicals to reach. Fry hatched from virus contaminated eggs may readily undergo infection and disease contraction. The virus is discharged from such diseased fish in water with excretions and infects other fish. There is a great possibility that the gills are the route of infection. There is also a possibility of oral infection.

Oral administration of polyvinyl pyrolidone iodine agent seems to be effective in treatment of IPN. The primary important preventive measure would be to select adult fish which do not carry IPN virus and eggs and semen should be stripped from such fish. Contaminated ponds should be dried (about 3 weeks) and then disinfect with 200ppm of active chlorine.

Viral Renal Disease in Eels

This disease which first occurred in many eel farms in Shizuoka Prefecture in the winter of 1968-1970, with great damage was characterized by pathological changes in the gills and kidneys and was termed "Gill-Kidney Inflammation". Other characters of the disease were abnormalities in blood chemical properties and among them, a conspicuous decline in the plasma chlorine ion concentration was characteristic. The virus isolated from infected eels morphologically resembled IPN virus. This virus is stable in glycerin and is solvents such as ether. Histological observations of the dead fish revealed hyaline droplet degeneration of the uriniferous tubules, collapse of the auriferous tubular epithelium, interstitial necrosis, glomerulus's collapse, exudation of protein-like material into Bowman's capsules and focal necrosis of the liver and spleen. This virus was termed EVE since it was initially isolated from European eels.

Channel Catfish Virus Disease (CCVD)

This disease first appeared in 1968 with a high mortality rate in cultivated fry and fingerlings and the pathogenic virus of this disease is characterized by the fact that it does not infect fish other than channel catfish.This disorder affects primarily fingerlings less than 4 months old. The disease outbreak occurs when the water temperature exceeds 23°C. The affected fish exhibit abnormal swimming patterns, circling on their sides, exhibiting convulsions sinking to the pond bottom, and finally floating perpendicularly at the water surface, where they die. Fish exhibits abdominal distension and protrusion of the anus in the terminal stages. This disease is also characterized by eye protrusion and extreme fading of the gills. Hemorrhaging of the skin and the fins is occasionally visible. Hemorrhaging is evident in the liver, spleen, kidneys and skeletal muscles. Since hemorrhaging is general, the disease could be termed viral hemorrhagic septicemia.The causative virus is a DNA virus, belonging to the herpes virus group. The organ initially infected by virus I s the kidney, followed by spreading to the intestines and the liver. The central nervous system is attacked in the terminal stage. No pathology occurs in the brain, nerves and the gills.Fish, infected at the fingerling stage and survived may be the main source of infection. Infection can also occur via water and also there is possibility of oral infection.

Herpesvirus Disease in Kokanee Salmon and Rainbow Trout

This disease is prevalent since 1970 among Kokanee salmon hatched at hatcheries with mortality rates occasionally reaching 80 per cent.Infected fingerlings undergo darkening in body colour, and eye protrusion. Anemia is extreme. Ascetic fluid accumulates in the abdominal cavity and the abdomen distends. Anemia and edema are conspicuous in the visceral organs and the liver, pancreas and digestive tract all become flauid, while the vascular organs are mottled with areas of hyperemia. The kidneys fade in colour. Skeletal muscles and cardiac muscles exhibit edema. The liver seems to be the organ principally attacked. The causative virus is a species of the herpes virus group, *Herpesvirus salmonis.* It is sensitive to ether and acid, but it exhibits resistance at pH 10.

Since the virus has been isolated from ovarian fluid in adult rainbow trout, there is a possibility that virus bearing adult fish are the source of infection. Since the kidneys of afflicted fish are also attacked, the virus is believed to be excreted in water with urine and is believed to be spread to fingerlings one by one.

Pox Disease in Cyprinids (Papilloma Disease)

This disease is characterized by the dense formation of papilloma in various sites on the trunk, head and fins of cyprinids, such as carp and goldfish. The papillomas are white or pink and have a flat gloss. They are fairly hard, and are growths which appear to be of wax at first glance. The shape and expanse of the growths vary. Abnormal propagation of epithelial cells initiates the formations of these papillomas, and connective tissues with blood vessels develop as the nutritional base. The growths are difficult to peel off, and hemorrhaging occurs when they are forcibly removed. There are cases, in which they spontaneously peel off, and healing seems to have occurred, but they frequently redevelop. Virtually no deaths occur, but when the grow they have formed over an extensive area of the body, fish growth becomes retarded, and they become weak, where they are liable to die from secondary disease. The causative virus is classified as a herpes virus and they have thick membrane and are about 110nm in size. There is a strong possibility that environmental factors participate in the disease contraction.

Lymphocystis Disease

This disease was first reported in the English flounders, *Pleuronectes flesus* and *P. platessa*.

Small blister-like matters appear scatteringly or are clusters on the surface of the trunk, head and fins and other parts. Such blisters are enlarged dermal connective tissue cells. Their size is from 100μm to 500μm, and occasionally they are even larger. These abnormally developed cells are termed lymphocystis cells. They appear also in the muscles, liver. ovaries, intestines, etc. The causative virus is classified as an Icosahedra cytoplasmic DNA virus (ICDvirus). Lymphocystis virus can withstand freezing at -20°C for prolonged periods, but it is sensitive to ether and glycerin.

The source of infection is probably diseased fish. Infection via water is also considered a principal infection route. Moreover, wounds on the skin, etc are believed to provide a portal to infection. Various parasites produce infection portals and the possibility of their serving as virus vectors have been suggested. Disease occurrence is usually seen at high water temperatures. Lymphocystis disease generally is cured in a few months after its outbreak. Presumably due to specific antibody production and other immunity reactions.

Piscine Erythrocyte Necrosis (PEN)

It is a type of disorder in which the myclin of erythrocytes fuse. Usually salt-water fish such as Cod (*Gadus morhua*), blenny (*Blennius pholis*) and Atlantic herring (*Clupea harengus*) are affected.

The causative pathogen is believed to be an ICD virus-like particle which is icosahedral.

Chapter 21

Impact of Pollution on the Outbreak of Microbial Diseases

Introduction

The vast continental shelf area, tidal flats, backwaters, estuaries and freshwater of India have rich resource potential. These resources can be better managed and exploited for meeting the growing demand of food for the increasing human population. But the current scenario of fish production from coastal and fresh waters is highly discouraging owing to various factors. Population intensity along the coastal region increases pressure on the utilization of resources leading to habitat destruction, degradation and fragmentation. Among others, the factors that seriously damage or reduce fish production in coastal and fresh water environment are reclamation of the area for the establishment of industries, pesticide leaching from agricultural fields, oil spills from oil tankers and boats, sewage and waste disposal and retting of coconut husk along estuaries and backwaters and pollution through various sources. According to reports more than 90 per cent of the industries in India do not have adequate facilities to treat the effluents and as a result huge quantities of untreated effluents containing chemicals and wastes are discharged into the aquatic environment leading to serious consequences in the environment. Many externally visible diseases are prevalent in fish population and their incidences have been proposed as an index of pollution. Comparative toxicity of various pollutants on aquatic organisms under controlled laboratory conditions has been assessed and it is reported that at lower concentrations, the toxic materials may not cause the death of the organisms. However, they may cause stress to the organisms by interfering with the normal physiological functions. There are several experimental studies on the sub-lethal effects of pollutants. A wide array of responses due to sub-lethal effects have been reported and they are changes in swimming performance, respiration, phototaxis, equilibrium, decreased heart rate, delayed embryonic development, disrupted liver function, biomagnifications, blood chemistry alteration, behavioural changes and tissue damage (Natarajan and James,1977). During the past two decades, the public has become increasingly concerned with hazardous and toxic materials discharged into the aquatic environment. The toxic materials may accelerate detrimental effects either directly or indirectly on the organisms in question. Due to the development of more industries, increased urbanization and reclamation of areas, the environment is under severe stress and these create an unfavourable situation for aquatic organisms to live. The pollutants can weaken the organisms making it susceptible to disease or they cause disease directly. Toxicants or other stressful situations may enhance the disease. It is also indicated fish diseases do not occur as single caused events but are end results of interactions between host, parasite and environment (Snieszko, 1973). Several biological agents including bacteria, fungi, viruses, protozoa and metazoan parasites have been implicated to cause disease in wild fish population (Austin and Austin, 1987; Dykstra *et al.*, 1986). In the present study an attempt was made on epizootic ulcerative syndrome, haemorrhagic septicaemia and tail and fin rot diseases of fresh water and brackish water fishes along the southern coast of Kerala.

Study Sites

Veli and Vellayani lakes were selected as study sites. Fish species collected from these water bodies were transferred live within minimum stress and stocked separately in cement cisterns whenever necessary.

Examination of Body Lesions

The fishes were examined for body lesions(haemorrahages,ulcers,erosion and penetrating wonds). Detailed post-mortem examination was carried out by naked eye and under a dissection microscope to observe diseases if any externally and internally.

Bacteriology

For bacteriological investigation the infected fishes were narcotized and brought to the laboratory in a sterile polythene bag. All microbiological samples were taken from live fishes. Standard plating technique was followed (Collins *et al.*, 2001) serial dilutions were made and plated on Zobell agar22 16e,Triptic soya agar(TSA) and TCBS agar. All inculcated media were incubated at 37°C for 24 hrs following incubation, the colonies that developed on the plates were enumerated.

Bacterial population was determined by calculating the total number of colonies divided by weight of sample multiplied by respective dilution factor. Colonies were taken then randomly selected from each plate, isolated and restreaked on fresh agar plate to get pure culture and slants were maintained.

Biochemical tests, growth in different concentration as well as carbohydrate fermentation tests were carried out to identify the organisms at the species level. The isolated cultures were identified according to Bergey's classification (Staley *et al.*, 1989).

Observations

Epizoatic Ulcervative Syndrome (EUS)

Prevalence in Relation to Different Species of Fish

The prevalence of Epizoatic Ulcervative Syndrome in relation to different species of fish is presented in Table 21.1 and Figure 21.1. A total of 1000 individuals belonging to three species collected for a period of one year from January to December, 1993 were examined for the evidence of diseases. Of the three species collected two species were found to be affected and showed the sympotoms of EUS disease. The species of fishes that showed the symptoms of disease include *Channa striatus* and *Etroplus suratensis*. Species such as *Mystus gulio* did not show symptoms of the disease. The prevalence of the disease was maximum in *Channa striatus* (76.1 per cent).

Table 21.1: Prevalence of Epizoatic Ulcervative Syndrome (EUS) in Different Species of Fish

Sl.No.	*Species*	*No of Fish Examined*	*No of Fish Infected*	*Infection Percentage*
1.	*Channa striatus*	376	286	76.1
2.	*Etroplus suratensis*	300	205	68.3
3.	*Mystus gulio*	324	–	–

Haemorrhagic Septicaemia

The prevalence of Haemorrhagic septicaemia in relation to different species of fish is presented in Table 21.2 and Figure 21.3. Of the three species collected two species were found to be affected and showed the sympotoms of Haemorrhagic septicaemia disease. The species of fish that showed the symptoms of disease include *Etroplus suratensis*. Species such as *Channa striatus* and *Mystus gulio* did not show symptoms of the disease. The prevalence of the disease was maximum in *Etroplus suratensis*.

Table 21.2: Prevalence of Haemorrhagic Septicaemia in Different Species of Fishes

Sl.No.	*Species*	*No of Fish Examined*	*No of Fish Infected*
1.	*Channa striatus*	–	–
2.	*Etroplus suratensis.*	240	75
3.	*Mystus gulio*	–	–

Tail and Fin Rot

The prevalence of Tail and Fin rot in relation to different species of fish is presented in Table 21.3 and Figure 21.2. Of the three species collected two species were found to be affected and showed the sympotoms of Tail and Fin rot. The species of fish that showed the symptoms of disease include *Mystus gulios*. Species such as *Channa striatus*,

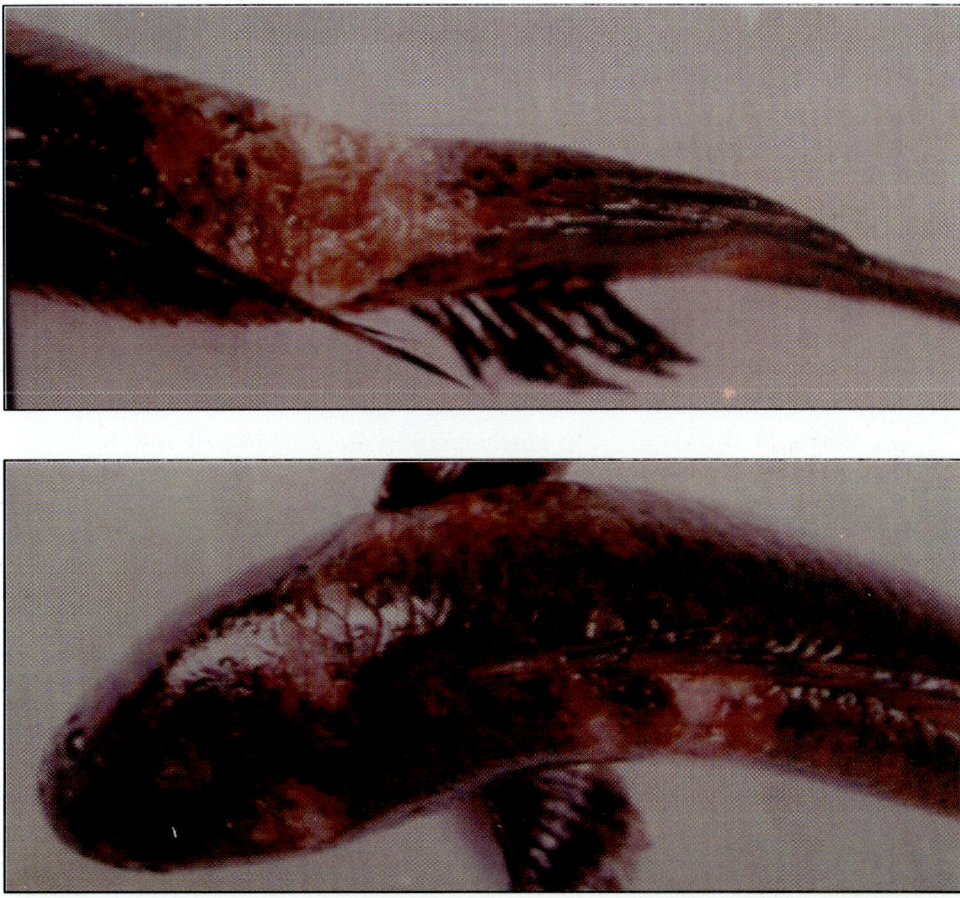

Figure 21.1: Epizoatic Ulcervative Syndrome Observed in *Channa striatus*

Figure 21.2: Tail and Fin Rot Observed in *Mystus gulio*

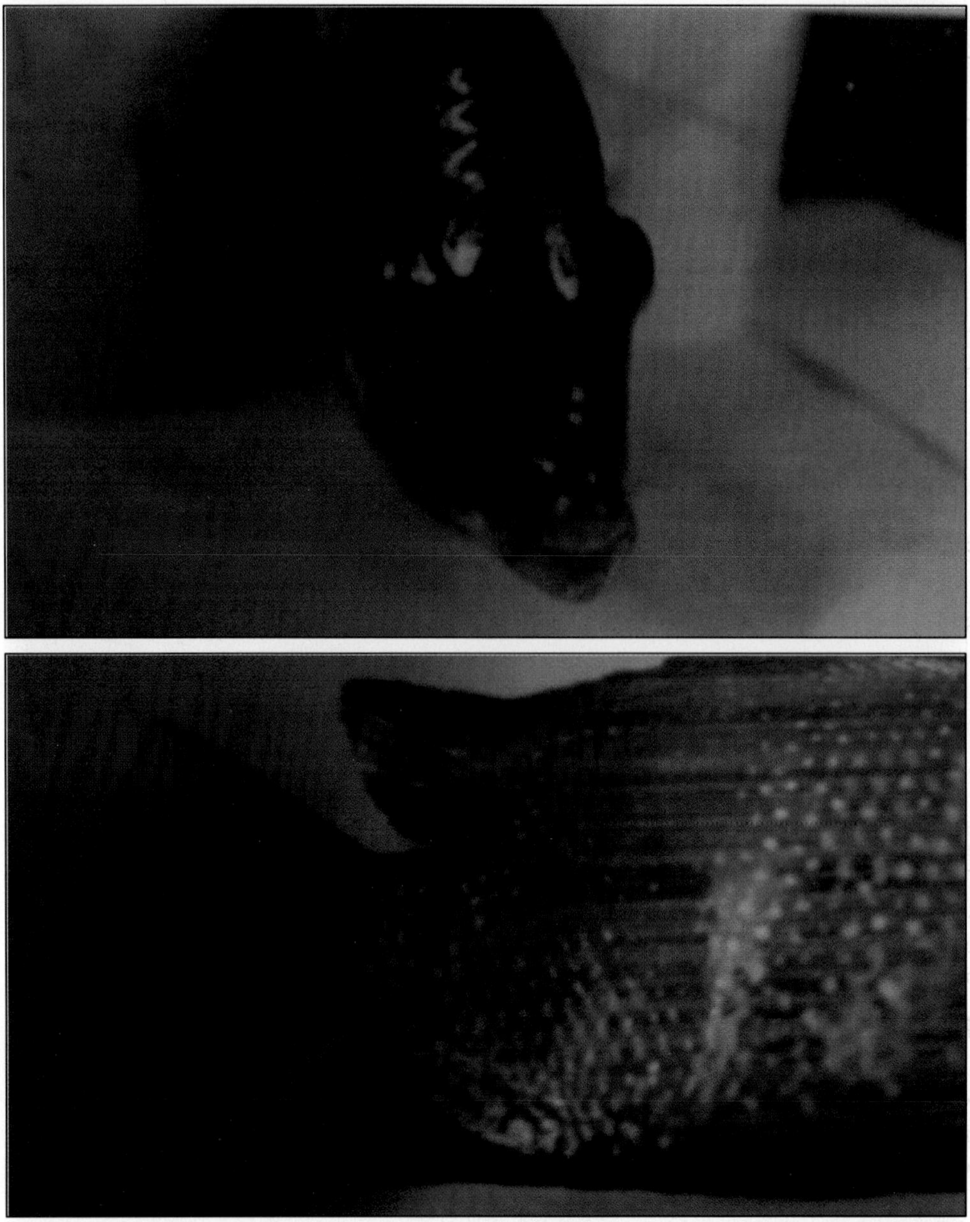

Figure 21.3: Haemorrhagic Septicaemia Observed in *Etroplus suratensis*

Etroplus suratensi did not show symptoms of the disease. The prevalence of the disease was maximum in *Mystus gulios*.

Bacterial Load

The total heterotropic bacterial (THB) population isolated in the study revealed that the infected fish harboured comparatively higher bacterial population than the uninfected fish (Table 21.4).

Table 21.3: Prevalence of Tail and Fin Rot in Different Species of Fishes

Sl.No.	*Species*	*No of Fish Examined*	*No of Fish Infected*	*Infection Percentage*
1.	*Channa striatus*	436	56	12.8
2.	*Etroplus suratensis*	329	53	16.1
3.	*Mystus gulio*	324	76	23.5

Table 21.4: Total Bacterial Population ($\times 10^6$/g) in Different Species of Fish

Sl.No.	*Species*	*Uninfected*	*No of Fish Infected*
1.	*Channa striatus*	45.5	65.5 (20.0)
2.	*Etroplus suratensis*	50.5	79.5 (29.0)
3.	*Mystus gulio*	30.0	60.1(30.1)

Epizoatic Ulcervative Syndrome (EUS)

The prevalence of the disease was maximum in *Channa striatus* (76.1 per cent). Similar observation was reported by Roberts *et al* (1989). Diseased snakeheads had visible signs of the EUS because they were robust and survived over a prolonged period to a very advanced state of ulceration whereas other species succumbed more readily.

Haemorrhagic Septicaemia

The prevalence of the disease was maximum in *Etroplus suratensis*. As early as 1936, Kluyer and Van Niel reported that microbes such as *Aeromonas, Pseudomonas,Bacillus,Proteus* and *Aerobacter* were associated with haemorrahagic septicaemia in fish. Lewis and Bender (1960 and 1961) and Meyer (1970) inferred that haemorrahagic septicaemia caused by *Aeromonas liquefaciens* in the pond fishes of United States culminated in very high mortalites.

Tail and Fin Rot

The prevalence of the disease was maximum in *Mystus gulio*. Similar results were reported in the case of fin and tail rot disease in *Mystus gulio* by Beena (1993). Incidence and severity of fin rot in Atlantic salmon have been related to rearing temperature (Schneider and Nicholson, 1980; Mahesh Kumar, 1985).

Conclusion

From the above findings it is concluded that the impact of pollution in freshwater and brackish water vastly affect the edible fishes. Epidemic surveillance and remedial actions are inevitable as epidemic can wipe-out species from earth. Controlling factors are to be brought to immediate action for the prevention of diseases.

Chapter 22
Dilution of the Pathological Samples

Introduction

The majority of marine microorganisms require sea water or its equivalent for their optimal activity. So, aged and filtered sterile sea water was used to dilute the marine pathological samples.

In the present investigation, procedures for the dilution of the samples as given by Dhevendaran and Maya (1998) were adopted with slight modifications.

The pathological samples were ground in a clean and sterile 25 ml mortar using a sterile pestle inside the surgery cell, in 1-3 ml of the suitable suspension fluid. The properly ground and diluted samples were aseptically collected in convenient sterile test tubes for timely inoculation on to a suitable media.

Media

The wet mount of the pathological sample was prepared and an examination of it under the microscope was made to select the suitable media for culturing bacteria or fungi (Dhevendaran and Salley, 2000). The pathological samples were cultured on appropriate media based on sea water or habitat water as the base. Fresh water with 5 per cent Sodium chloride was also tried as the base.

In addition to culturing in solid media, samples were also cultured in the corresponding liquid media, sea water peptone or peptone water or in both according to the nature and source of the samples. The following media were used for culture:

1. Nutrient broth with sea water base

2. Nutrient broth
3. Nutrient agar
4. Nutrient sea water agar
5. ZoBell's medium, 2216
6. Fish infusion agar
7. Tripticase soy agar
8. Blood agar
9. MacConkey's agar
10. Sabouraud's glucose agar
11. Leetown medium

Nutrient Sea Water Agar

1. Peptone (Bacto)	1.00gm
2. Ferric phosphate (BDH, AR)	0.01gm
3. Beef extract (Oxoid)	0.30gm
4. Agar (Bacto)	1.50gm
5. Aged and filtered sea water	100 ml

Fish Infusion Agar

1. Peptone (Bacto)	1.00gm
2. Agar (Bacto)	1.50gm
3. Fish infusion	100 ml

Preparation of Fish Infusion

Fresh fish muscle weighing 454 gm was mixed into one liter of water (aged and filtered sea water or habitat water). The minced fish muscle overnight stored in a refrigerator at about 4°C and later boiling was done in a water bath for 30 minutes. The precipitated and coagulable proteins were removed by filtering through lint and finally using whatman 1 filter paper.

Fish muscle consists of a high amount of protein, sugar and amino acids. So, fish infusion medium is nutritionally rich and favours the growth of most forms of bacteria (Maya and Dhevendaran, 1997). This medium was found to be one of the best media giving luxuriant growth.

All the media were in regular routine use except ZoBell's agar as the purpose of this medium was satisfied by the rest of the media. MacConkey's media did not give satisfactory growth. Tripticase soy agar, Lee town medium and Sabouraud's agar were used only when the need arose.

pH of the Media and Autoclaving

The available information regarding the pH of fresh fish muscle shows that it is approximately neutral in reaction. Usually the pH of the sea water is 7-8. The maximum growth of marine microbes is in the pH range between 7-8. Normally, high alkaline and acid ranges are not favourable for the growth of marine and terrestrial microbes respectively. In order to obtain the maximum counts and number of species, the pH of the media was adjusted to 7.2 to 7.5 unless and otherwise mentioned.All the media were sterilized at 121°C for 10-15 minutes, if not otherwise mentioned.

Culturing of the Samples

A 0.1 ml aliquot of the pathological sample, was taken with a sterile pipette and aseptically spread over the surface of the media according to the modified methods of Dhevendaran and Salley (2000). The time lag between capture of the fish and the commencement of culture of the samples was generally within one to two hours.

Incubation Temperature

The most suitable temperature for the culture and cultivation of the microbes from the fish and shellfish depends on the ambient water temperature as the microbes associated with various types of fish whether pelagic or deep sea, will have corresponding varying optimum growth temperatures. However, the optimum growth temperature for marine microbes varies from 18-22°C, for the development of the maximum plate counts. The growth of the marine forms at 37°C is very poor because most of the marine bacteria are thermo sensitive (Dhevendaran *et al.*, 1992).

The surface temperature of the tropical waters in the open sea ranges generally between 28-30°C and a little higher in certain areas of the littoral zone and the same is the case with fresh water systems. Even though the temperature of the marine environment is relatively low, quite surprisingly enough, many of the marine bacteria grow at 20-25°C than at12-15°C. In order to isolate the pathogenic forms from the diseased specimen, from both marine and fresh water, the media were incubated at aerobic temperature, 28 ± 2°C, to 72 hours.

The predominating bacteria represented by most numerous colonies similar in colonial characteristics were selected and considered as the causative agents. Representatives of the isolated colonies, after studying the colonial morphology-size, shape, margin, elevation, consistency, capacity and pigment were transferred into sea water nutrient or nutrient broth and each isolate was re streaked to check for its purity. After ascertaining the purity of the isolates, they were streaked twice on to the corresponding fish infusion agar slants and stocked at 4°C for future studies, since Dhevendaran and Annie (1999) observed that refrigerator temperatures favour the constancy of the species characteristics.The organisms were sub cultured in fish infusion agar slants once for every 30 days.

Identification of microorganisms is useful and essential in many ways as they form the indicator organisms in several respects (Staley *et al.*, 1989). In nature, many of the bacteria autochthonous to the sea from the indigenous flora of the finfish and shellfishes. It is also important to take into consideration the fact that a few microbes of terrestrial origin also find a place and survive in the sea. Even though there are no imfallible criteria to differentiate marine bacteria from the terrestrial forms, it is an established fact that marine bacteria indigenous to the do exist, and it has been calculated that they form atlest 12 per cent of the existing forms of the total bacteria (Staley *et al.*, 1989).

It is well established fact that most of the bacteria of marine fish are Gram negative rods (Staley *et al.*, 1989) but, there are considerable difficulties in identifying the Gram negative rods other than the species belonging to the family Enterobacteriaceae.

To tentatively identify the isolates obtained in the present study, two systems of identification, a modified scheme of Staley *et al.* (1989), were used to identify the bacterial isolates.

Characters Examined

Unless specially mentioned, all the tests were carried out in freshly prepared media, inculated at room temperature 28 ± 2°C for 24 hours to 14days.

The following characters were studied

1. Colonial morphology on fish infusion agar.
2. Cell morphology and Gram's staining of the isolates from fish infusion agar according to Hucker's modification.
3. Motility in hanging drop preparations from cultures following incubation for 6-18 hours in peptone water or sea water peptone (Bacto peptone, 1 per cent in aged and filtered sea water pH 7.2-7.5).
4. Indole production in tryptone water.
5. Methyl Red and Voges Proskauer reactions.
6. Utilization of citrate in Koser's citrate medium.
7. Nitrate reduction in peptone water containing 0.2 per cent Potassium nitrate, and testing at 24-72 hour culture with Dimethyl alpha naphthylamine sulphanilic acid reagent. Zinc dust was added to the cultures showing negative reactions, to detect if nitrate remained and decomposition had not proceeded beyond the nitrite stage.
8. Ammonia from peptone in peptone water cultures aged 18-22 hr with Nessler's reagent.
9. Hydrogen sulphide production in nutrient broth with lead acetate papers suspended over the medium.
10. Gelatin liquefaction.
11. Starch hydrolysis in starch media using Lugol's iodine as an indicator.
12. Catalase by emulsifying an an 18-24 hr culture, from fish infusion agar, with a few drops of 20 vol Hydrogen peroxide.
13. Oxidase according to Kovacs.
14. Urease by inoculating the organism in Christensen's agar.
15. Metabolism of glucose.
16. Reaction in litmus milk.
17. Haemolysis in blood agar (cooled sterile nutrient agar plus 5 per cent sterile defibrinated blood).
18. Casein hydrolysis and pigmentation observed in milk agar.

19. Lypolytic activity in plates of tributyrin agar.
20. Sensitivity to antibiotics. The vibriostatic compound, '0/129' (2,4 Diamino 6,7 diisopropylpteridine) was also tested by using filter paper discs.
21. Luminescence in Yeast peptone medium.
22. Peptone water including 1 per cent of the respective sugars with Andrade's indicator to observe the production of acid and gas.

Each organism was studied individually for its characteristics. The cases are recorded according to the clinical symptoms noted at the time of documentation of diseases.

Chapter 23
Characterization, Classification and Identification

Introduction

Morphology

Cocci, Rods and Spiral

Size

0.1 to 5 um. Chemical characteristics: When cell was broken, their cell ingredients are subjected to chemical analysis.

1. Lipopolysaccharide(LPS) in the cell wall is characteristic of Gram Negative bacteria
2. Teichoic acid is for Gram Positive bacteria
3. Virus has either DNA or RNA

Culture Characteristic

Media, Temperature, and other physical characteristics.

Metabolic Characteristics

Microbes derive energy from light, or oxidising organic compounds or inorganic compounds. Chemical reaction like starch, protein and lipids are disintegrated by enzymes

Antigenic Characteristics

Toxins of microbial compounds are antigen. Animal produces specific blood serum protein called antibodies. The reaction between antigen and antibody like a lock and key system.

If we take typhoid bacterium antibody and mix with a suspension of unknown bacterial cells and a positive reaction.

Agglutination occurs we can conclude that the cells are those of the typhoid organism.

Genetic Characteristics

The double stranded chromosomal DNA of each kind of microorganism has certain features that are constant and characteristic for that organism and useful for it's classification.

1. *DNA-Base composition*: Nuceotide Base Pair: Guanine-Cytosine and Adenine and Thymine.

   Percentage expressed as mole per cent G+C. It ranged from 23 to 75.
2. *Sequence of nucleotide base in DNA*: It is unique for each kind of organism; significant for classification in addition to chromosomal DNAP. lasmid DNA may sometimes be present in microbial cells, having autonomous replication within the bacterial cells. They have the ability to make toxins (Toxigenicity). It becomes resistant to various antibiotics to use unusual chemical compounds as nutrients.

Pathogenicity

The ability to cause disease is called pathogenicity. A few microbes cause disease in plants and animals. Wilt disease in plants and typhoid fever inhuman beings. Further, Bdellovibrio is parasitic to other bacteria.

Ecology

Halophilic bacteria from marine sources and differs from fresh water *Bacillus.*

Classification: Nomenclature and Identification

In microbiology, taxa are initially constructed from strains. For example, strain ATCC 19554 is a strain of spirilla, isolated from pond Virginia in 1965 and this strain is maintained at the American Type Culture Collection(ATCC), Maryland, USA.

Cultures of the same species that were isolated from other sources would be considered different strains.

The basic taxonomic group (Taxon) is the species. *i.e.*, collection of strains having similar characteristics. Bacterial species consists of a special strain called the Type strain together with all other strains that are considered sufficiently similar to the type strain as to warrant inclusion in the species. Special attention is to be given to their maintenance.

In England National Collection of Type culture is established. Similar to species genus is composed of collection of different species.(*e.g. Vibrio* is the genus and *cholerae* is the species of *Vibrio.*

Taxonomic groups of higher rank than genus are familiy, order, class division and kingdom.

Goals of Classification

Classification has two qualities.

1. Stability without confusion
2. Predictability assuming that one member of the taxonomic has similar characteristics.

Those strains having a high per cent to each other are placed into groups and also placed in large groups. The degree of similarity needed to rank a group as a species, genus, or other taxon. This is of practical use and is being relatively unbiased in it's approach. It has high degree of stability and predictability.

Genetic Relatedness

Important method of classification. In 1960 the development of molecular biology used for the DNA of two organisms are compared. Crude comparison of mole per cent G+C values was taken. However, to realise that organisms that are completely unrelated may have similar mole per cent G+C values.

Therefore, more precise methods of comparison were needed. DNA molecules from various organisms could be compared with respect to the sequence of their component nucleotides. This sequence is the most fundamental characteristic of an organism. Modern techniques have now made it possible to make such comparision. It is as follows:

1. DNA homology experiments: The double stranded molecules from two organisms are heated to convert them into single strands. The single strands from one organism are mixed with that of other organism.

   If they are closely related hetero duplexes are formed (Figure 23.1).

   If not hetero duplexes are not formed.
2. Ribosomal RNA homology experiments and ribosomal RNA oligonucleotide cataloging: Two organisms may not be so closely related as to give high level of DNA homology, yet they may still have some degree of relatedness. Ribosomes are granular structures in the cell which synthesise protein, composed of protein and RNA. This ribosomal RNA cloned (rRNA) by only a fraction of the DNA molecules, the rRNA cistrons. In all bacteria the nucleotide sequence of these rRNA genes has been found to be highly conserved. During evolution the nucleotide sequence has changed more slowly than that of bulk of DNA molecule.

That means that even if two organisms are only distantly related and show no significant DNA homology, there still may be considerable similarity in the nucleotide sequence of their r RNA cistrons. The degree of similarity that exists can therefore be used as a measure of relatedness between organisms, but at a level beyond that of species (at the level of genus, families, order, etc.). Classifications based on the genetic relatedness come the closest to achieving the taxonomic goals of stability and predictability. Such classification is the basis for bacteria that infer the evolutionary significance so that the present day bacterial genera and species can be arranged in a hierarchy that reflects their ancestral relationships *i.e.* in a phylogenitic classification.

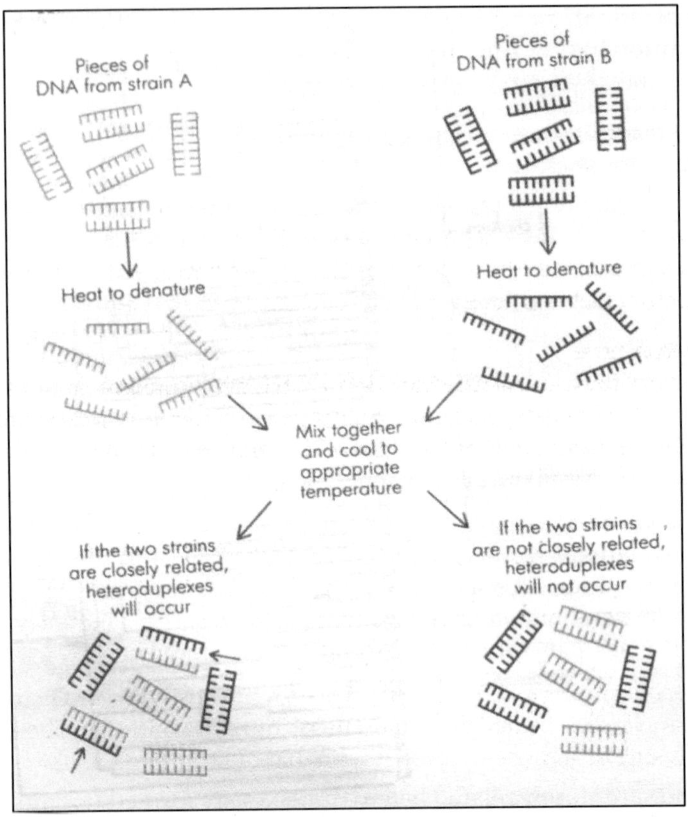

Figure 23.1: DNA Homology Modeling

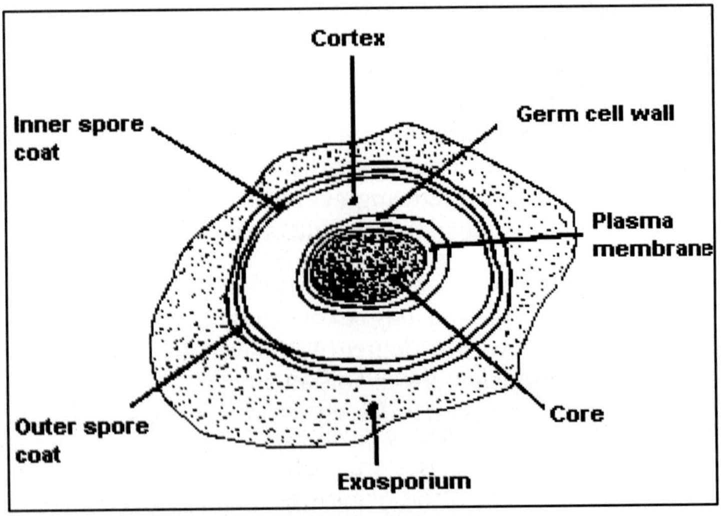

Figure 23.2: Bacterial Spore Structure

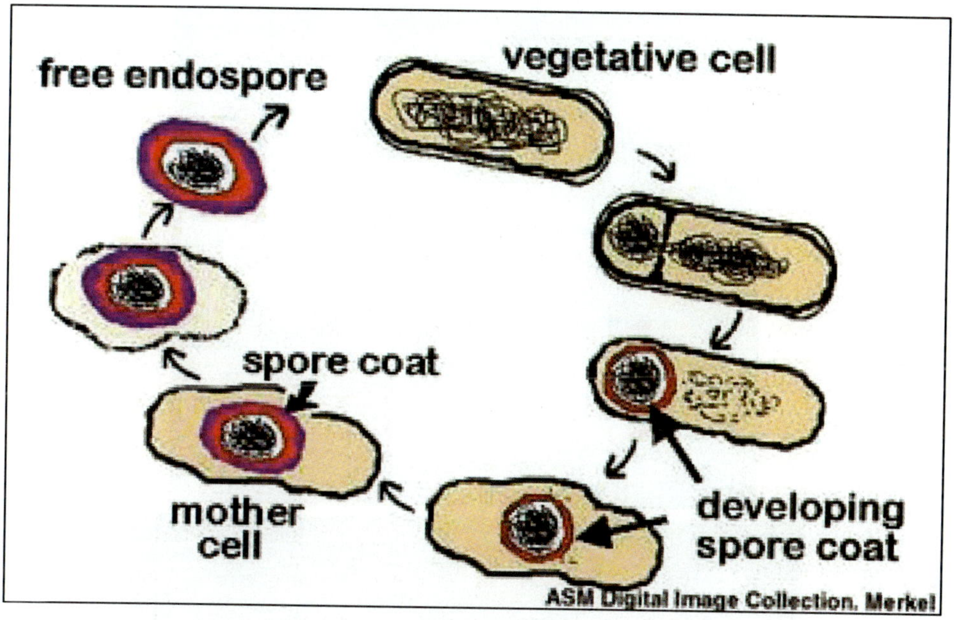

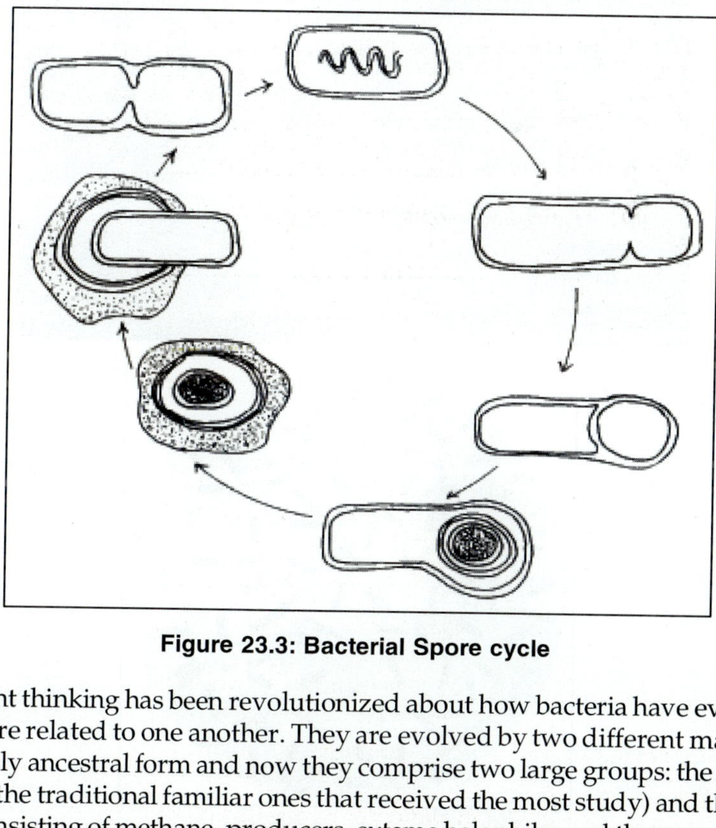

Figure 23.3: Bacterial Spore cycle

Current thinking has been revolutionized about how bacteria have evolved and how they are related to one another. They are evolved by two different major routes from an early ancestral form and now they comprise two large groups: the eubacteria (which are the traditional familiar ones that received the most study) and the archaeo bacteria(consisting of methane- producers, exteme halophiles and thermoacidophiles.

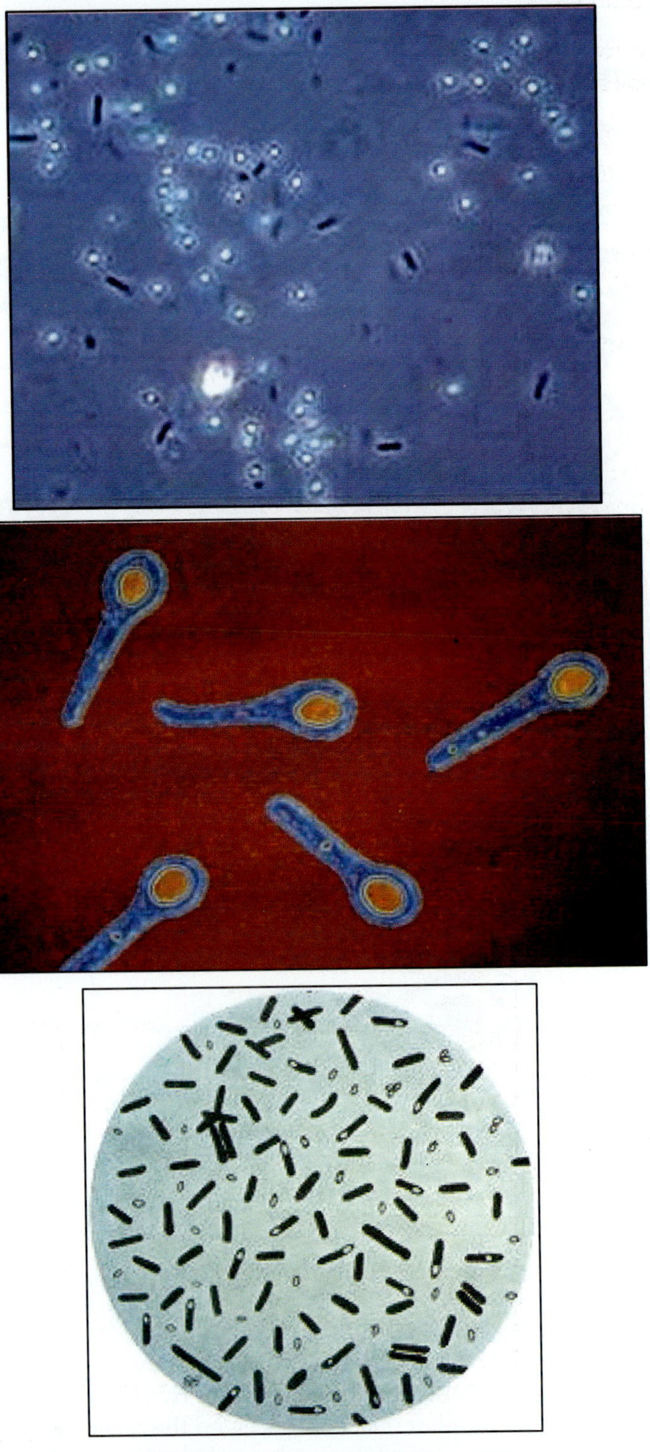

Figure 23.4: Bacterial Spores

They distinct each other. Invariably archaeobacteria are separated from other bacteria by a phylogenitic gulf.

Nomenclature

Each species of microorganisms has only one officially accepted name.

E.g: Micrococcus luteus -means yellow berry

Proteus vulgaris in Latin" common organism of many shapes

Binominal

Two words First –Genus: Second-species

E.g: Escherichia coli (Figure 23.23).

Identification

An organism must be classified before it can be identified.

E.g: Legionella pneumophila which causes pneumonia disease.

Past and Present of Bacteria

First classification in 1773. The valid one is Bergey's Manual of Determinative Bacteriology. It increases the coverage of bacterial characterisation.

Bergey's Manual of Systematic Bacteriology

It is universally used reference for classification of microbes preferably bacteria.

The Spirochetes (Figures 23.4A and 25)

The distinguishing features are

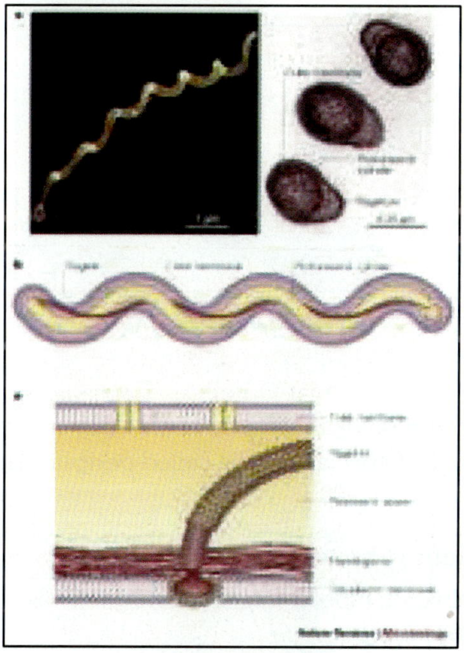

Figure 23.4A

Order Spirochaetales

Fam. Spirochaetaceae — Fam. Leptospiraceae

E.g. *Treponema pallidum* (Causing syphilis in human animals and being) — E.g. *Leptospira* (Microaeropsilic) *interrog* (pathogenic to human beings.)

Figure 23.5

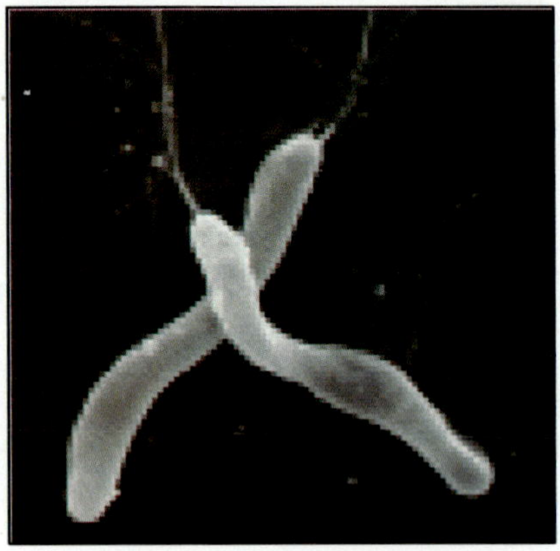

Figure 23.6: Aquaspirillum

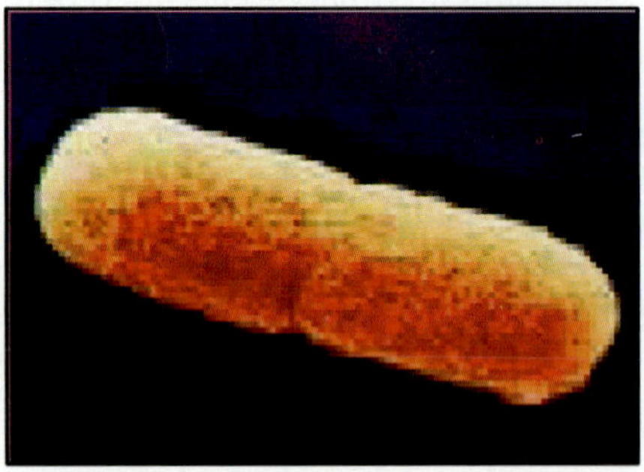

Figure 23.7: Azospirillum

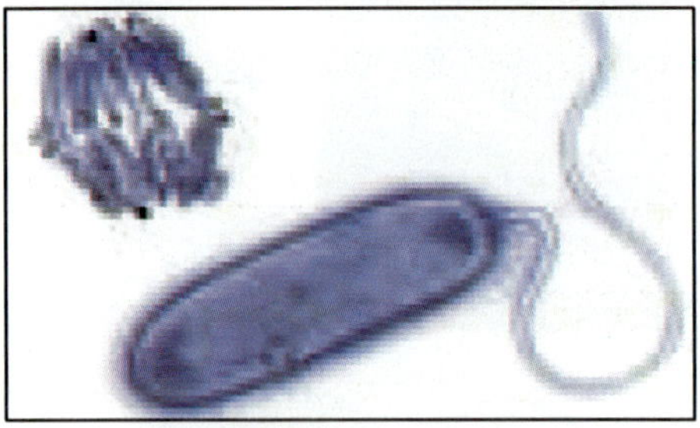

Figure 23.8: *Bdellovibrio*

1. A helical shape.
2. An ability to twist their shape (*i.e.* flexibility).
3. The occurrence of special flagella called periplasmic flagella (*Axial fibril*).

Spirochaetes

Aerobic, motile Gram-ve bacteria:

Aquaspirillum is vibrioid, bipolar flagella motile, harmless, aerobic, occurring in freshwater ponds (Figure 23.6).

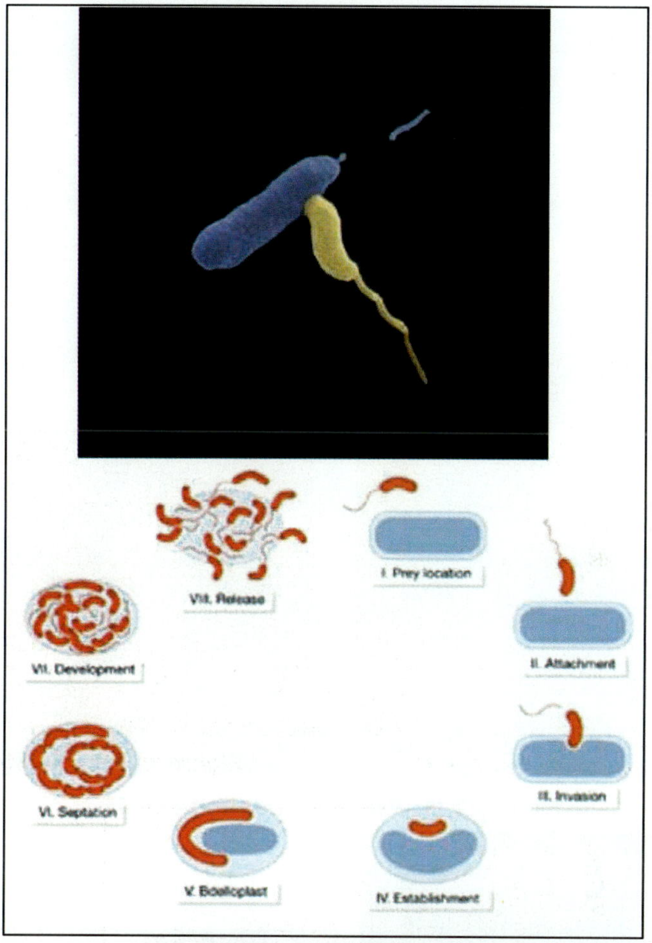

Figure 23.9: Life Cycle in Bacteria

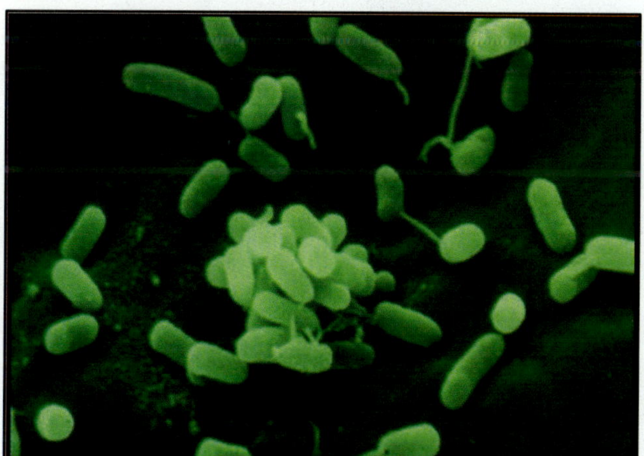

Figure 23.10: *Pseudomonas fluorescens*

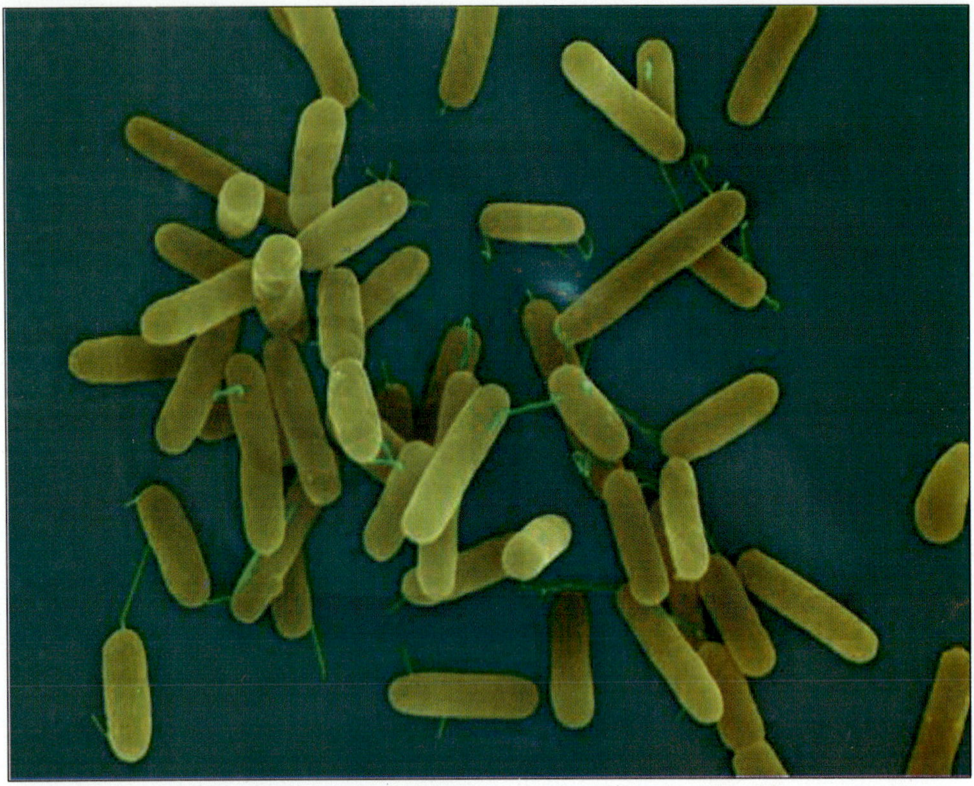

Figure 23.11: *Pseudomonas aeruginosa*

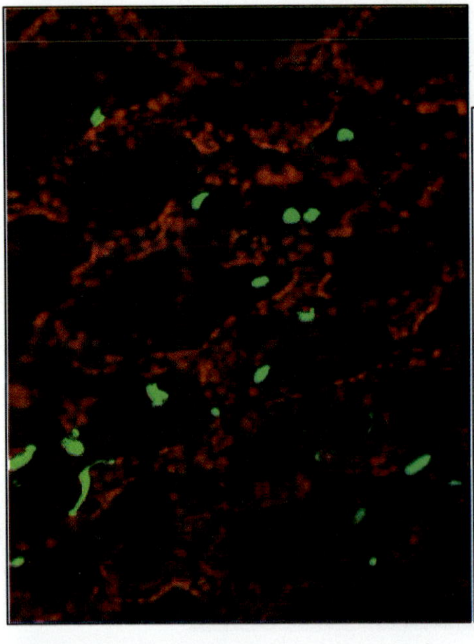

Figure 23.12: *Pseudomonas syringae*

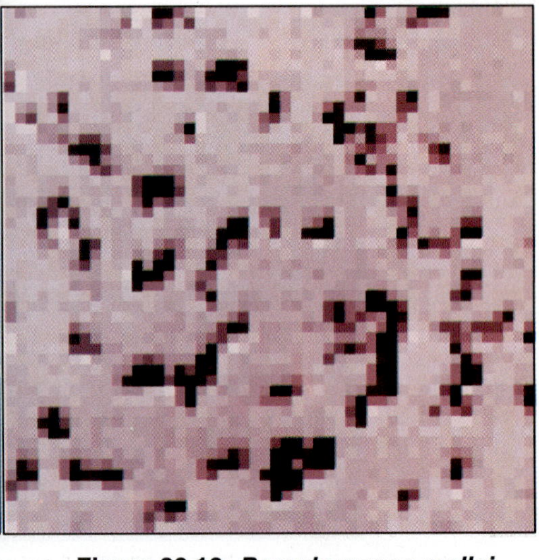

Figure 23.13: *Pseudomonas mallei*

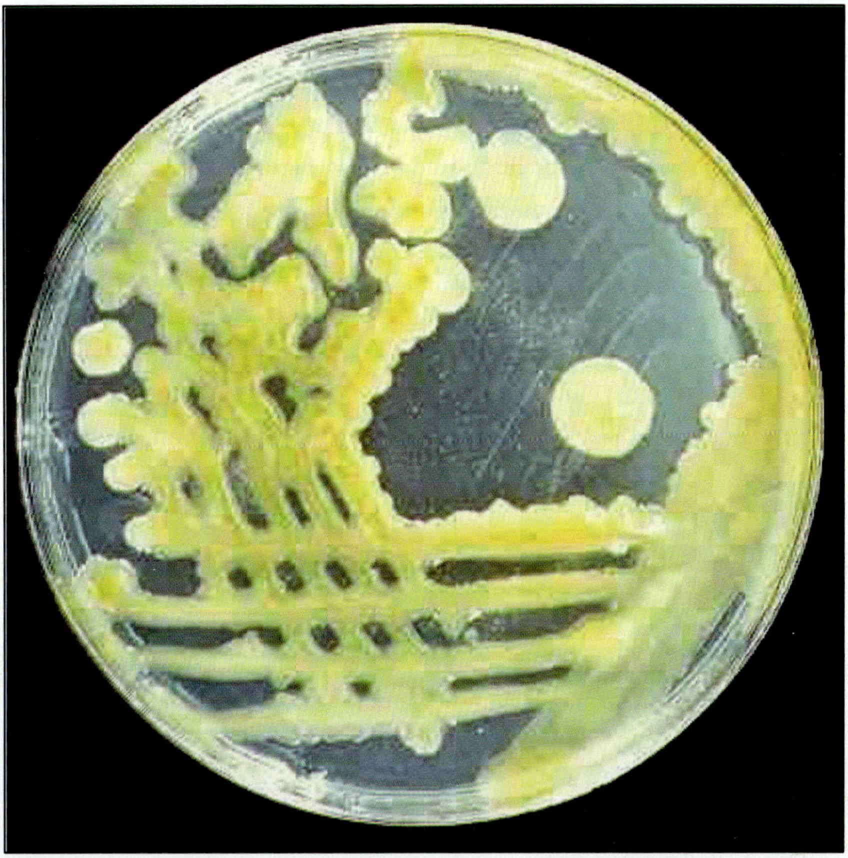

Figure 23.14: ***Xanthomonas*** **Colonies on Plate**

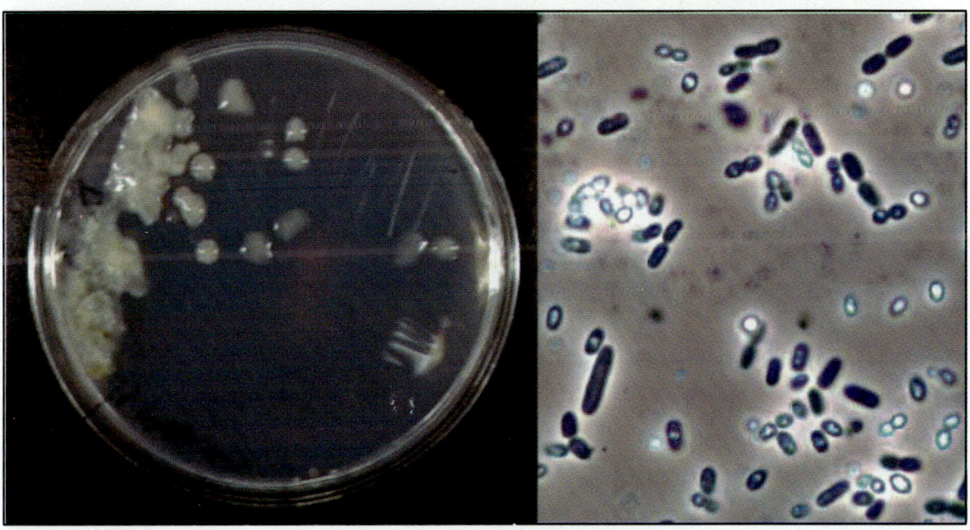

Figure 23.15: ***Azotobacter*** **Colonies**

Azospirillum is vibrioid with a single polar flagellum, fixing N2 within plant root or in laboratory culture (Figure 23.7).

A. brasilense fix N2.

Bdellovibrio, aerobic, vibriod, single polar flagellum present, parasitic on other bacteria. Life cycle in bacteria is illustrated below (Figures 23.8 and 23.9).

Vampiro vibrio has similar characteristics to *Bdellovibrio* but attack algae but not bacteria.

Family. Pseudomonadaceae. *Pseudomonas*. Majority are aerobic and a few are anaerobic.

P. aeruginosa an opportunistic pathogen producing pyocyanin pigment (Figure 23.11).

P. fluorescens, a saprophylic and and produces fluorescent pigment (Figure 23.10).

P. syringae – plant pathogen causing leaf-spot, leaf-stripe, wilt and necrosis (Figure 23.12).

P. mallei causes diseases in horses and donkeys that are transmissible to human (Figure 23.13).

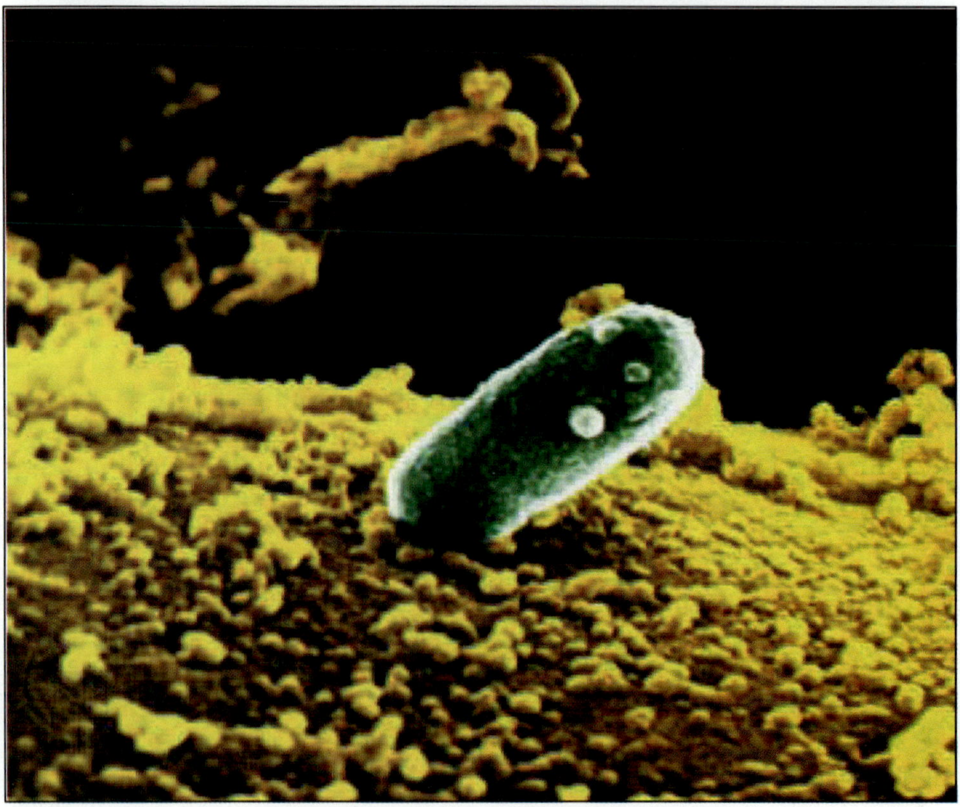

Figure 23.16: *Rhizobium* sp.

Xanthomonas

It produces yellow xanthomondin pigment, causing diseases such as spots, streaks, cankers, wilt and rots. It can produce Xanthan germs which can be used in food as stabilizer and antridrips in paints (Figure 23.14).

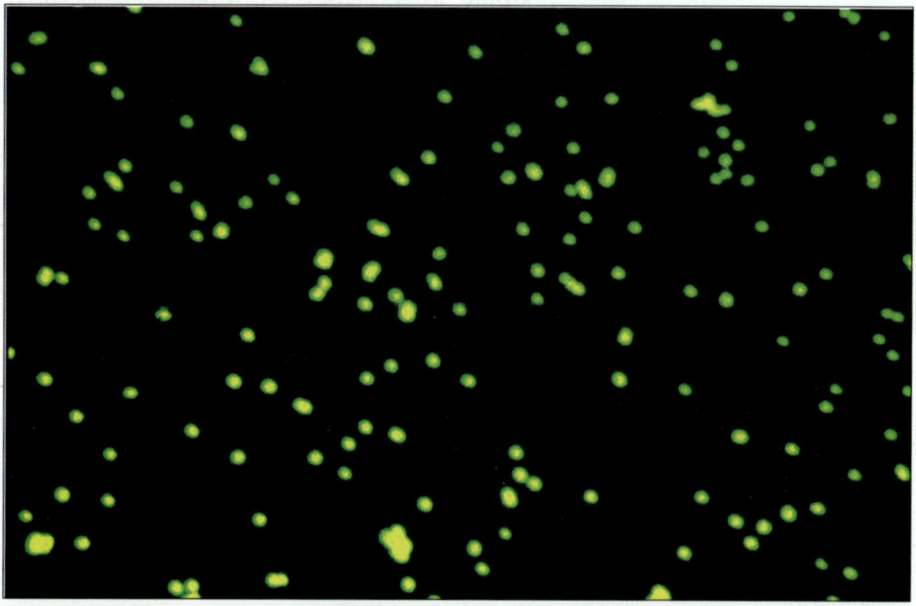

Figure 23.17: ***Methanococcus***

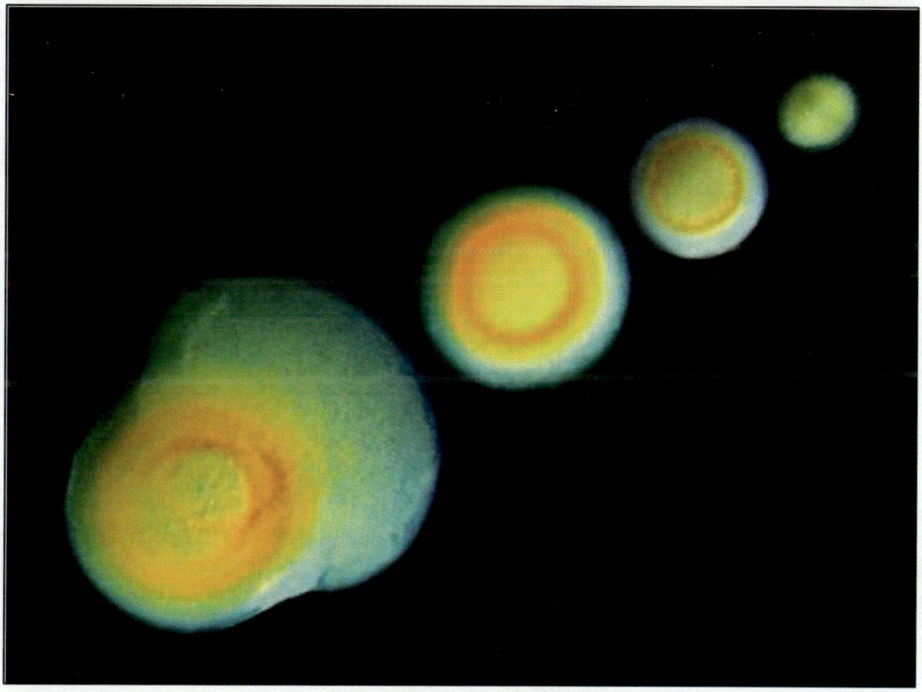

Figure 23.18: ***Acetobacter***

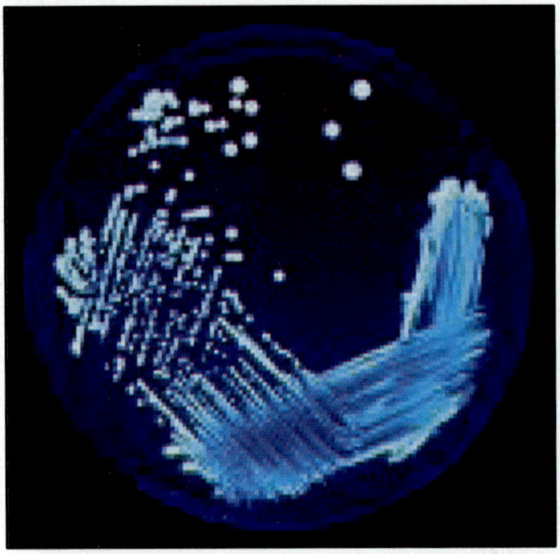

Figure 23.19: *Legionella* colonies

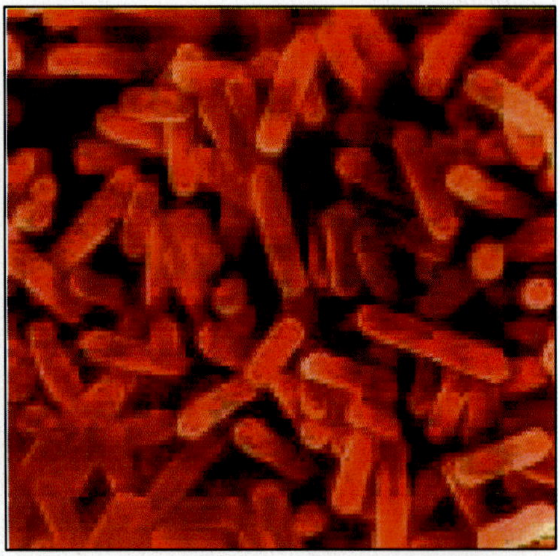

Figure 23.20: *Legionella* spp.

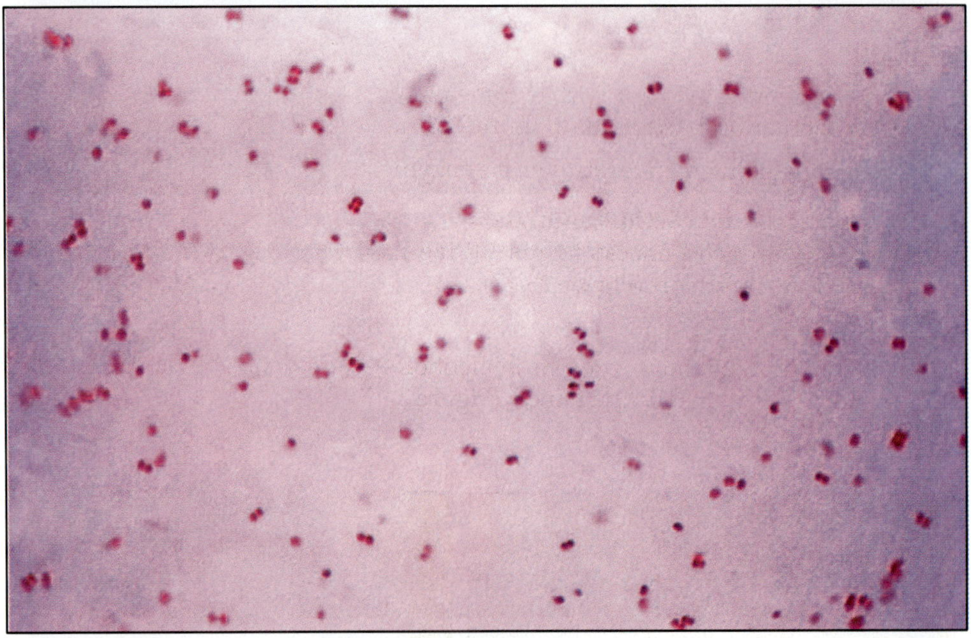

Figure 23.21: *Neisseria* spp.

Azotobacter

It is an aerobic N2 fixer, blunt nod, oval or cocci, present in soils. The unusual high respiratory rate of the cells serve to use up 02 rapidly at the cell surface and maintain an anaerobic cell interior (Figure 23.15).

This, inturn, protects the oxygen sensitive nitrogenase enzyme complex. It has desiccation-resistant cyst.

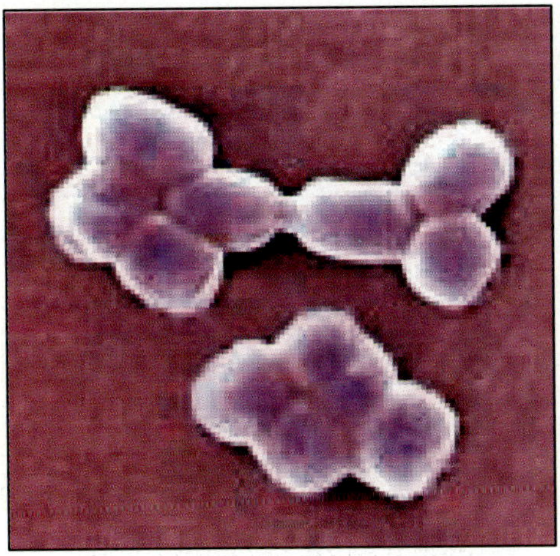

Figure 23.22: *Acinetobacter* spp.

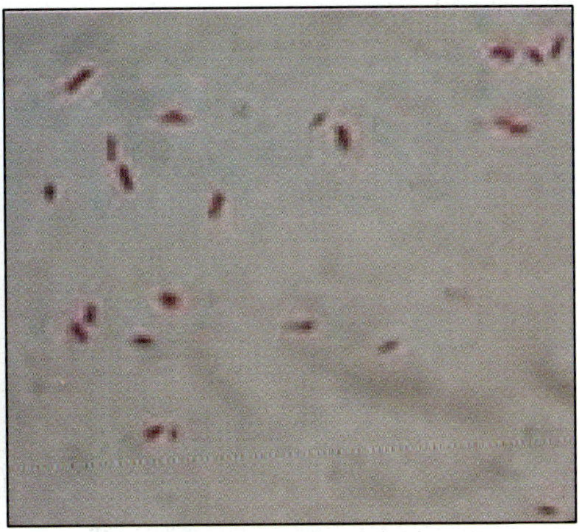

Figure 23.23: *E. coli*

Rhizobium

It is a symbiotic N2 fixer with legume plants. It is aerobic forming nodules in plants as pleomophic bacterioids (Figure 23.16).

Leghemoglobin protects nitrogenase enzyme in nodules.

Methonococcus and Methanomonas

They are red and coccal forms utilize methane as carbon source under microaerophilic condition (Figure 23.17).

Acetobacter

It is rod shaped and in present in alcohol-enriched acidic environments like flowers, beer, wine,vinegar and honey (Figure 23.18).

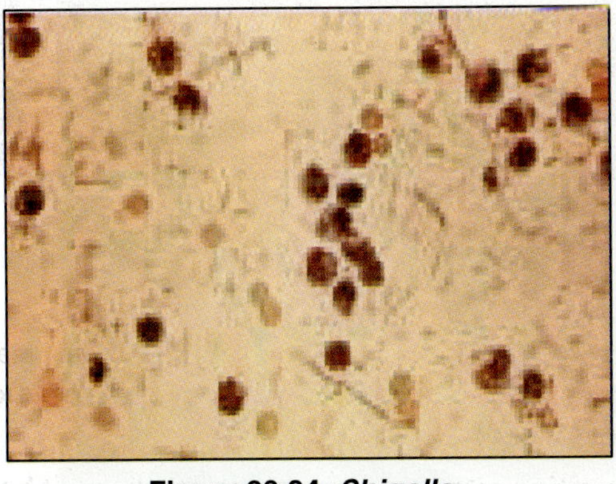

Figure 23.24: *Shigella*

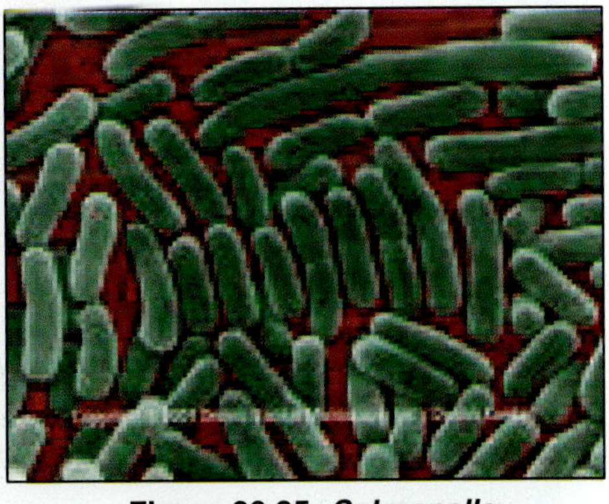

Figure 23.25: *Salmonella*

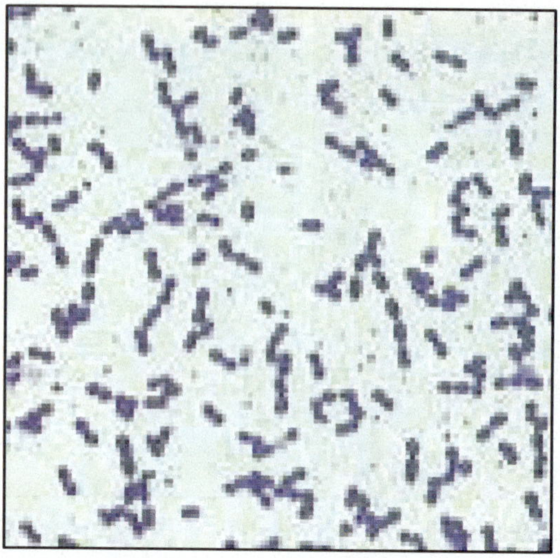

Figure 23.26: *Enterobacter*

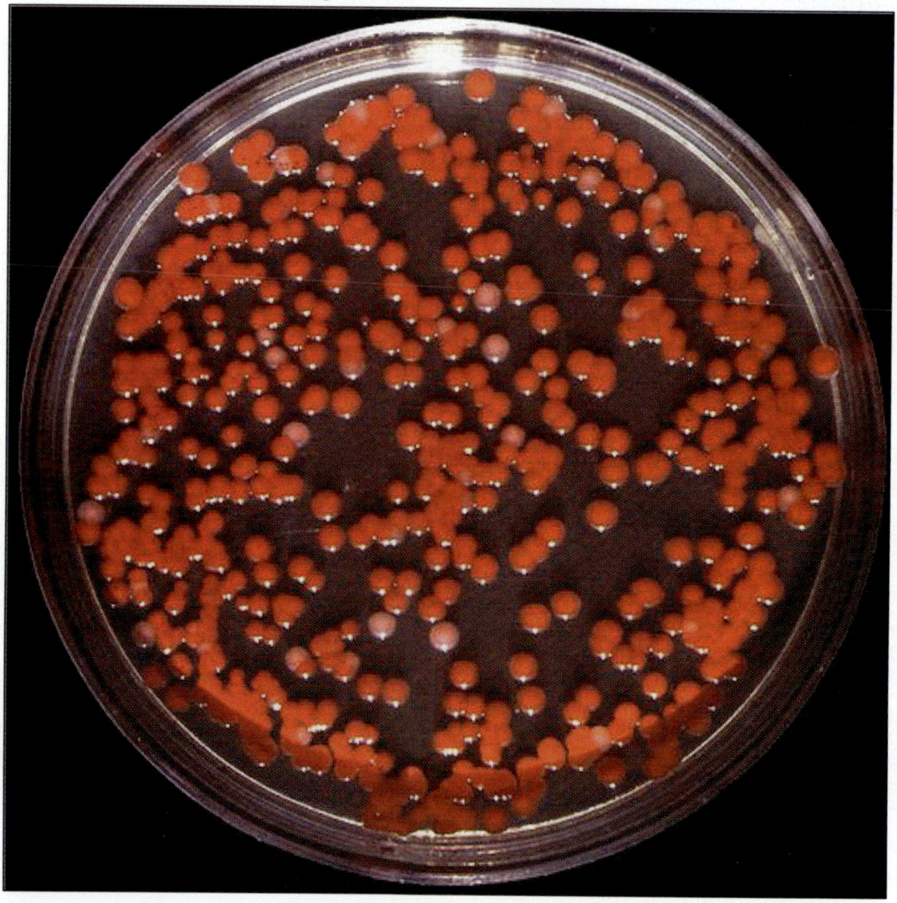

Figure 23.27: *Serratia* Colonies

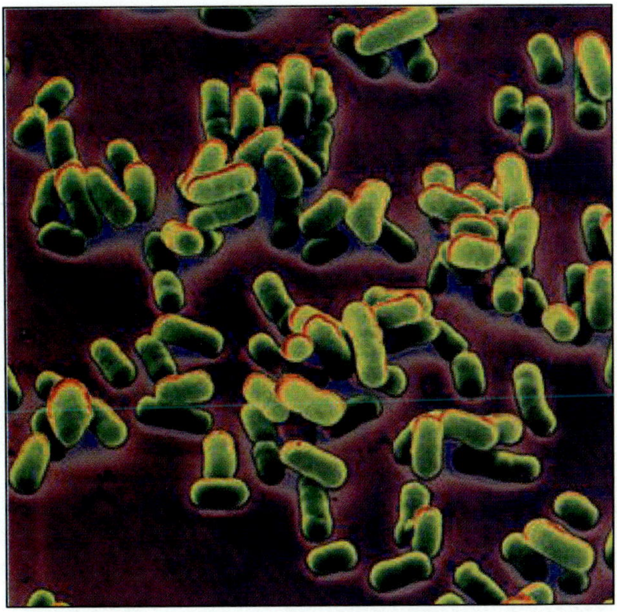

Figure 23.28: *Yersinea*

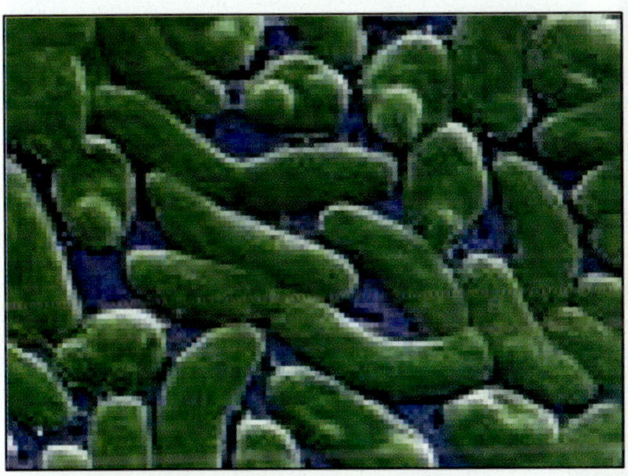

Figure 23.29: *V. vulnificus*

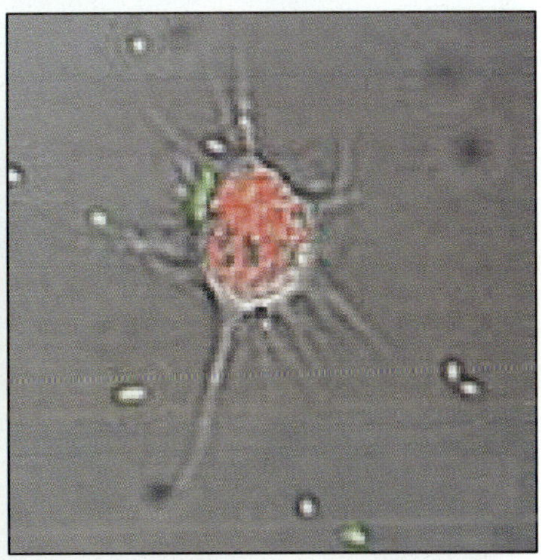

Figure 23.30: *V. fischeri*

Legionella

It is usually rod shaped, requiring L-cysteine, iron salt and activated charcoal for growth. It is present in thermally polluted lakes, air –conditioning cooling towers and evaporative condenser (Figures 23.19 and 23.20).

Neisseria (Figure 23.21)

It is oxidase and catalase positive coccus. *N.elongata* is parasitic, present in mucous membrane of human and animals.

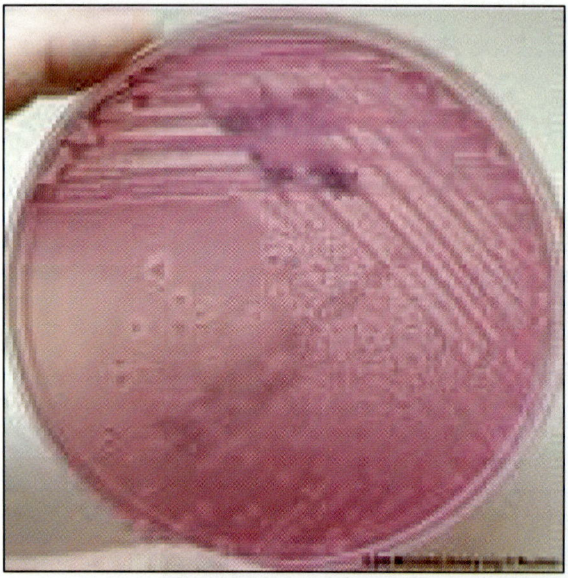

Figure 23.31: ***V. alginolyticus*** **Colonies**

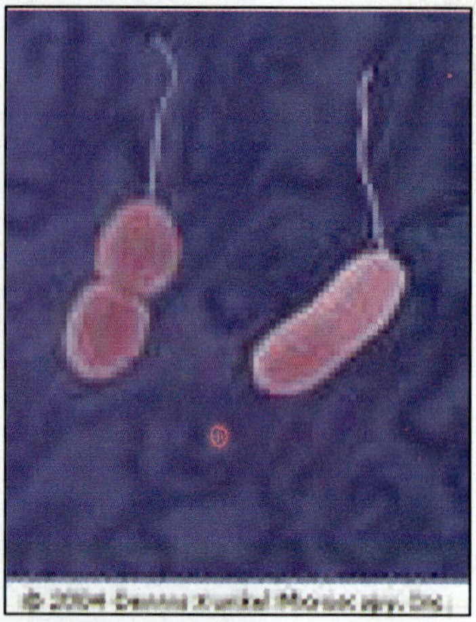

Figure 23.32: ***V. parahaemolyticus***

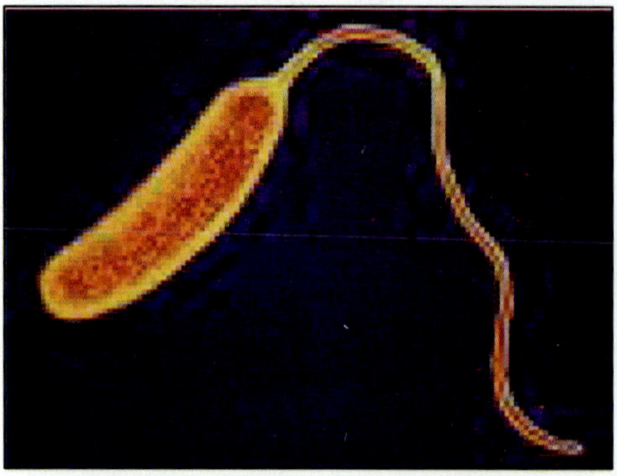

Figure 23.33: ***V. cholerae***

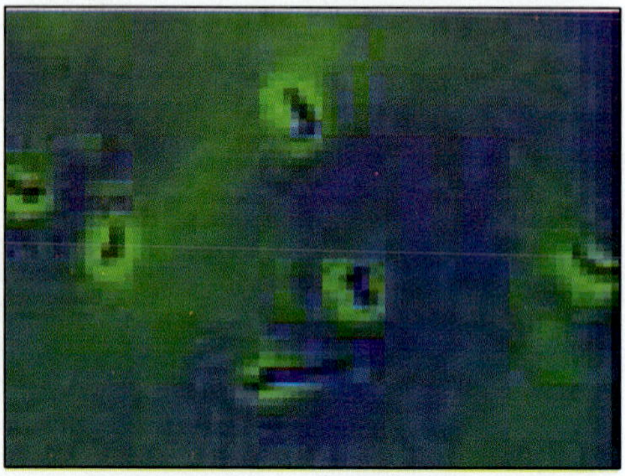

Figure 23.34: ***V. anguillarum***

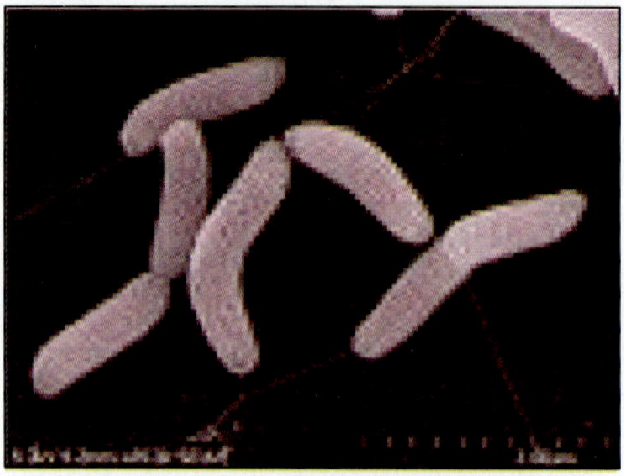

Figure 23.35: ***Desulfovibrio*** **spp.**

Figure 23.36: ***Desulfococcus*** **spp.**

Figure 23.37: ***Mycolplasma* spp.**

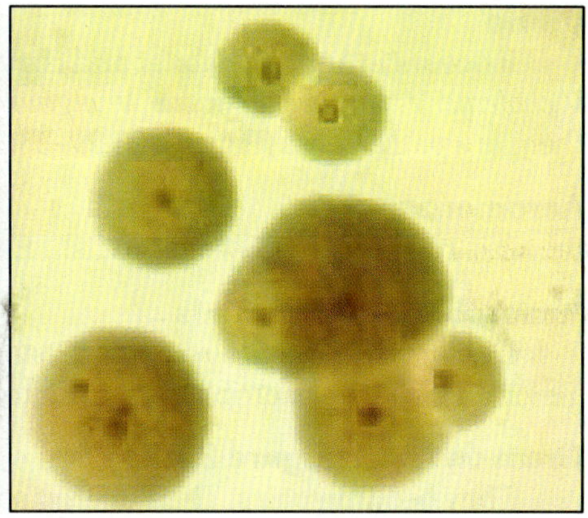

Figure 23.38: ***Mycoplasma* spp.**

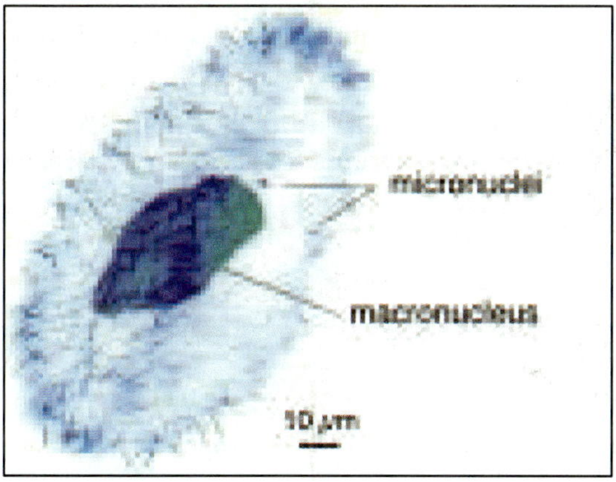

Figure 23.39: ***Paramecium tetraurelia***

Acinetobacter

It is opportunitic pathogens to human, diplococci in shape.

Family Enterobacterraceae

Members are *E.coli, Shigella, Salmonella Enterobacter, Serratia,Proteus and Yersinia.* They can be differentiated by their motility, VP test, indole production, Gas and acid production with carbon sources, enzymes such as Lysine decarboxylase, urease, phenylalanine deaminase and deoxyribo-nuclease (Figures 23.22, 23.24–23.26).

Yersinia pestis

It causes plague in human (Figure 23.28).

Erwinia caratovora

It is a plant pathogen causing leaf-spot and wilt diseases.

Vibrio

It occurs in aquatic habitat and curved rod certain species are *Vibrioanguillarum, V.cholerae, V. parahaemolyticus, V. alginolyticus, V.fischeri, V.vulnificus* and majority of them are pathogenic (Figures 23.29–23.34).

Aeromonas

It is rod causing diseases in fish. Eg. *Aeromonas hydrophila,* and *A.salmonicida.*

Anaerobic Gram-ve Rods

Desulfovibrio, Desulfuromonas, Desulfatomaculum and *Desulfococcus* are important genera. They are anaerobic. Dissimilatory sulfate –reduction is due to these groups.

Gram-ve cocci (Figure 23.27)

They occur in pairs. Their energy source from anaerobic environment. They are ibhabitants of oral cavity, respiratory tract of human, rodents and pigs.

Rickettsias and Chlamydia

Rickettsias are associated with arthropods and provide essential factors for the growth.

Rickettsias are pleomorphic and pathogenic to human causing diseases like mountain spotted fever, classical typhus fever and rickettsial pox. Chlamydias do not synthesise ATP. They are present in mucous membrane and pneumonia disease is caused by chlamydia (Figure 23.29 and Table 23.1).

Table 23.1

Property	*Bacteria*	*Rickettsia*	*Chlamydia*	*Viruses*
Nucleic acid DNA or RNA	Both	Both	Both	One or the other
Multiplication by binary fission	Yes	Yes	Yes	No
Cell wall with muramic acid	Yes	Yes	No	No
Active enzymes	Yes	Yes	Yes	No
Inhibition by Drugs	Yes	Yes	Yes	No
Synthesis ATP	Yes	Yes	No	No

Mycoplasma

Mycoplasma has no cellwall and *Mycoplasma pneumonie* caused pneumonia disease in human (Figures 23.37 and 23.38).

Endo-symbionts

Lyticum flagellatum, a symbiotic bacterium is associated with the protozoa, *Paramecium tetraurelia.* Symbiotic bacteria synthesize vitamin -folic acid for it's host. Symbiont free protozoa are killed by a strain, producing toxins. However, the endo - symbiont –bearing strains are resistant to this toxin.

Chapter 24
Fungi: Molds and Yeast

Introduction

Characteristics of Fungi

They are eucaryotic chemoorganotrophic without chlorophylls. Some fungi are dimorphic- when they are pathogens they are unicellular. When they are saprophytes they are filamentous. It has the size of 1 to 5 μm width and 5 to 30μm length.

Thallus of the mold consists of two parts. They are;

1. Mycelium and
2. Spores.

The mycelium is a complex of several filaments called hyphae (Figure 24.1).

Hyphae

Occurs in 3 forms. They are:

1. Non-septate or coenocytic
2. Septate with uni nucleate cells and
3. Septate with multi nucleate cells. Growth is distal, near the tip (Figure 24.2).

Asexual (Somatic or Vegetative) Reproduction

It does not involve the union of nuclei, sexual organs sexu cells. It can be accomplished by:

1. Fission of somatic cells yielding two similar cells.
2. Budding of somatic cell and
3. Fragementation of hyphae and 4. spore formation

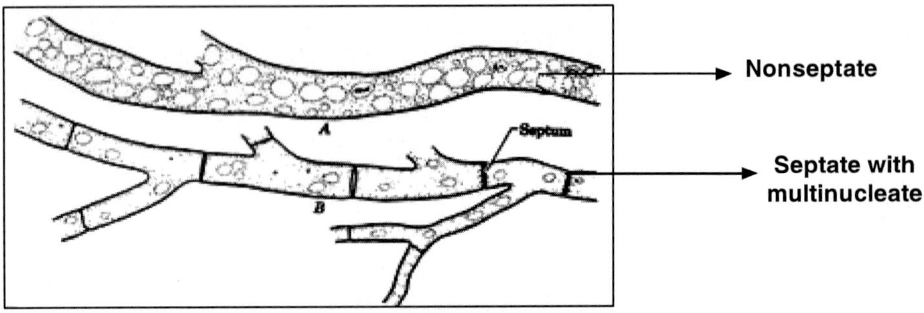

Figure 24.1: Hyphae Forms

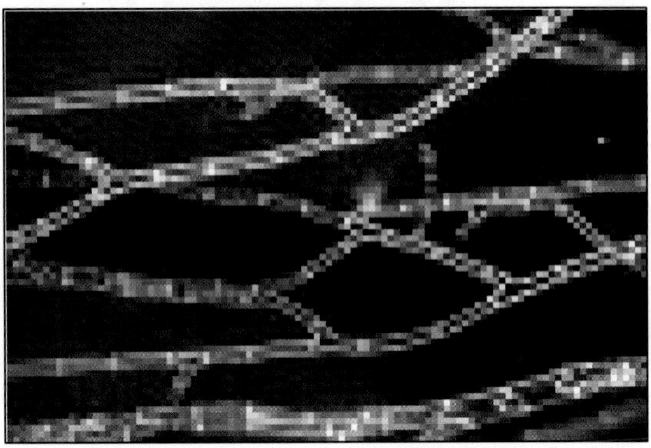

Figure 24.2: Septate with Uninucleate

There are many kinds of asexual spores

1. *Sporangiospores*: Single celled within a sac called sporangia at the end of sporangiospore.
2. *Conidiaspore*: Conidia are formed at the tip or side of hyphae
3. *Oidia*: These single celled spores are formed by disjoining of hyphal cells.
4. *Chlamydospores*: Thick walled, single celled spores are highly resistant to adverse conditions, formed from vegetative hyphae (Figure 24.3).

Sexual Reproduction

Fusion of compatible nuclei of two parent cells. The fusion of two protoplasts of parents is called plasmogamy. Fusion of two parents nuclei is called karyogamy. Sex organelles are called as gametangium.Sex cells are called as gametes. Male gametangium is called antheridium and that of female is called oogonium (Figure 24.4).

Methods of Sexual Reproduction

1. *Gametic couplation*: Fusion of naked gametes; one or both of which are motile.

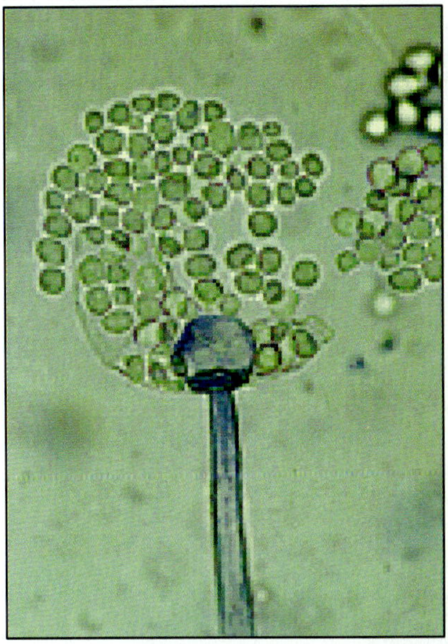

Sporangiospore

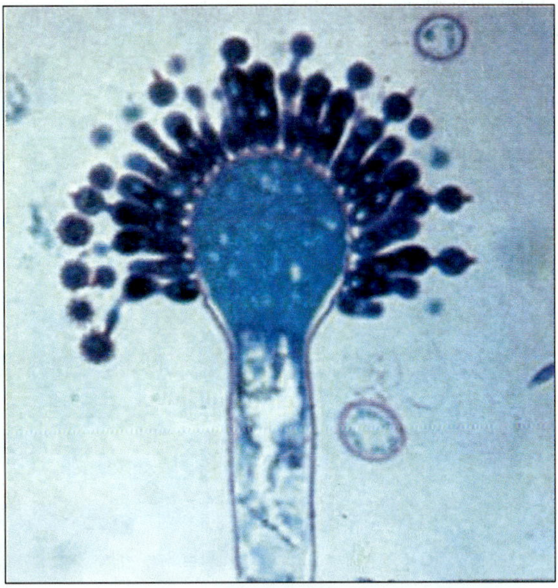

Conidiospore

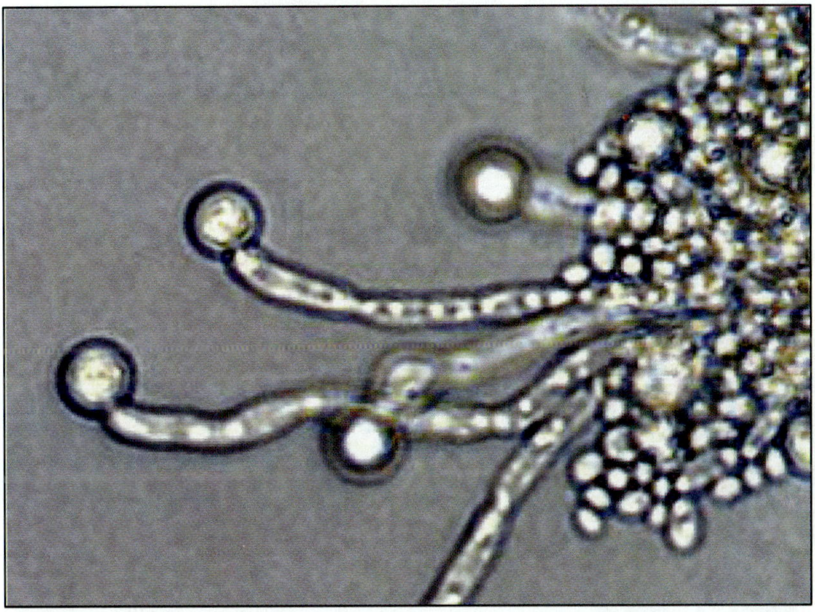

Clamydiospore

Figure 24.3: Asexual Reproduction of Fungi

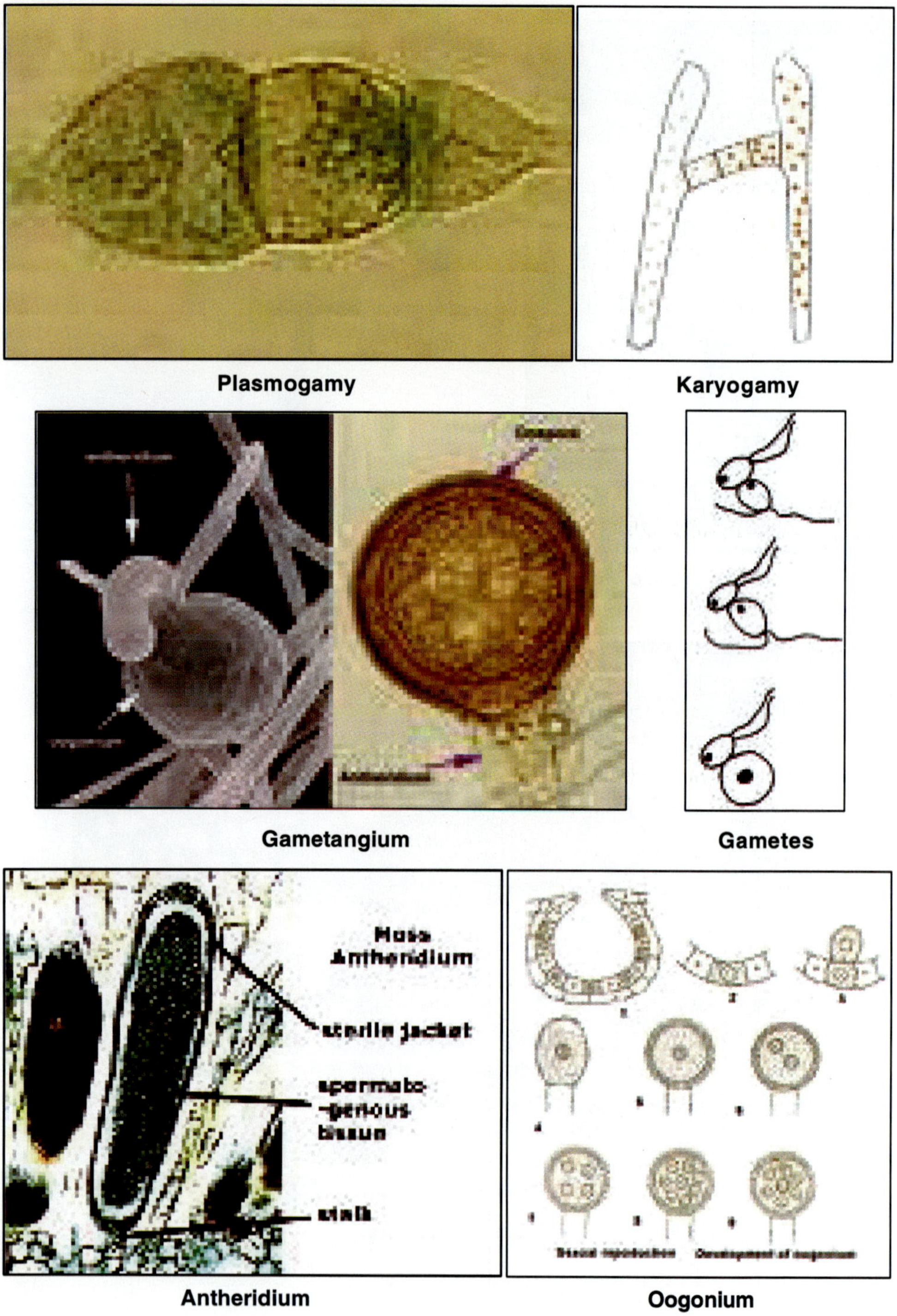

Figure 24.4: Sexual Reproduction

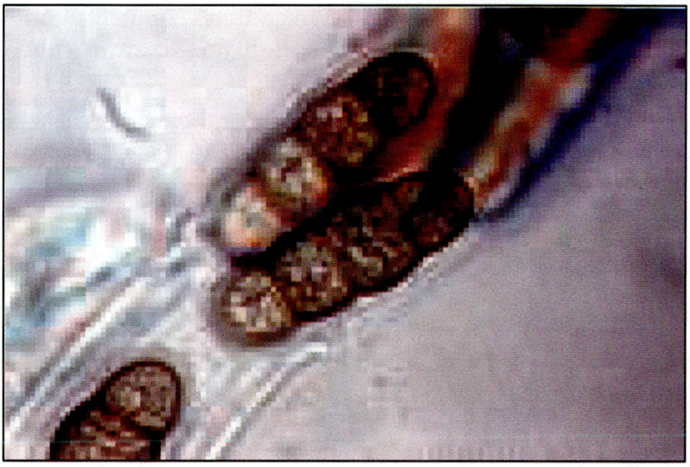

Ascospore

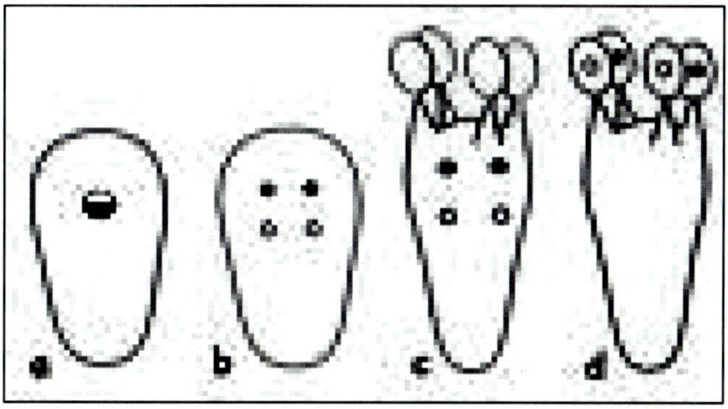

Basidiospore

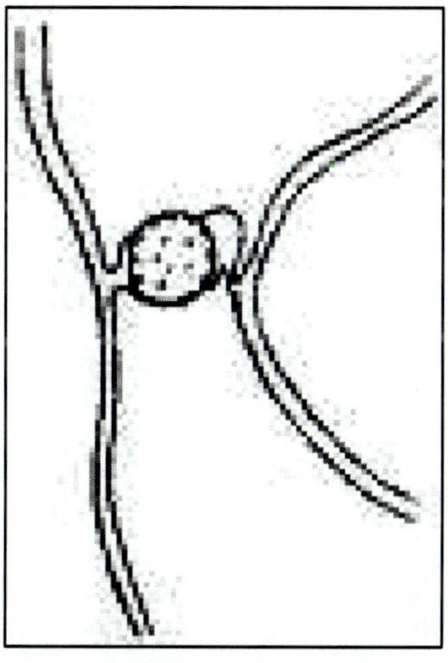

Zygospore

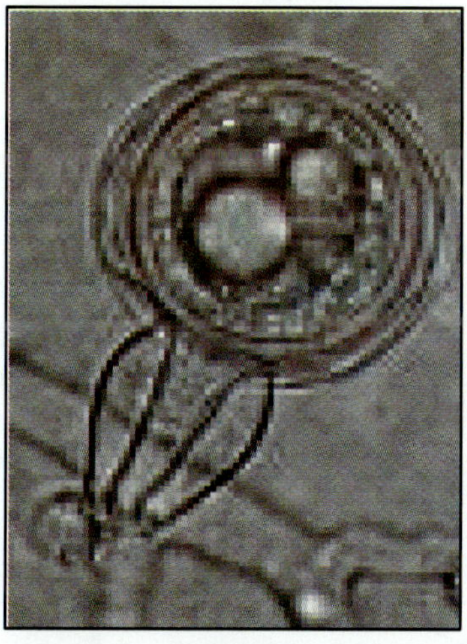

Oospore

Figure 24.5: Sexual Spore

2. *Gamete–Gametangial copulation*: Two gametes come into contact but donot fuse; male nucleus migrates through fertilization tube into female gametangium.
3. *Gametangial copulation*: Two gametangia or their protoplasts fuse and give rise to a zygote that develops into a resting spore.
4. *Somatic copulation*: Fusion of vegetative cells.
5. *Spermatization*: Union of aspecial male structure called "spermatium" with female receptive structure.

There are several types of sexual spores

1. *Ascopspores*: Singles celled spores are produced in a sac called ascus.
2. *Basidiospores*: Single celled spores are borne on a cub-shaped structure called a basidium.
3. *Zoospores*: Thich walled spores formed when the tips of two sexually compatible gametangia fuse together.
4. *Oospore*: This is formed in female structure called oogonium. Fertilization of eggs by male gamete that gives rise to oospore. A single fungus (mold) can produce asexual and sexual spores at different times and conditions (Figure 24.5).

Classification of Fungi

It is based on the characterisation of the sexual spores and fruting bodies (asexual) present during the sexual stages of their life cycles. However, many fungi

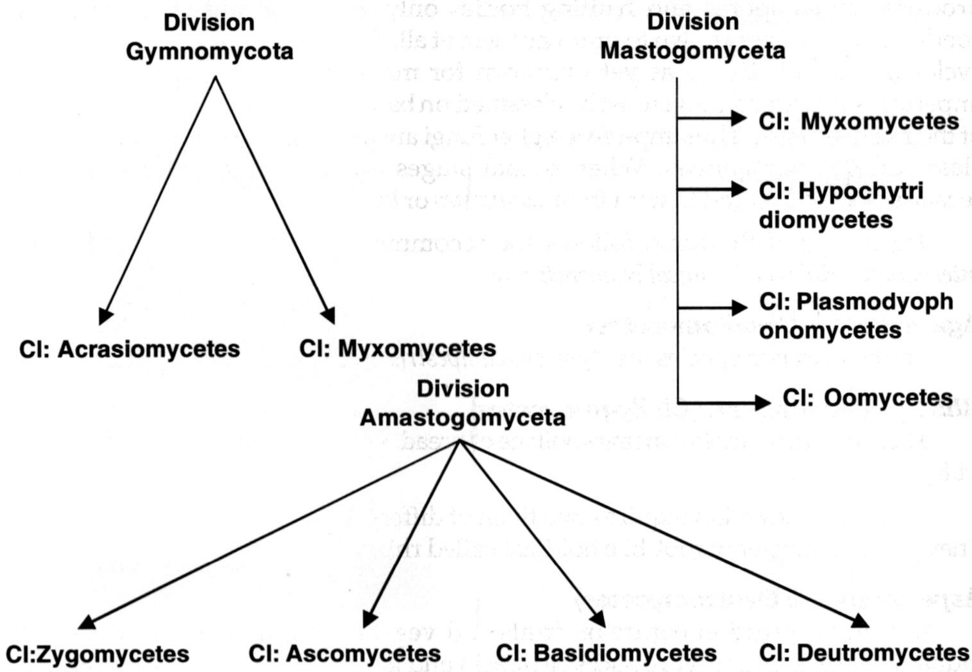

Figure 24.6A: Classification Scheme Proposed by Alexopoulos (Modified after 1979)

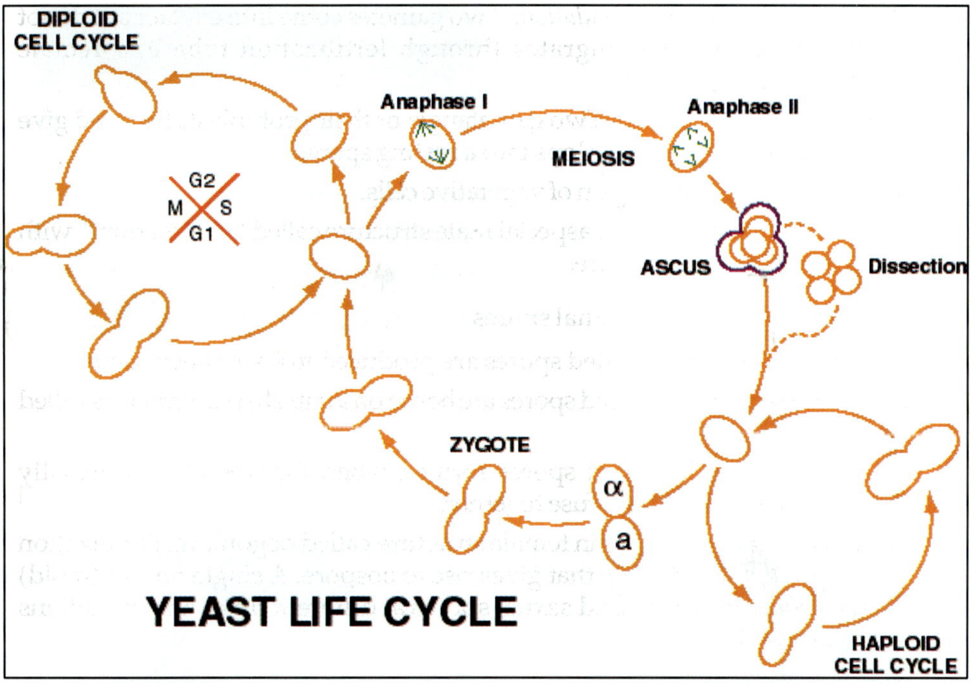

Figure 24.6B: Life Cycle of *Saccharomyces cerevisiae*

produce sexual spores and fruiting bodies only under certain environmental conditions, if they are known to produce them at all. Thus, the complete or perfect life cycles (Figure 24.6B) are as yet unknown for many fungi. Consequently, these imperfectly described fungi must be classified on bases other than the characteristics of their sexual stages.Thus imperfect higher fungi are provisionally placed in a special class called *Deutromycetes.* When sexual stages are found, the organisms are recalassified and placed in with the *ascomycetes* or *basidiomycetes.*

Taxonomy of the fungi follows the recommendations of the *Committee on International Rules of Botanical Nomenclature.*

Agaricus: (*Cl: Basidiomycetes*)

The best known species are *Agaricus campestris* and *A. bisporus* (Figure 24.7).

Rhizopus stolonifer L (*Cl: Zygomycetes*)

These are bread mold causing spoilage of bread, vegetable, fruits and food (Figure 24.8).

Sexual reproduction requires two thalli of different mating (Heterothalli) types. They produce cluster of root-like holdfast called rhizoids.

Aspergillus (*Cl:* Deutromycetes)

It is wide spread in nature on fruits and vegetables and other substrates. In industrial fermentation, it produces citric acid and gluconic acid. Eg: *A. niger* (Figure 24.9).

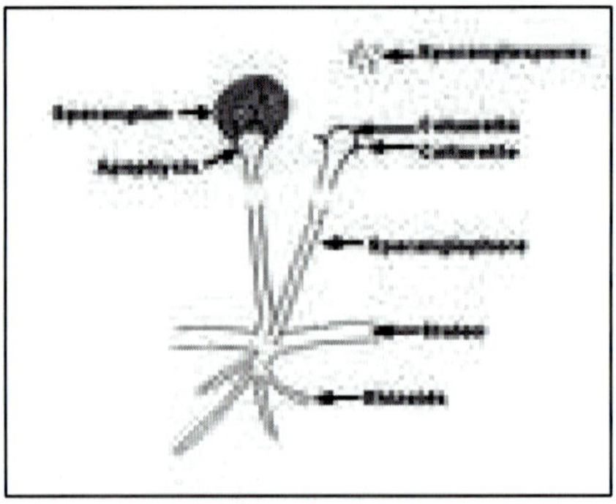

Figure 24.7: *Agaricus*

Figure 24.8: *Rhizopus*

Figure 24.9: *Aspergillus*

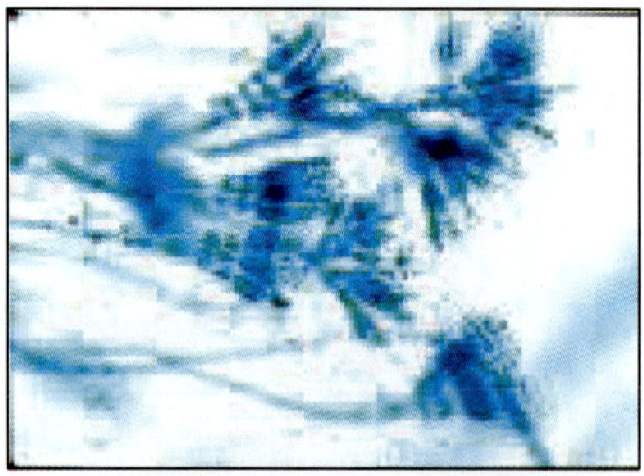

Figure 24.10: *Penicillium*

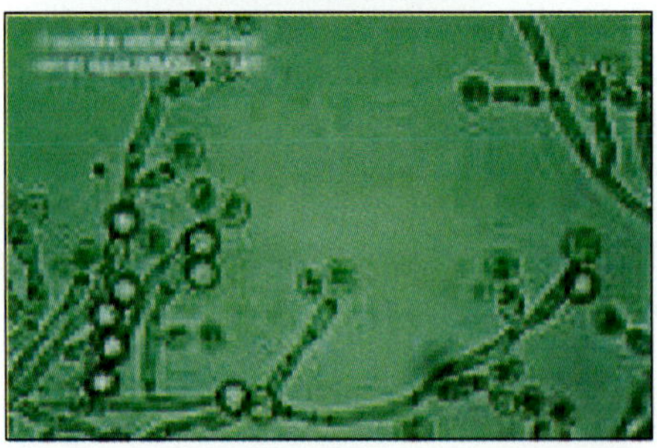

Candida

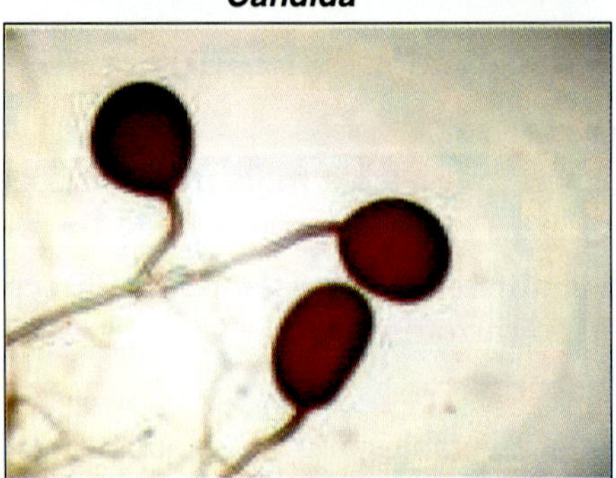

Figure 24.11: Mycorrhizae

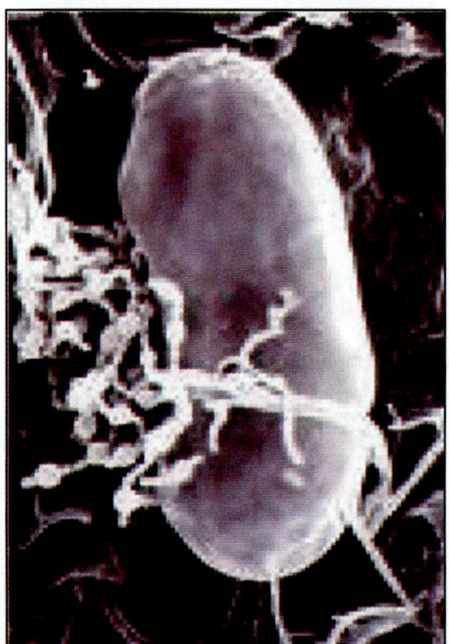

Figure 24.12: *Verticillium chlamydosporium* on *Heterodera avenae*

Penicillium (*Cl: Deutromycetes*)

P.notatum and *P.chrysoginum* produce penicillin (Figure 24.10).

They grow best at temperature from 15 to 30°C. They have septate mycelia.

Candida (*Cl: Deutromycetes*)

C. albicans is isolated from warm-blooded animals, causing disease called candidiasis in mucus membrane, in blood (septicemia) and in brain (meningitis).

Fungi as Parasites of Nematodes

Nematodes are small, roundworms, pathogenic to plants. Parasitic fungus, *Arthrobotrys conides* parasitises nematode and kill it (Figure 24.12) and *Verticillium* on *H. avenae.*

Mycorrhizas in an infected root fungus arising from rootlets.It enhances meneral absorption by the green plants. In roothairless Orchids the fungus serves the absorption of nutrients (Figure 24.11).

Chapter 25

Characterization of Microbes with Special Reference to *Streptomyces* spp.

Introduction

The first object of study was to determine the usefulness of many of the criteria which are employed in descriptions of these microbes. Were the characteristics that had been used to define as streptomycete and photosynthetic bacteria sufficiently stable, clear, reproducible and readily capable of interpretation. If these criteria could not be fulfilled, then their inclusion in a key would only lead to confusion and misidentification of species. Better characteristics or tests would be necessary for the definition of the species. SEM analyses for *Streptomyces* are done to describe the spore morphology and spore surface ornamentation. As photosynthetic bacteria concerned SEM was done to determine structural morphology. The biochemical parameters were done for confirming *Streptomyces* and photosynthetic bacteria up to genus level.

Biochemical Characterization

The taxonomic properties of the strains were determined according to the International Streptomycetes Project (ISP) (Shirling and Gottlieb, 1972). The genus *Streptomyces* was proposed by Waksman and Henrici (1943) and classified in the family *Streptomycetaceae* on the basis of morphology and subsequently cell wall chemotype. Streptomycetes are Gram-positive aerobic members of the order *Actinomycetales* within the classis *Actinobacteria* (Stackebrandt *et al.*, 1997) and have a DNA G-C content of 69±78 mol per cent (Korn-Wendisch and Kutzner, 1992). Streptomycetes produce an extensive branching substrate and aerial mycelium. The substrate hyphae are approximately 0.5±1.0 μm in diameter and often lack cross-walls during the vegetative phase. Growth occurs at the hyphal apices and is accompanied by branching, thus producing a complex tightly woven matrix of hyphae during the vegetative growth phase. As the colony ages, aerial mycelia (sporophores) are produced which develop into chains of spores (conidia) by the formation of crosswalls in the multinucleate aerial flaments. This is followed by separation of individual cells directly into spores (Wildermuth and Hopwood, 1970). The morphological features of the aerial mycelium are regarded as more signifcant for characterization than the aerial mycelium and they include the mode of branching, the configuration of the spore chains and the surface of the spores. The surface of the conidial wall often has convoluted projections which, together with the shape and the arrangement of the spore-bearing structures, are characteristic of each species and were often used for the separation of *Streptomyces* species (Pridham *et al.*, 1958; Korn-Wendisch and Kutzner, 1992). Shirling and Gottlieb (1968a and b, 1969 and 1972) described the standard criteria that involved the use of spore chain morphology, spore surface ornamentation, colour of spores, substrate mycelium and soluble pigments, production of melanin pigment and the utilization of a range of carbon sources. More than 450 *Streptomyces* species were redescribed and type strains were selected and deposited in internationally recognized culture collections.

The application of chemical methods influenced the development of the actinomyces systematics at the genus level and even at higher taxonomic levels (Goodfellow and Cross, 1984, Kroppenstedt,1985, Stackebrandt, 986). The separation of genus *Streptomyces* from other *Streptomycetes* groups (on the same taxonomic level) is not a problem [Sneath, 1963]. The streptomycetes can be distinguished from all the rest actinomycetes in morphological and chemotaxonomical (aminoacids, diaminopimelic acid, fatty acids, lipids, phospholipids, menaquinones, mycolic acids, sugars) characteristics (Soput, *et al.*, 1967, Lechevalier and Lechevalier, 1970 and 1980). The use of the above mentioned chemotaxonomic methods leads to more profound and precise characterization of the genera.

The biochemical parameters of photosynthetic bacteria include cell size, Vitamin B12 utilization and grams staining. The present arrangement of genera and species of *Chlorobium* and *Chromatium* entirely based on simple phenotyphic characteristics as used in the eighth section of Bergeys Manual (Pfennig and Truper, 1974).

SEM (Scanning Electron Microscope)

The original proposal to create a new generic name *Streptomyces* was made by Waksman and Henrici in a manuscript submitted to the Journal of Bacteriology in April, 1943, and to Breed in May, 1943, for publication in the sixth edition of Bergey's Manual. This name first appeared in the Journal of Bacteriology in October, 1943; unfortunately, the sixth edition of Bergey's Manual, which contained a detailed description of *Streptomyces griseus* was not published until 1948. Pridham, Hesseltine, and Benedict (1958) assigned prime importance to the morphology of *Streptomyces* sporophores and grouped species into primary morphological sections on this basis. In the system of Ettlinger, Corbaz, and Hultter (1958) the type of sporophore (spore chain) was also one of the characteristics given major emphasis but the morphology of the spores themselves, as observed by electron microscopy, was given equal weight. In recent years several other investigators (Flaig *et al.*, 1952; Kuister, 1953; Flaig and Kutzner, 1954; Flaig, Kuister, and Beutelspacher, 1955; Baldacci and Grein, 1955; Baldacci, Gilardi, and Amici, 1956; Kutzner, 1956; and Preobrajenskaya *et al.*, 1959) using the electron microscope have studied the details of *Streptomyces* spore structure and have shown that an apparently distinctive spore surface configuration exists in many of the species. Provided with this encouraging prospect, it seemed highly desirable to expand these studies to include mass collections of strains of individual *Streptomyces* species, for it is only by demonstration of constancy from strain to strain within a species that the true value of any criterion for species differentiation can be firmly established.

The present report concerns an investigation of the chromatophore arrangement in several photosynthetic bacteria by Scanning electron microscopy of ultrathin sections and supports the observations of Pardee *et al.* (1952) which discrete chromatophores are a basic structure in the photosynthetic bacterial cell.

Protein Profiling

Bacterial whole-cell protein profiles can be analysed by polyacrylamide gel electrophoresis (PAGE) to generate complex banding patterns that can be used to delimit species and subspecies groups. Proteins can be analysed either by using one-dimensional (1-D) or two-dimensional (2-D) protein electrophoresis. 1-D protein electrophoresis allows many bacterial strains to be compared rapidly and reproducibly (Goodfellow and O'Donnell, 1993). Manchester *et al.* (1990) investigated the use of SDS-PAGE to analyse and compare the cellular protein profiles of 32 *Streptomyces* strains and five *Streptoverticillium* strains. Taxonomic correlations were found between the profiles obtained and the phenotypic groupings observed by Williams *et al.* (1983a). Correlations between protein profiles and DNA hybridization groupings have also been demonstrated. Paradis *et al.* (1994) used PAGE and DNA±DNA hybridizations to elucidate the taxonomy of streptomycete isolates that are the causal agent of common potato scab. Isolates obtained from potato tubers were divided into two groups with a correlation coefficient of 0±75 using SDS-PAGE analysis. The same two groups were resolved at approximately 44 per cent similarity using DNA±DNA hybridization analysis. The protein profiles also enabled further differentiation within the two groups; one group contained two subgroups which clustered at a level of 0±82. In the same study, fatty acid analysis results did not correlate with the SDS-PAGE and the DNA±DNA hybridization groupings. This is partly due to the influence of growth conditions on the profiles obtained (Saddler *et al.*, 1986, 1987). Despite the success of these tests to taxonomically group *Streptomyces scabie,* they could not differentiate between pathogenic and non-pathogenic strains.

Taxonomic studies of *Streptomyces* can also be done as strains, by isolating specific proteins from cell extracts and determining the protein sequence so that divergence between strains can be evaluated. Taguchi *et al.* (1996) used *Streptomyces* subtilisin inhibitor protein (SSI), which plays unidentified role(s) in physiological and/or morphological regulation, to investigate the taxonomic status of the *Streptomyces coelicolor* strains. The SSI-like proteins have molecular masses of 9±12 kDa and show significant amino acid sequence homology in the conserved regions that are required for maintaining the tertiary structure of the SSI. The amino acid sequence of SSI from *'Streptomyces lividans'* 66, *Streptomyces coelicolor* Mu$ ller ISP5233T and *'Streptomyces coelicolor'* A3(2) were compared. The alignments supported ribosomal sequence comparisons, indicating that *'Streptomyces coelicolor'* A(3) is more closely related to *'Streptomyces lividans'* 66 (cluster 21) than to the type strain, *Streptomyces coelicolor* Mu$ ller ISP5233T (cluster 1).

Observations with Tables and Figures

Among 271 *Streptomyces* spp isolated only seven strains (AQB-T, AQB-E, AQB-S, AQB-Gr, AQB-C, AQB-G, AQB-CA) are selected. This selection is mainly based on more pigment production and efficient antibacterial activity. The seven strains isolated were tested biochemically and scanning electron microscope was done to determine spore morphology. Various biochemical parameters include colour determination, utilization of carbon sources, amino acid, melanoid pigment production, spore morphology and sodium chloride tolerance.

The characterization of *Streptomyces* spp should be studied following the method recommended by International *Streptomyces* Project. The aerial and substrate mycelial colours of the seven stains AQB-T, AQB-E, AQB-S, AQB-Gr, AQB-C, AQB-G, AQB-CA) grown in Glycerol asparagines agar was shown in Table 25.1 and Figure 25.1. The Aerial mycelial colour expressed was White, Green, Grey and cream. The Mycelial colour characteristics in microplates of selected strains of *Streptomyces* in different media like Nutrient agar, Actinomycetes agar, Glucose yeast agar, Tryptophan medium, Streptomyces agar medium (Table 25.2, Figure 25.2). The colouration patterns of aerial and substrate mycelia were totally different in each medium with enrichment. This may be due to that the enriched medium provides certain nutrients for triggering of genes for the conversion as well as expression of other metabolic products, which in turn, leads to different aerial and substrate mycelial colouration. This colouration difference may be due to other primary as well as secondary metabolites production provided by the enriched media. Carbon utilization showed better growth when glucose is added and poor growth in Sucrose (Table 25.3, Figure 25.3). The effects of amino acids showed maximum in strains with Tryptophan and Glycine showed minimum in strains (Table 25.4). At 2.5 per cent sodium chloride concentration profused growth of strains (Table 25.5). Melanoid pigmentation concerned the strains (AQB-E, AQB-C, AQB-G) showed positive (Table 25.6). The mass culture was done in Actinomyctes broth for these *Streptomyces* strains (Figure 25.4). Spore morphology concerned the strains (AQB-E, AQB-C, AQB-G) showed Rectiflexibiles(RF) and the rest of strain showed Spiral spores (Table 25.7, Figure 25.5). As far as the Scanning electron microscope results showed structural morphology (Figure 25.6). With regard to spore chains, the strains can be grouped into 'sections'. The species belonging to the genus *Streptomyces* are divided into three sections (Shirling and Gottlieb, 1972), namely rectiflexibiles (RF), retinaculiaperti (RA) and Spirales (S). When a strain forms two types of spore chains, both are noted (*e.g.* SRA).

Comparing the properties of the isolated strain with the representative species found in the key (Nonomura, 1974) and Bergey's Manual of Determinative Bacteriology (Buchanan and Gibbons, 1974) can help in the genus level identification.

Figure 25.1: Mycelial Colour Characteristics of Selected Strains of *Streptomyces* in Selective Media Glycerol Asparagine Agar

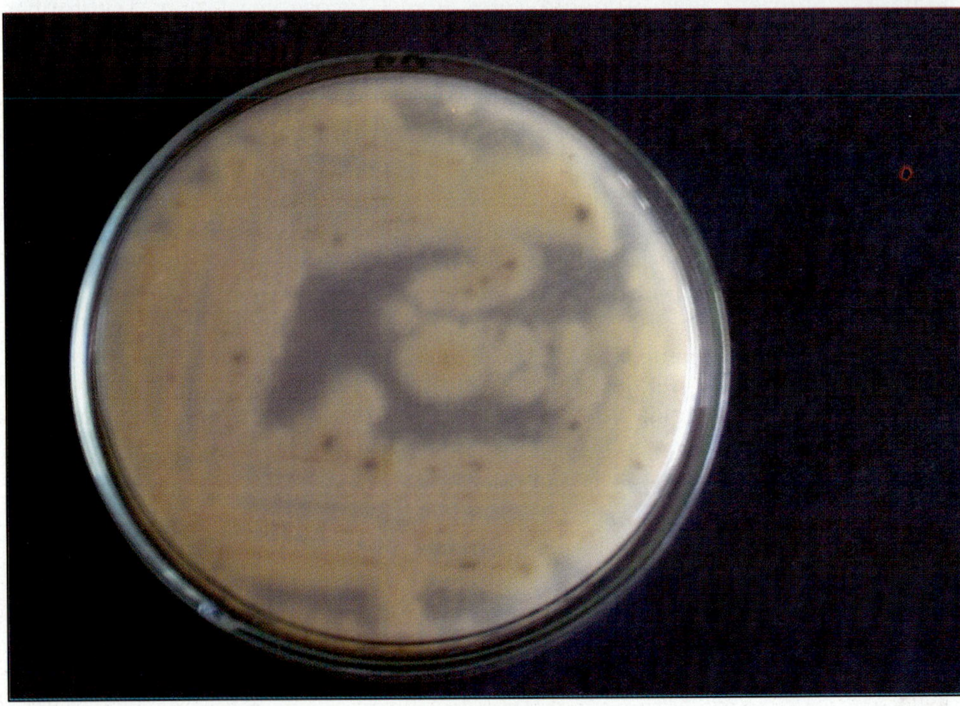

AQB-Gr

Figure 25.1–*Contd...*

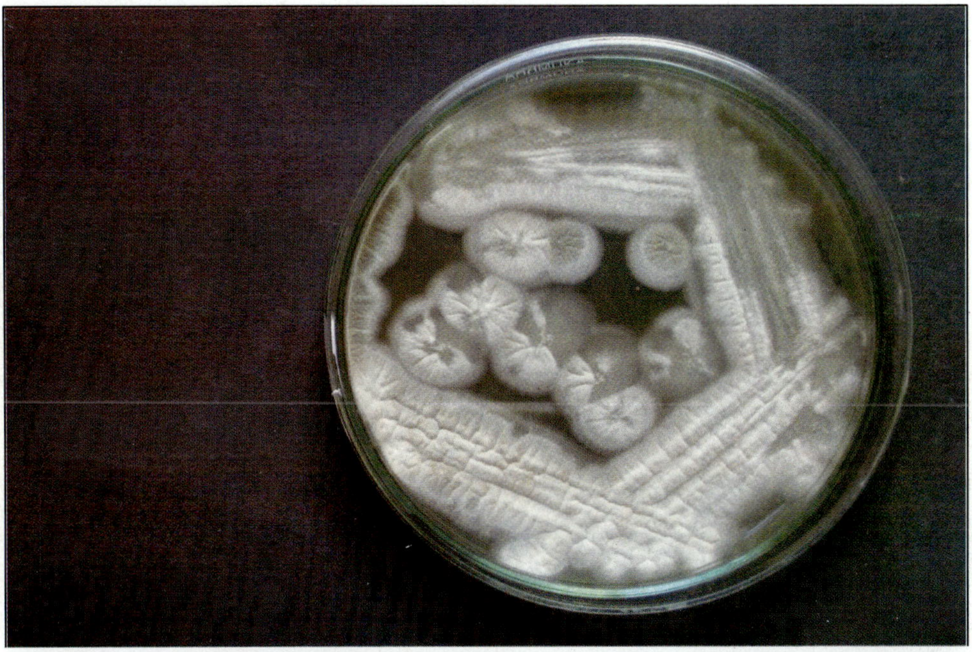

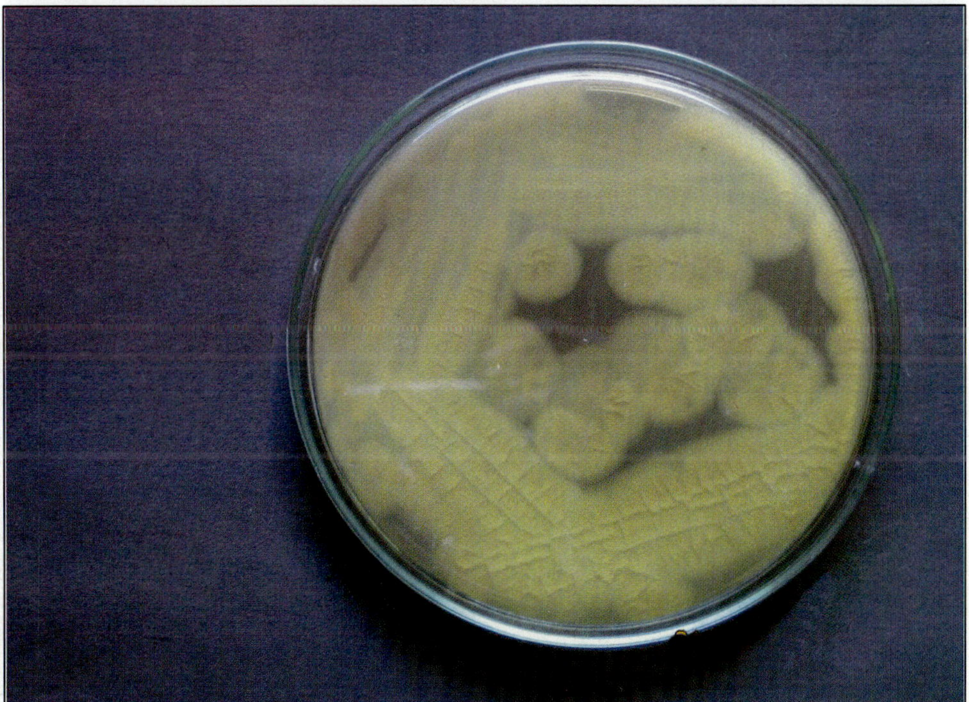

AQB-C

Figure 25.1–*Contd...*

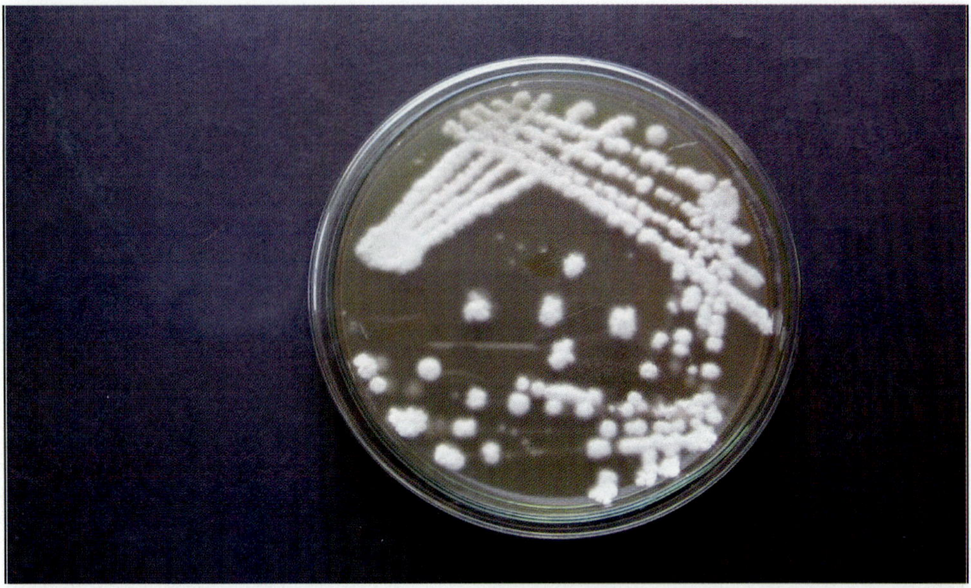

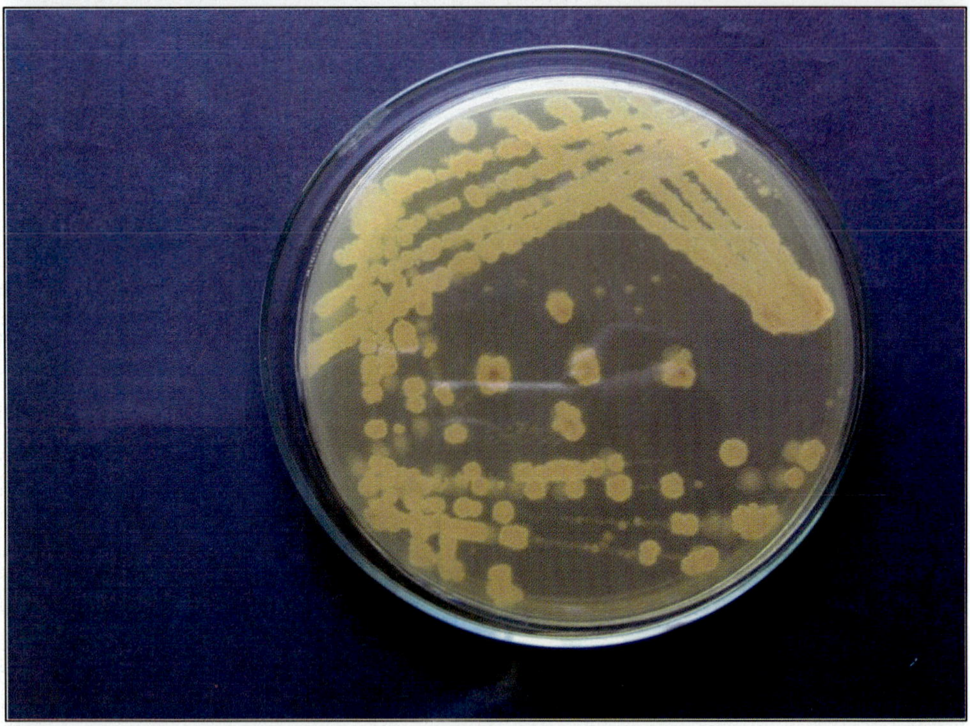

AQB-G

Figure 25.1–*Contd...*

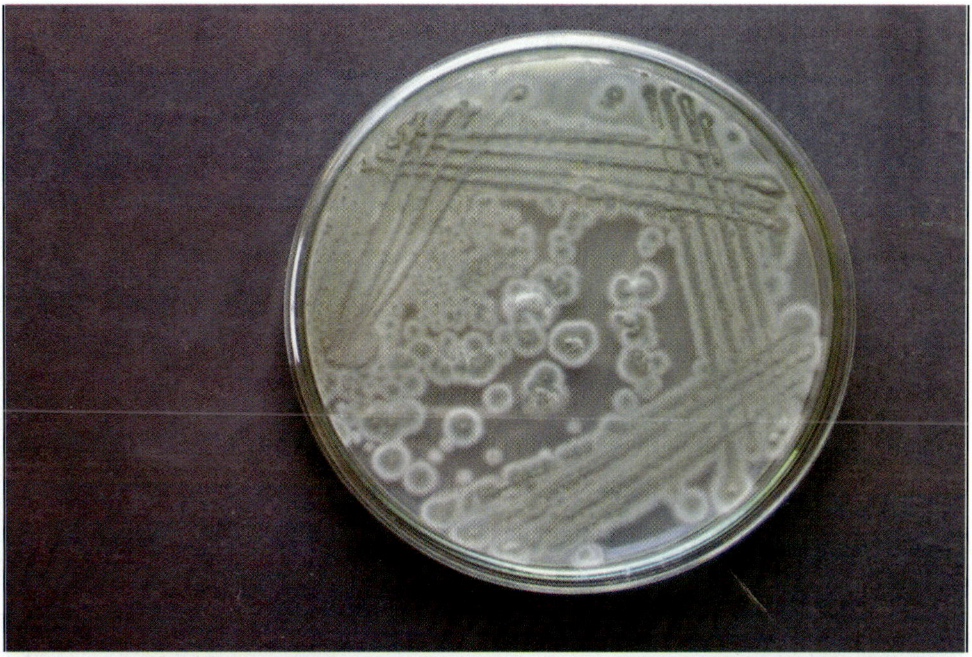

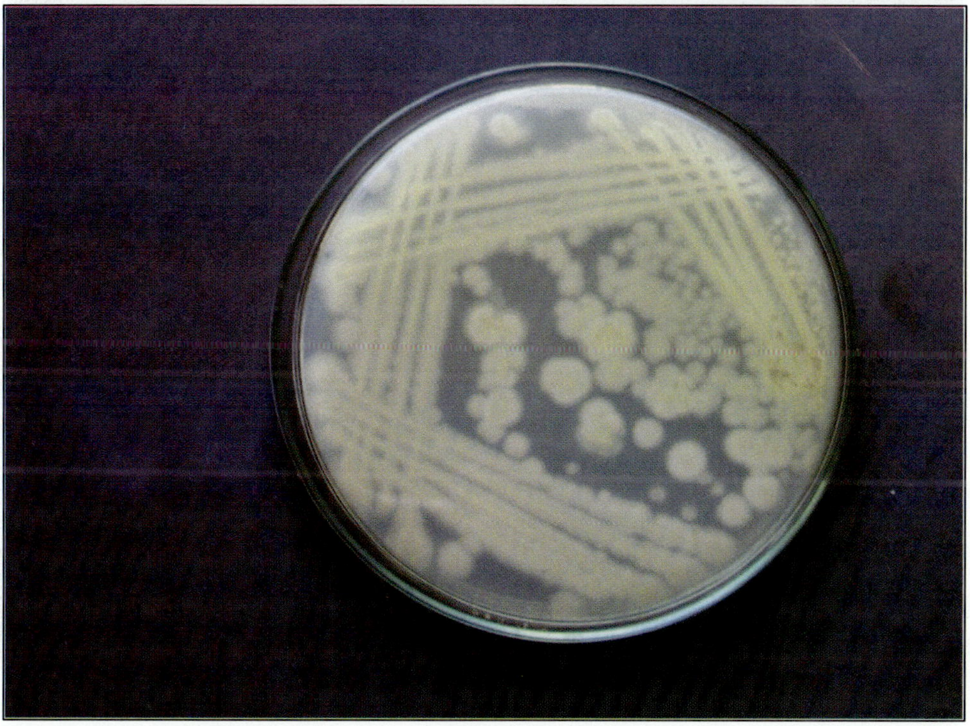

AQB-CA

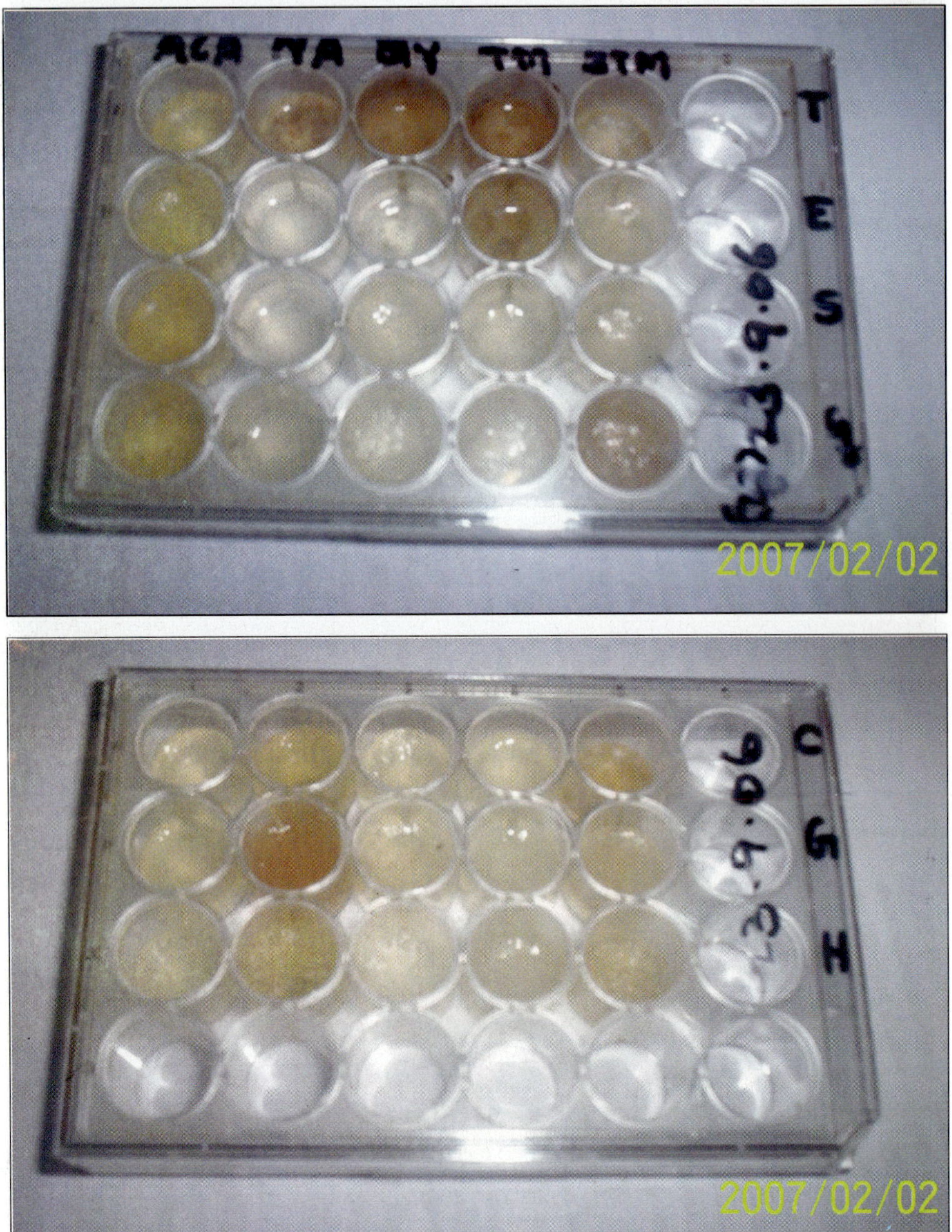

Figure 25.2: Mycelial Colour Characteristics of Selected Strains of *Streptomyces* in Different Media

Figure 25.3: Utilization and Colouration Expression of Selected Strains of *Streptomyces* in 10 Different Carbon Sources (Glucose, Xylose, Arabinose, Rhamnose, Fructose, Galactose, Raffinose, Mannnitol, Inositol and Sucrose)

Table 25.1: Mycelial Colour Characteristics of Selected Strains of *Streptomyces* in Selective Media Glycerol Asparagine Agar Medium

Sl.No.	*Strains*	*Glycerol Asparagine Agar Medium*	
		Aerial Mycelium	*Substrate Mycelium*
1.	AQB-T	Green	Yellow
2.	AQB-E	White	Reddish yellow
3.	AQB-S	Greenish ash	Yellow
4.	AQB-Gr	Creamy white	Yellow
5.	AQB-C	Grey white	Yellow
6.	AQB-G	White	Yellow
7.	AQB-CA	Grey green	Light yellow

Streptomyces isolates from T: Tilapia; E: Pearl spot; S: Sponges; Gr: Grouper; C: Common carp, G: Gourami; CA: Gold fish.

Table 25.2: Mycelial Colour Characteristics of Selected Strains of *Streptomyces* in Different Media

Strains	*Different Media*									
	Nutrient Agar		*Actinomycetes Agar*		*Glucose Yeast Agar*		*Tryptophan Medium*		*Streptomyces Agar Medium*	
	AM	*SM*	*AM*	*SM*	*AM*	*SM*	*AM*	*SM*	*AM*	*SM*
AQB-T	Brown	White	Yellow	Yellow	Brown	White	Brown	White	Green	White
AQB-E	White	White	Yellow	Yellow	White	White	Brown	White	White	White
AQB-S	White	White	Yellow	Yellow	White	White	White	White	White	White
AQB-Gr	White	White	Yellow	Yellow	White	White	White	White	Brown	White
AQB-C	White	White	Yellow	Yellow	White	White	White	White	White	White
AQB-G	Brown	White	Yellow	Yellow	Brown	Brown	Brown	Brown	White	White
AQB-CA	Brown	White	Yellow	Yellow	Brown	White	Brown	White	Green	White

Streptomyces isolates from T: Tilapia; E: Pearl spot; S: Sponges; Gr: Grouper; C: Common carp, G: Gourami; CA: Gold fish.

As far as the results of gel electrophoresis concerned the samples of tissue as well as broth cultured *Streptomyces* spp. of different sources expressed many new proteins of different molecular weight and R_f values in both 10 per cent TCA and 70 per cent Ethanol were observed. The molecular weight ranged from 14 to 80 kd and R_f values ranged from 0.1 to 0.9. But concerned with clear band formation 10 per cent TCA shows the results better. The newly discovered target protease SAM-P20 was found to have significant structural homology, in both its N-terminal amino acid and whole nucleotide sequences (Figure 25.7), with the already well-characterized proteins SGPA and SGPB from *Streptomyces griseus* (8). Also, to date, production proteins with molecular sizes ranging from 14 to 20 kDa from several *Streptomyces* species has been

Figure 25.4: Mass Culture of *Streptomyces*

AQB-G

AQB-C

AQB-E

AQB-CA

Figure 25.4–*Contd...*

AQB-Gr

AQB-T

AQB-S

Figure 25.5: Characteristics of the Spore Bearing Hyphae and Spore Chains Determined by Direct Microscopic Examination of the Culture Surface

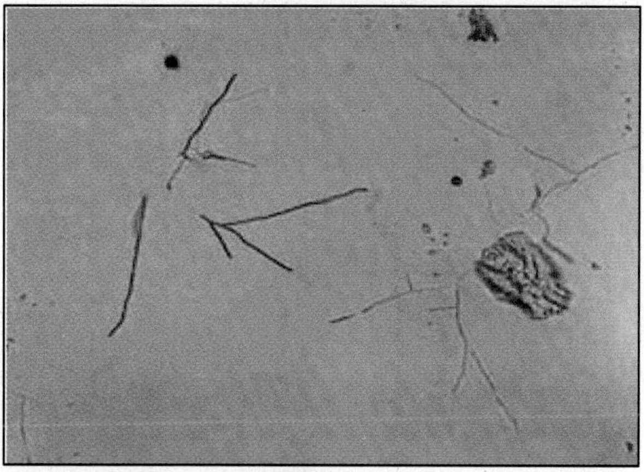

AQB-T

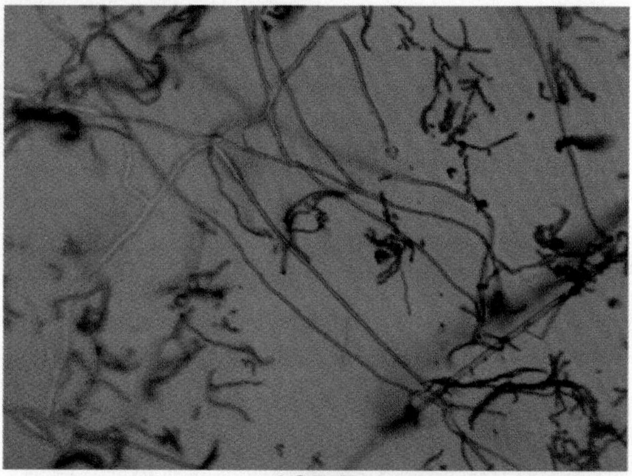

AQB-E

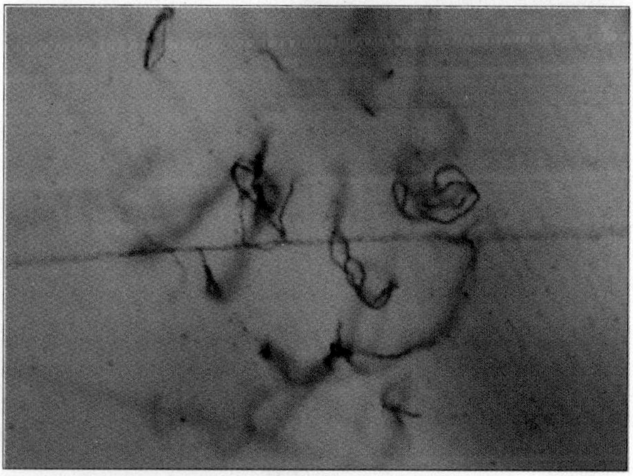

AQB-S

RF Spore chains (400X)

Figure 25.5–*Contd...*

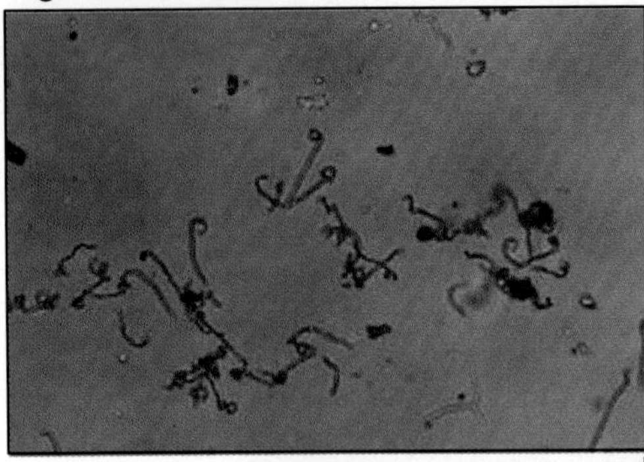

AQB-G

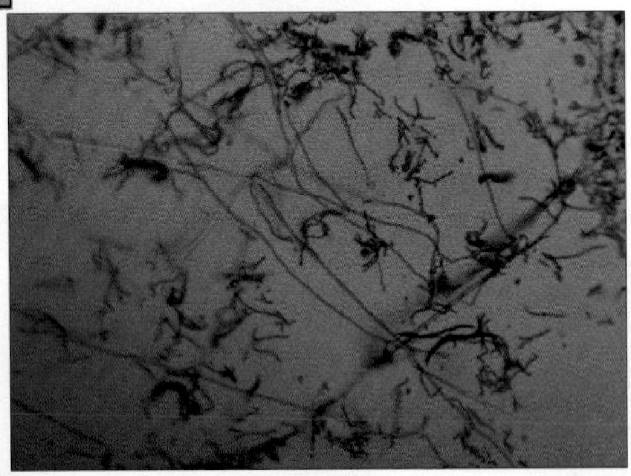

AQB-Gr

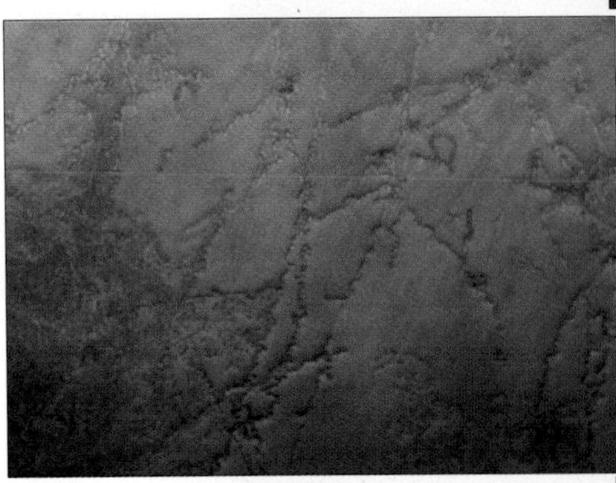

AQB-C

AQB-CA

Spiral Spore Chains (400X)

Figure 25.6: SEM Analysis Results of *Streptomyces* spp

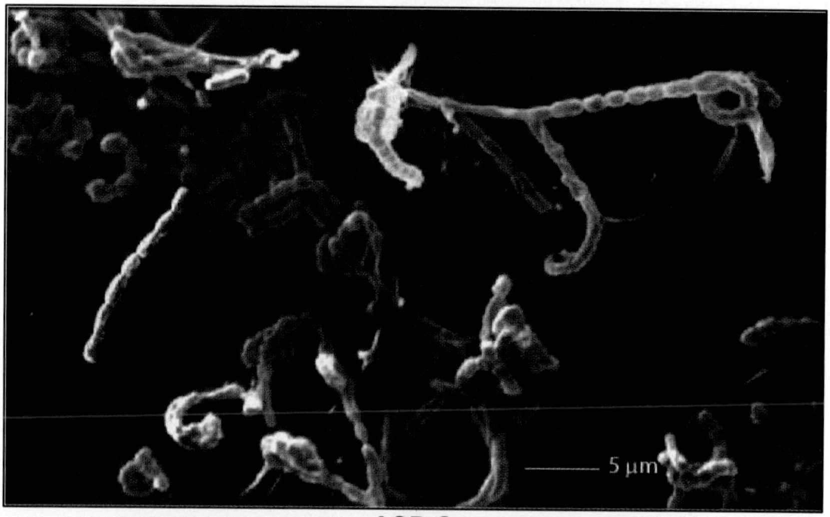

AQB-G

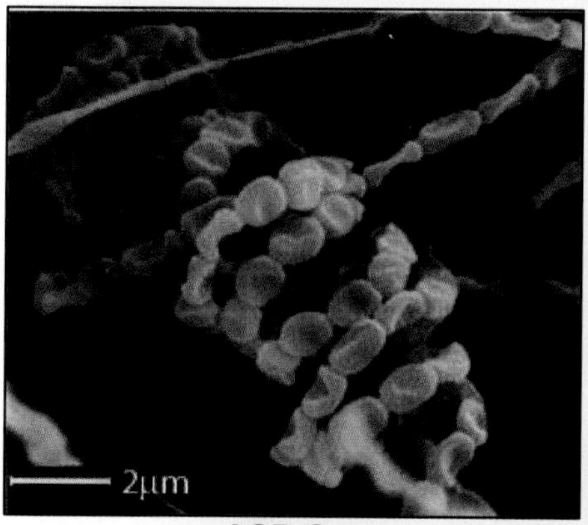

AQB-Gr

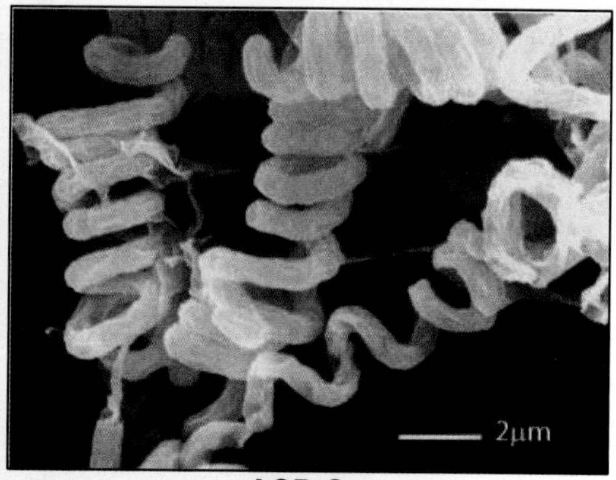

AQB-C

A

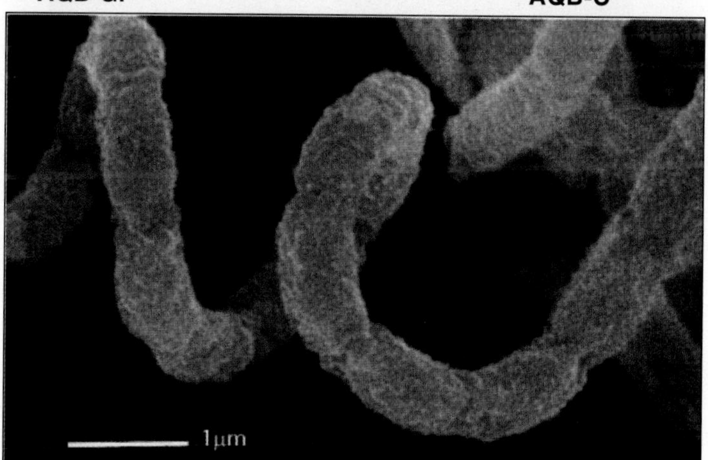

AQB-CA

Figure 25.6–*Contd...*

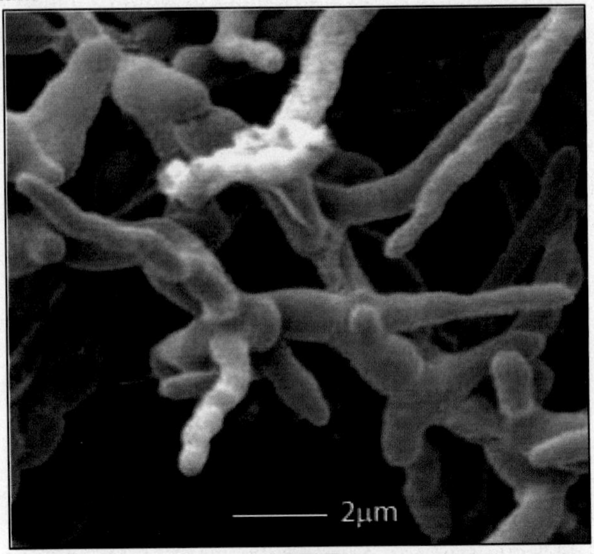

AQB-T

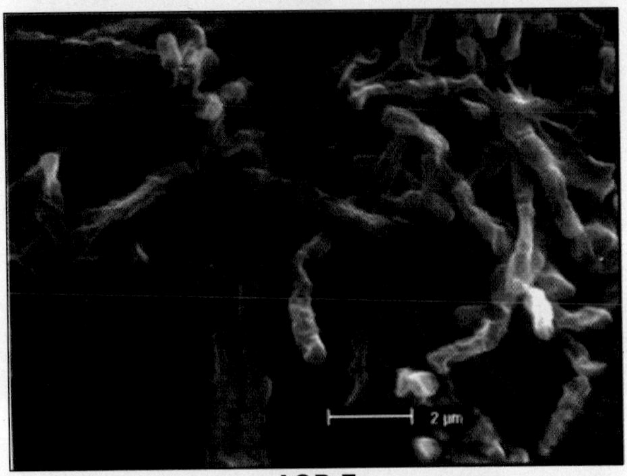

AQB-E

B

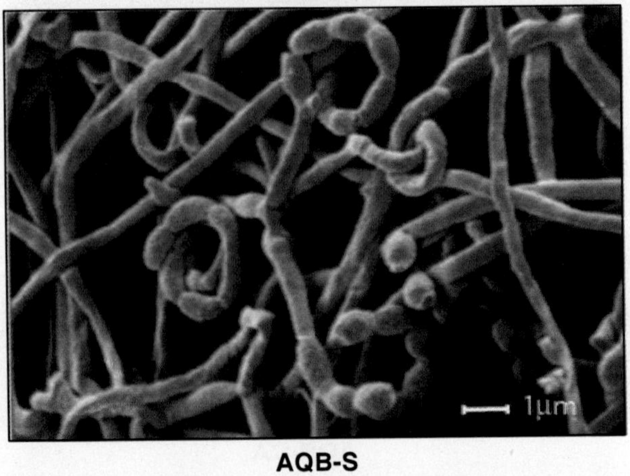

AQB-S

Table 25.3: Utilization of Carbon Sources (Glucose, Xylose, Arabinose, Rhamnose, Fructose, Glactose, Raffinose, Mannnitol, Inositol and Sucrose)

Strains	*Carbon Sources*									
	Glu-cose	*Xy-lose*	*Arabi-nose*	*Rham-nose*	*Fruc-tose*	*Galac-tose*	*Raffi-nose*	*Mann-nitol*	*Ino-sitol*	*Suc-rose*
AQB-T	+	+	+	+	+		–	+	+	–
AQB-E	+	+	+	–	–	+	+	–	–	–
AQB-S	+	+	+	+	+	+	–	–	+	±
AQB-Gr	+	+	–	–	±	+	–	+	–	
AQB-C	+	+	+	+	+	+	–	+	+	–
AQB-G		+		+			+	–	–	+
AQB-CA	+	+	+	+	+	+	+	+	+	

+: Indicates carbon utilization; –: Indicates not utilized, ± very slight utilization

Table 25.4: Influence of Amino Acids on the Growth of Selected *Streptomyces* spp. Strains

Strains	*Control*	*Glycine*	*Cystine*	*Alanine*	*Tryptophan*	*Valine*
AQB-T	0.022	0.048	0.058	0.070	0.047	0.080
AQB-E	0.017	0.067	0.056	0.059	0.064	0.057
AQB-S	0.020	0.071	0.085	0.064	0.081	0.072
AQB-Gr	0.015	0.049	0.051	0.054	0.061	0.062
AQB-C	0.024	0.056	0.059	0.061	0.064	0.074
AQB-G	0.024	0.038	0.043	0.054	0.067	0.057
AQB-CA	0.021	0.039	0.076	0.069	0.083	0.049

Table 25.5: Influence of Sodium Chloride on the Growth of Selected *Streptomyces* spp

Stains	*0.0%*	*1.0%*	*1.5%*	*2.0%*	*2.5%*	*3.0%*
AQB-T	0.094	0.052	0.068	0.068	0.050	0.046
AQB-E	0.047	0.057	0.035	0.070	0.079	0.048
AQB-S	0.033	0.043	0.055	0.062	0.102	0.071
AQB-Gr	0.032	0.047	0.049	0.078	0.072	0.063
AQB-C	0.038	0.053	0.047	0.049	0.070	0.053
AQB-G	0.044	0.049	0.026	0.057	0.065	0.044
AQB-CA	0.037	0.043	0.052	0.050	0.062	0.037

Table 25.6: Melanoid Pigmentation

Strains	*Peptone Yeast Extract Iron Agar Slants*
AQB-T	–
AQB-E	+
AQB-S	–
AQB-Gr	–
AQB-C	+
AQB-G	+
AQB-CA	–

+: Pigments produced, –: Not produced

Table 25.7: Spore Morphology

Strains	*Spore Morphology*
AQB-T	S
AQB-E	RF
AQB-S	S
AQB-Gr	S
AQB-C	RF
AQB-G	RF
AQB-CA	S

S: Spiral spores; RF: Rectus flexibilis.

Here A represents Spiral Spore and B represents Rectiflexibilies Spore chains chains

Figure 25.7: Gel Electrophoresis of *Streptomyces* strains Isolated from Gut of Marine [*Tendania anhelans* (Lieberkuhn) (Sponges), *Epinephelus diacanthus* (Grouper)], Estuarine [*Oreochromis mossambicus* (Tilapia), *Etroplus maculates* (Pearl spot)], Fresh-water [*Cyprinus carpio* (common carp), *Ospherenemous gourami* (Gourami)] and 4) *Carassius auratus* (Gold fish) Samples Using 10% Trichloroacetic Acid (Gel 1 and 2), 70 per cent Ethanol (Gel 3 and 4)

T: Tilapia; E: Pearl spot; S: Sponges; Gr: Grouper; C: common carp; G: Gourami; CA: Gold fish

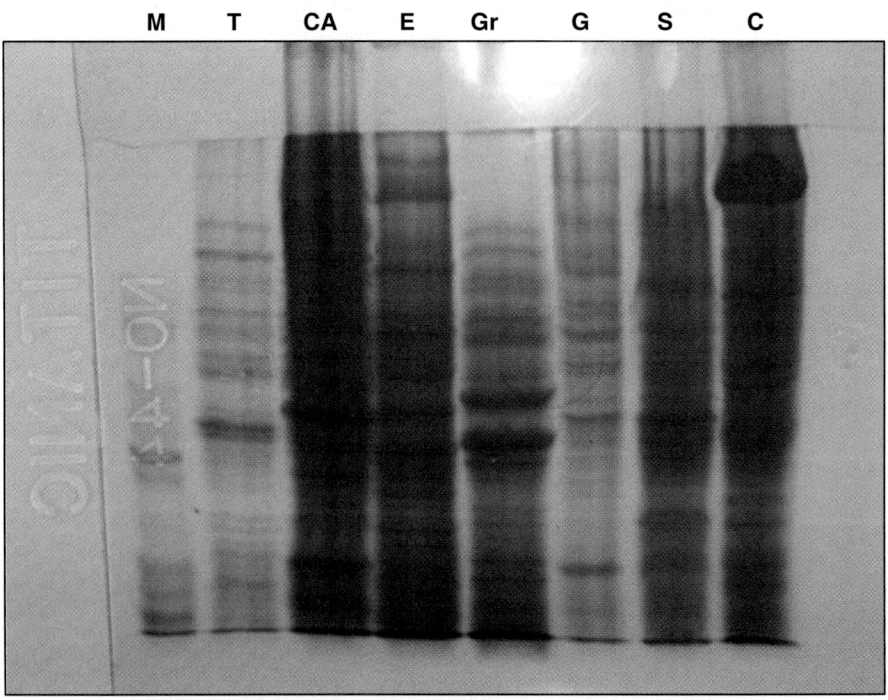

Gel 1

Figure 25.7–*Contd...*

M T CA E Gr G S C

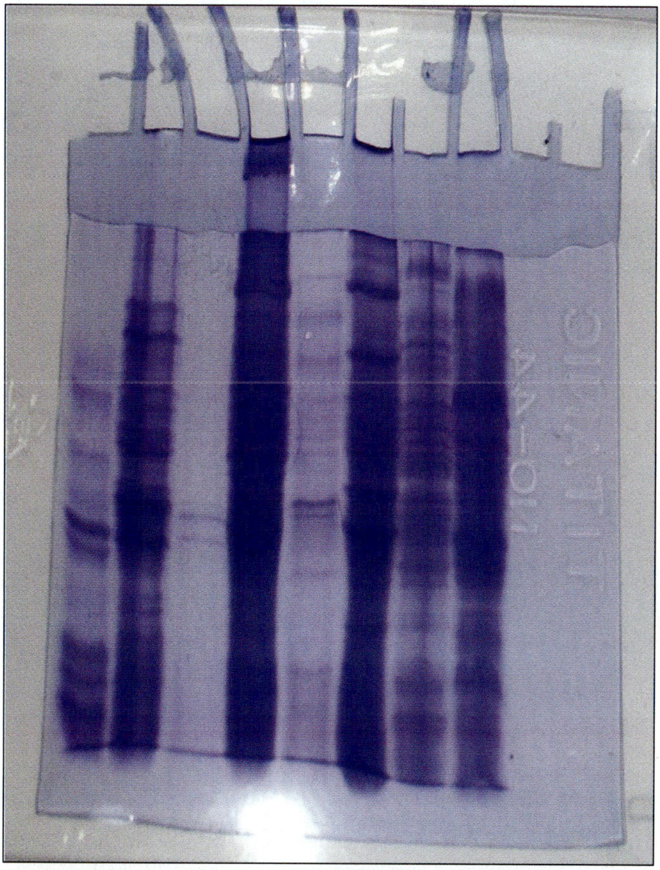

Gel 2

M T CA E Gr G S C

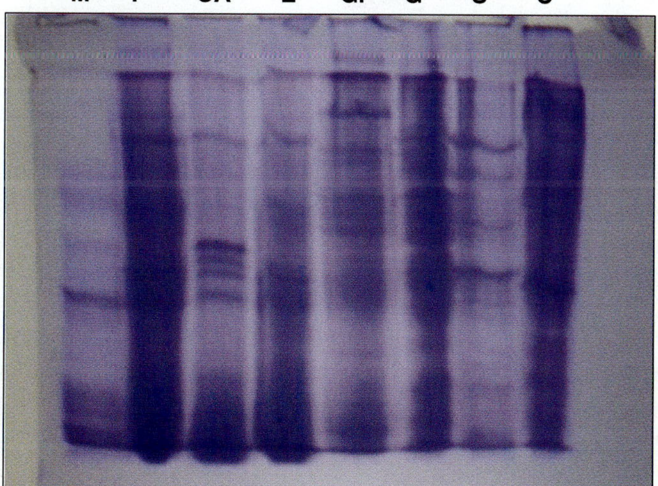

Gel 3

Figure 25.7–*Contd...*

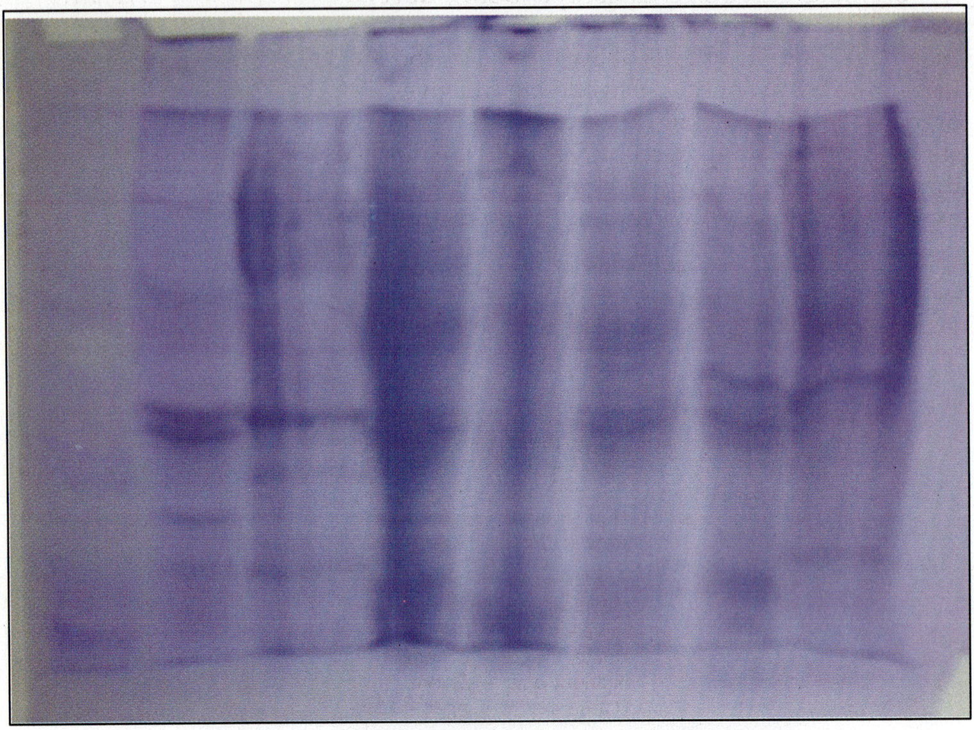

Gel 4

reported, although their structures have not been well characterized [Aretz *et al.* (1989), Henderson *et al.* (1987), Morihara and Tsuzuki (1969), Vinci *et al.* (1993)]. Activity gels were run as described as above.The regular native gels in anionic system did not resolve the clearing bands distinctly and the enzyme did not enter the separating gel. This is probably due to its net positive charge conferred by the bound metal ions and hence the cationic system of Gabriel (1979) was tried. A sharp clearing band corresponding to molecular weight marker trypsinogen was found both in *S. fradiae* and TK23 (pMKU1). Below this band diffuse bands were observed (data not shown). No clearing bands were observed for TK23 (pIJ699). The purified sample (Table 25.2) showed a single band of 18 kDa on SDS-PAGE both with Coomassie blue and silver staining. The extracellular protein profile of TK23 (pMKU1) (SDS-PAGE) showed that the number of protein bands are limited, unlike in *S. lividans* TK23 and TK23 (pIJ699). This was in contrast to the high concentration observed by *in vitro* protein assay of the TK23 (pMKU1) culture filtrate. It is most likely that the protease degrades the extracellular proteins secreted by the host. This could also explain the ease of purification of pMKU1 encoded protease, an 18 kDa protein. Based on these results the 8·8 kb *S. fradia* DNA insert in pMKU1 encodes a new protease protein of 18 kDa in *S. lividans*.

Systematics of Twleve Selected *Streptomyces* spp.

Scanning Electron Microscope (SEM), developed in the late 1960s, enables researchers to see the surface of objects in the natural state and without sectioning. Magnifications with the SEM are limited to about 75,000 to 100,000 times, but the instrument is relatively easy to operate and gives vivid and undistorted views of an organism's surface details. It is very useful for investigating the general shape and surface morphology of microbial cells and their interaction with each other as well as their environment (Edward Alcamo, 2001). Mycelial colours in selected media are expressed by *Streptomyces* spp. associated with seaweeds.

Systematics of *Streptomyces* AQB. BPJ.10

Spore chain morphology	Section Rectiflexibilis. Long aerial mycelium with 20 – 50 spores in each chain.
Spore surface	Warty
Aerial mycelial colour	Actinomyces agar – Mustard Yellow Glycerol asparagine agar – Pistha Green Inorganic salt – Starch agar – Greenish Yellow Kuster's agar – Green Nutrient agar – Light Peach Potato dextrose agar – Light Green
Substrate mycelial colour	Actinomycetes agar – Light Yellow Glycerol asparagine agar – Cream Inorganic salt – Starch agar – Cream Kuster's agar – Cream Nutrient agar – Cream Potato dextrose agar – Cream
Soluble pigment	Actinomyces agar – Nil Glycerol asparagine agar – Light Brown Inorganic salt – Starch agar – Light Brown Kuster's agar – Light Brown Nutrient agar – Yellow Potato dextrose agar – Yellow
Melanoid production	Negative
Carbon utilization	Profused growth – Maltose and Fructose Moderate growth – Control, Fructose and Glucose Poor growth – Dextrose
Sodium chloride tolerance	Profused growth – 6 per cent Moderate growth – 2 per cent, 3 per cent, 4 per cent, 5 per cent, 7 per cent, 8 per cent, 9 per cent and 10 per cent Poor growth – 0 per cent, 0.1 per cent, 0.2 per cent, 0.3 per cent, 0.4 per cent, 0.5 per cent, 0.6 per cent, 0.7 per cent, 0.8 per cent, 0.9 per cent and 1.0 per cent
Amino acid utilization	Profused growth – Alanin and Glycine Moderate growth – Control, Aspartic acid and Vanillin Poor growth – Not detected
Antibiotic activity	Activity against Streptococcus pyogenes and *Salmonella typhii*
Source	*Bryopsis plumose*
Habitat	Marine environment (Kovalam coast)

Remarks: The strain Streptomyces AQB. BPJ.10 does not resemble any one of the reported species of Streptomyces in the ISP and also from the estuarine sediment at Porto Novo along the east coast of India (Lakshmanaperumalsamy, 1978) and it may be a new taxon.

Colour of *Streptomyces* from Marine Seaweeds Sources

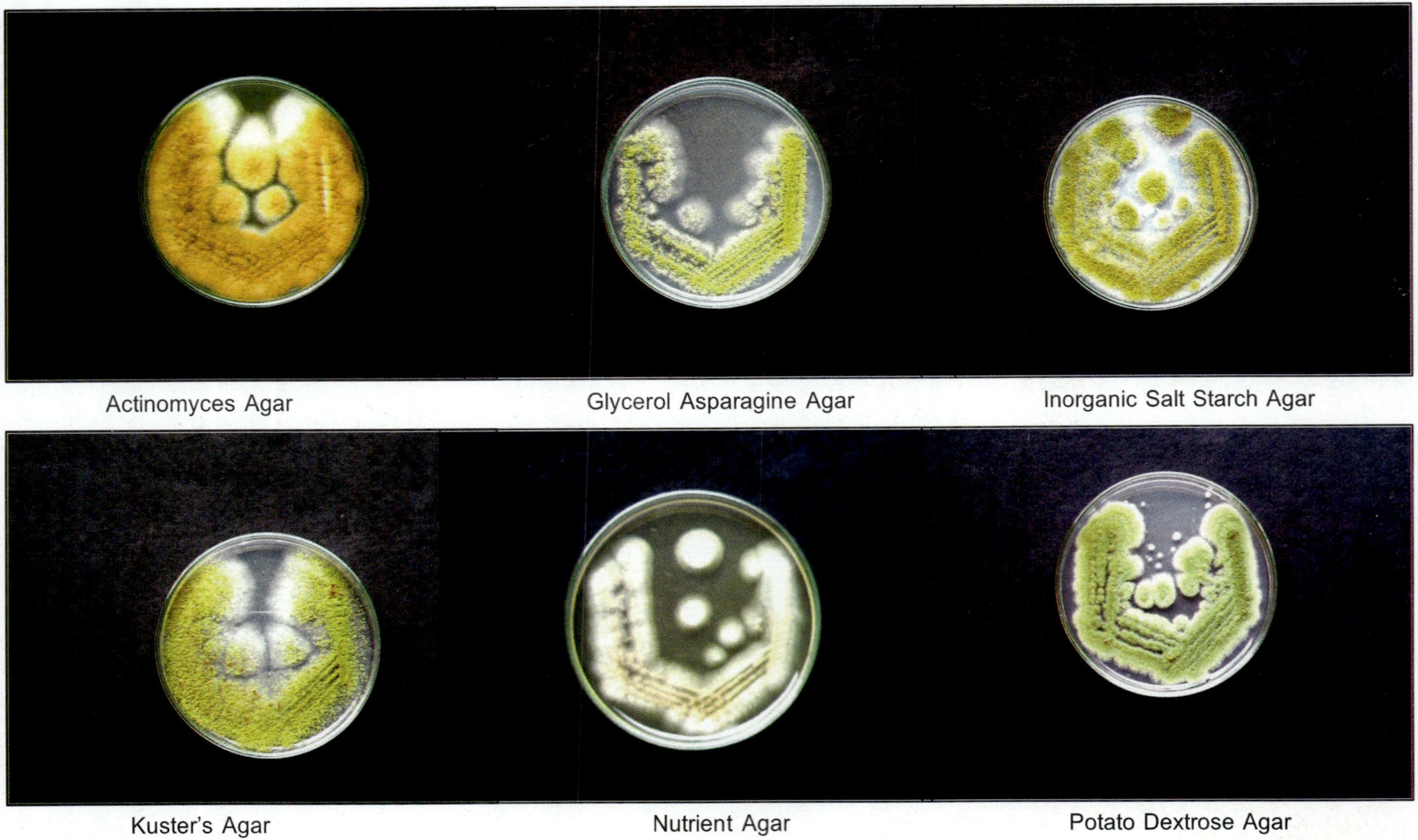

Figure 25.8: Aerial Mycelial Colors of *Streptomyces AQB.BPJ.10* in Six Different Media

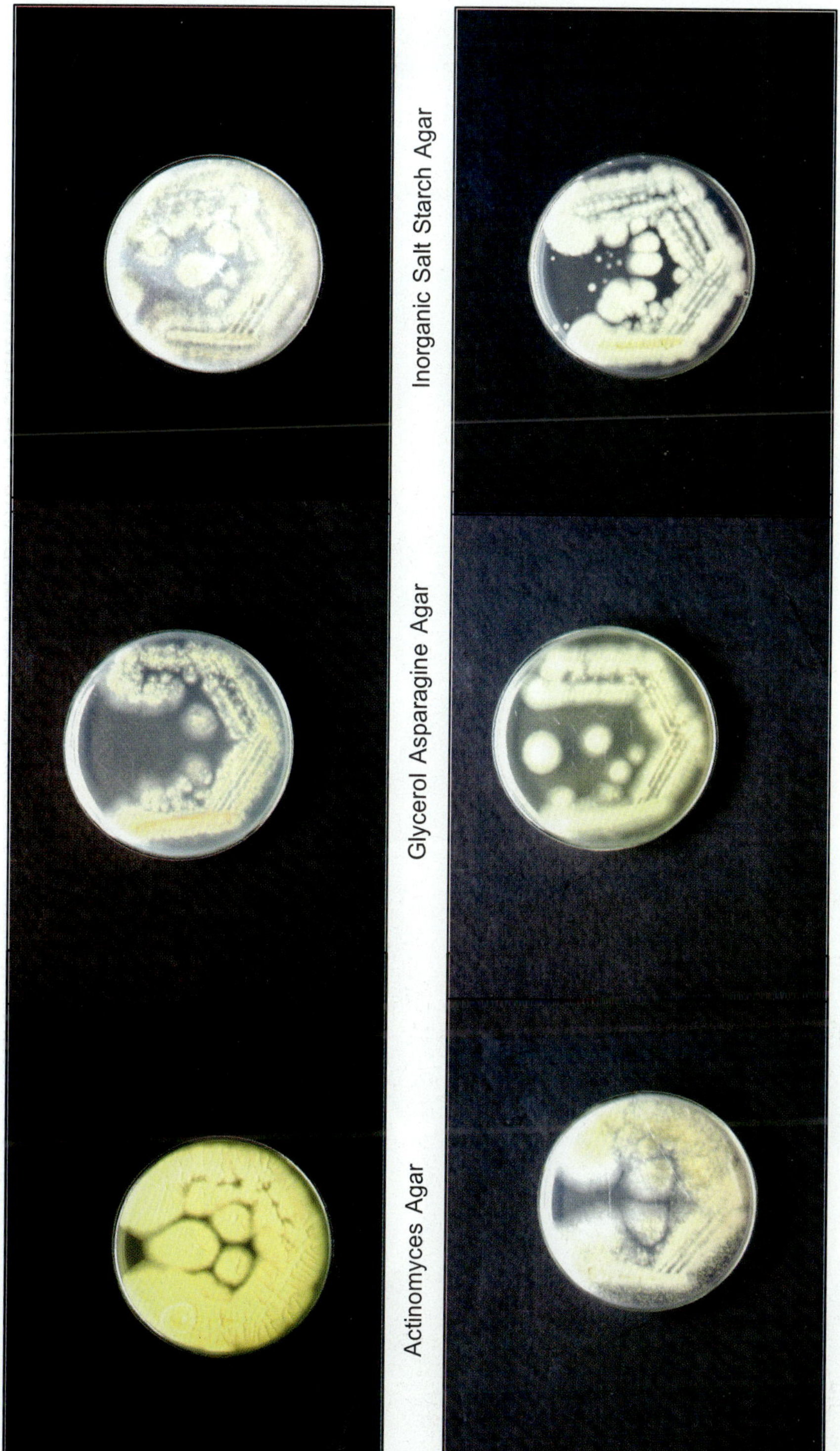

Figure 25.9: Substrate Mycelial Solors of *Streptomyces AQB.BPJ.10* in Six Different Media

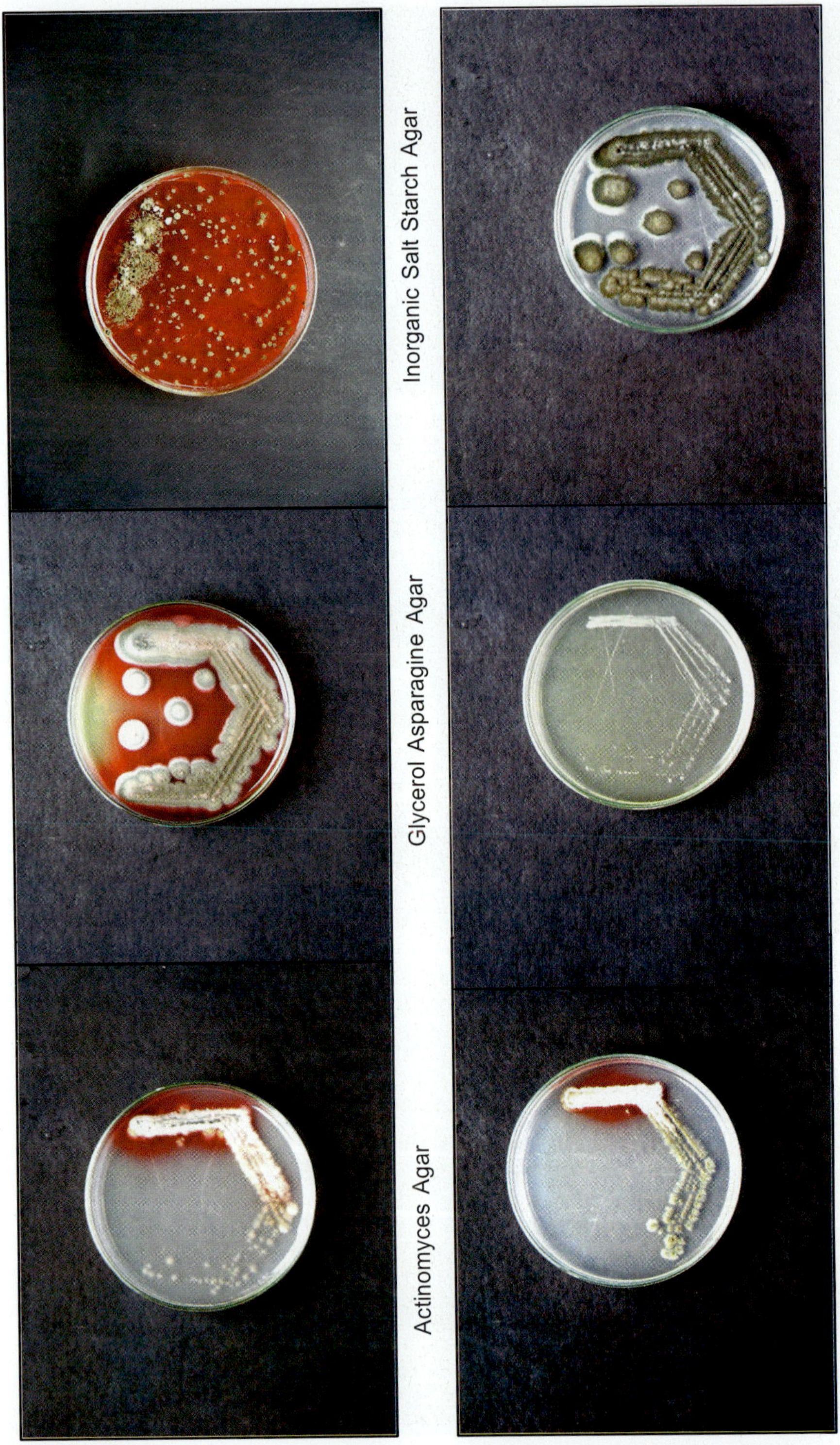

Figure 25.10: Aerial Mycelial Colors of *Streptomyces AQB.CRA.11* in Six Different Media

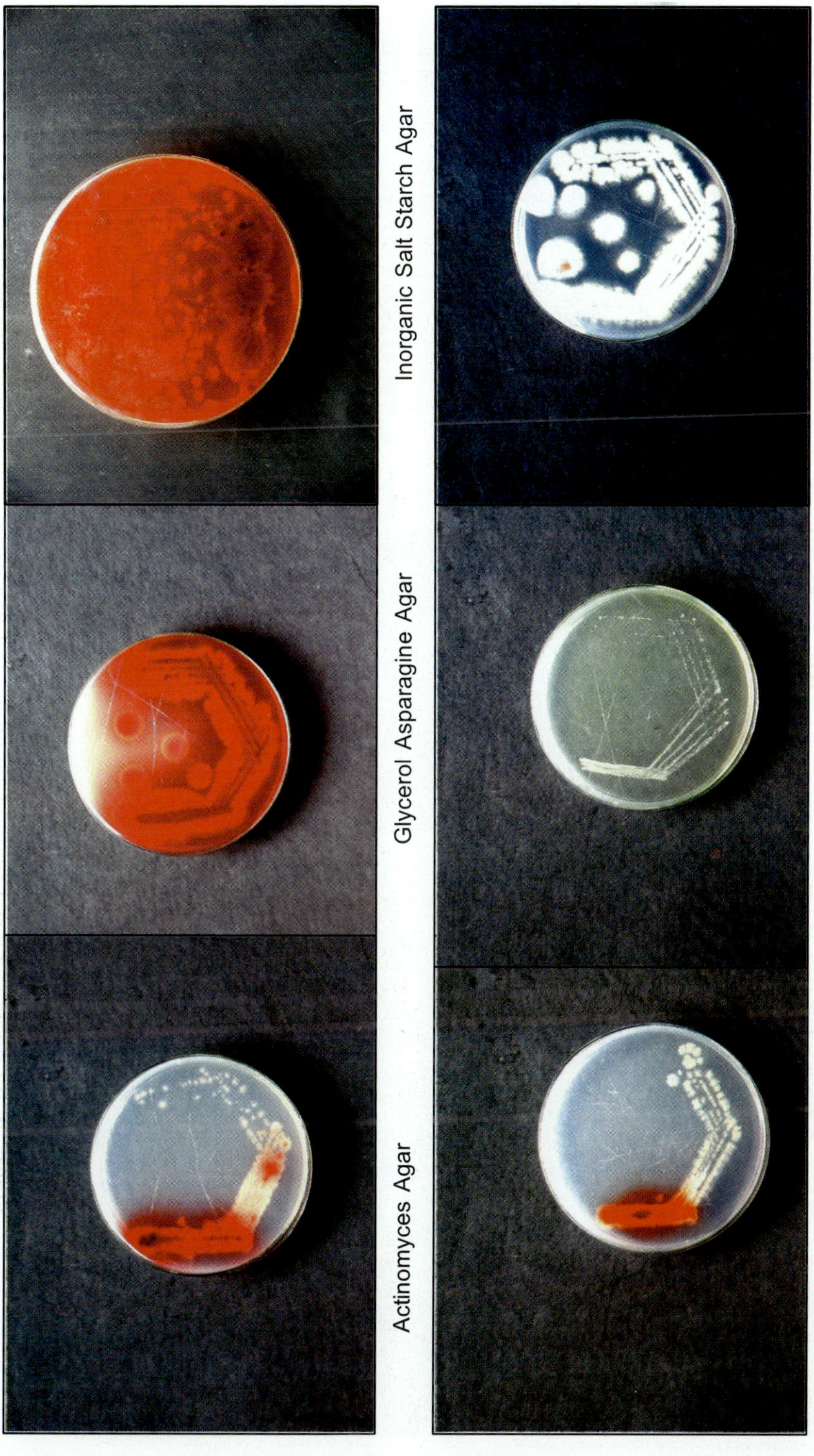

Figure 25.11: Substrate Mycelial Colors of *Streptomyces AQB.CRA.11* in Six Different Media

Figure 25.12: Aerial Mycelial Colors of *Streptomyces AQB.CAY.12* in Six Different Media

Figure 25.13: Substrate Mycelial Colors of *Streptomyces AQB.CAY.12* in Six Different Media

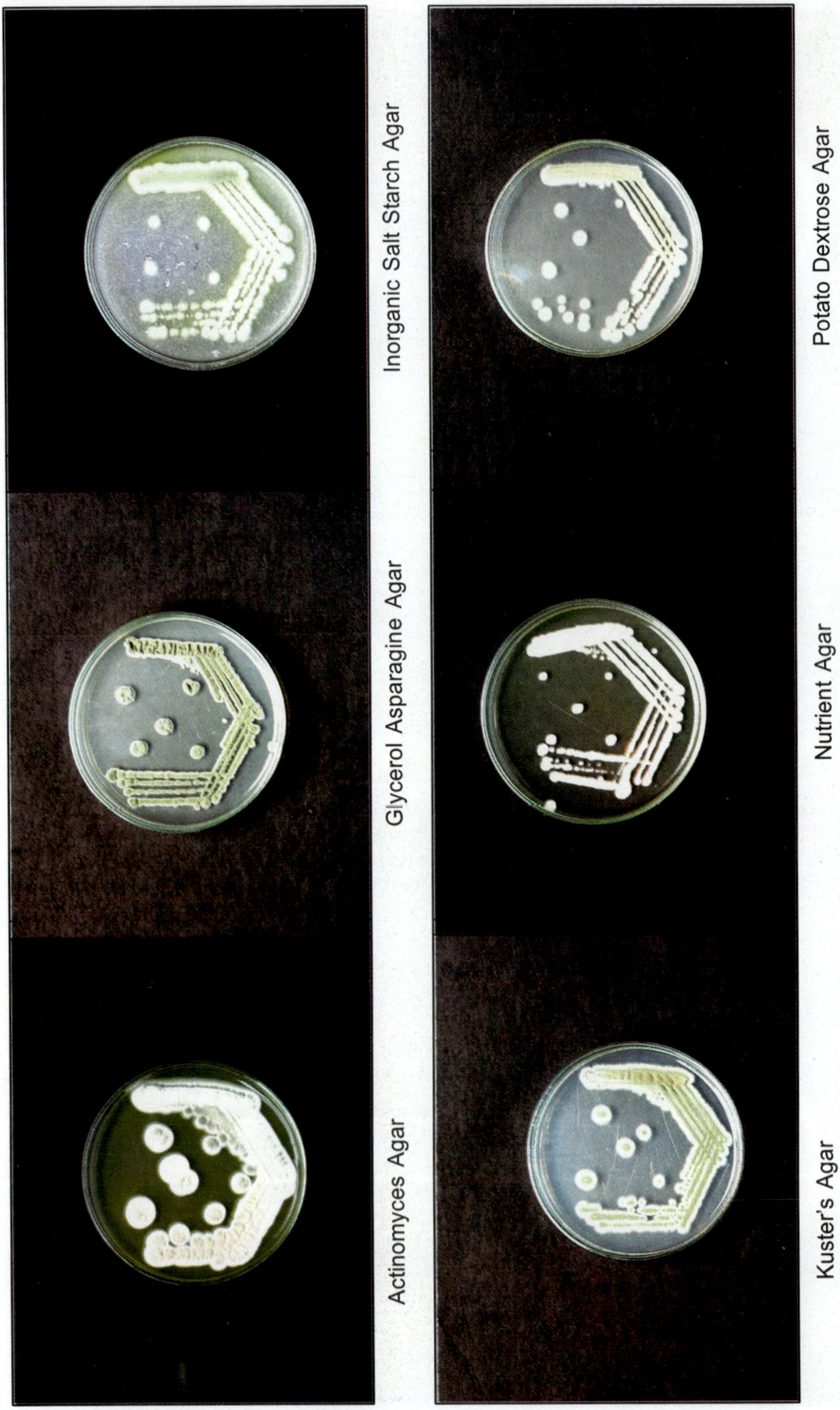

Figure 25.14: Aerial Mycelial Colors of *Streptomyces AQB.CSA.13* in Six Different Media

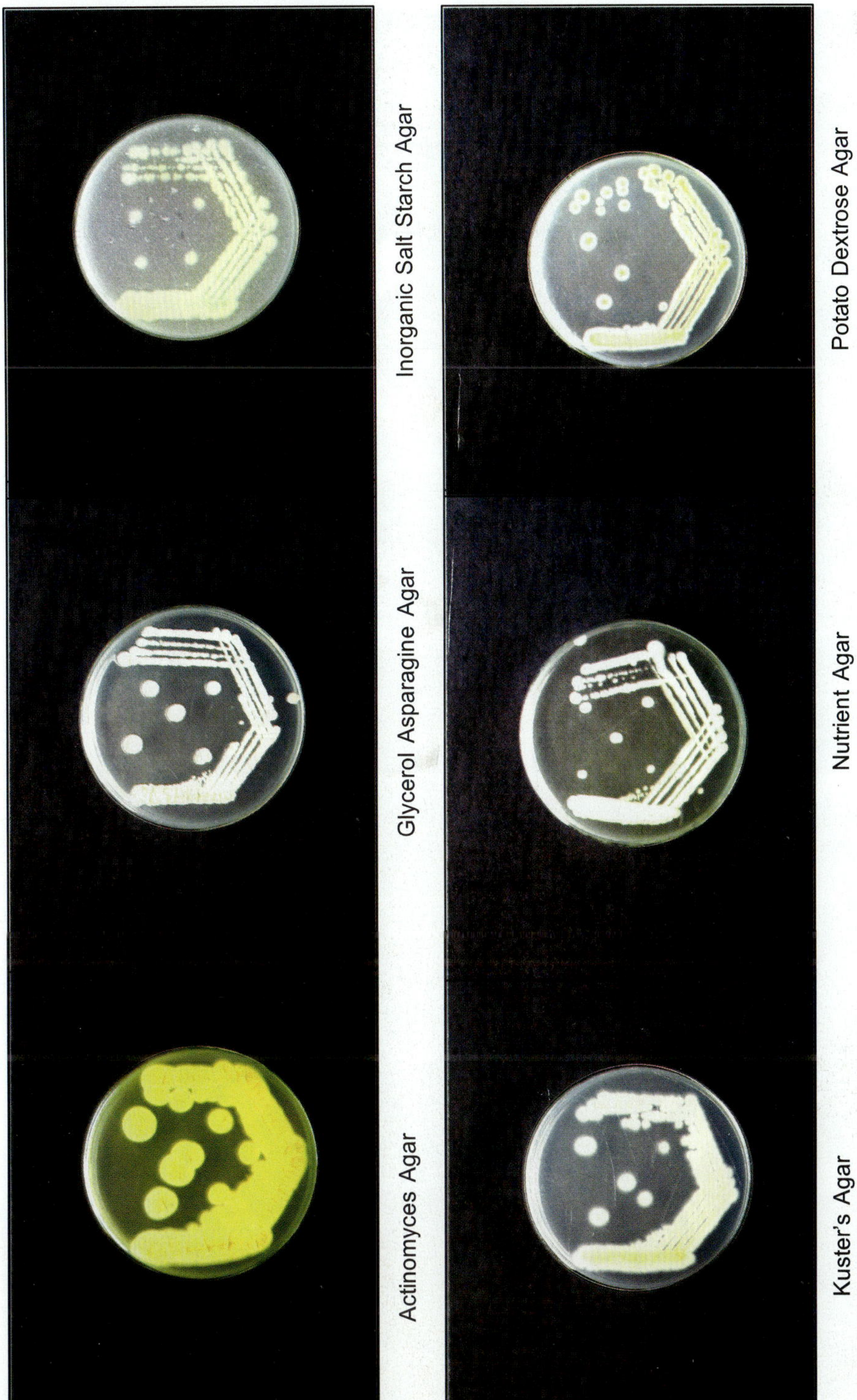

Figure 25.15: Substrate Mycelial Colors of *Streptomyces AQB.CSA.13* in Six Different Media

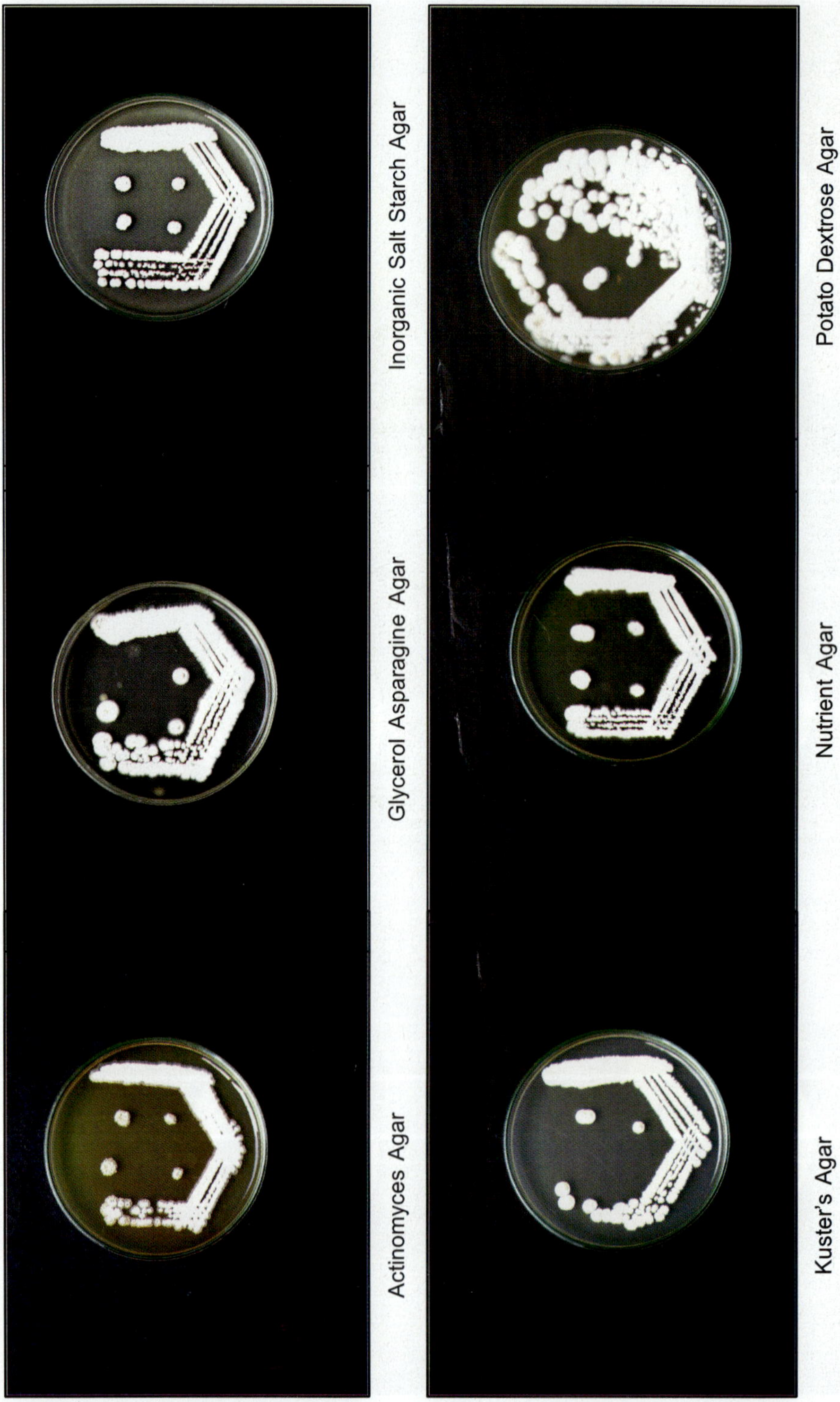

Figure 25.16: Aerial Mycelial Colors of *Streptomyces AQB.CMC.14* in Six Different Media

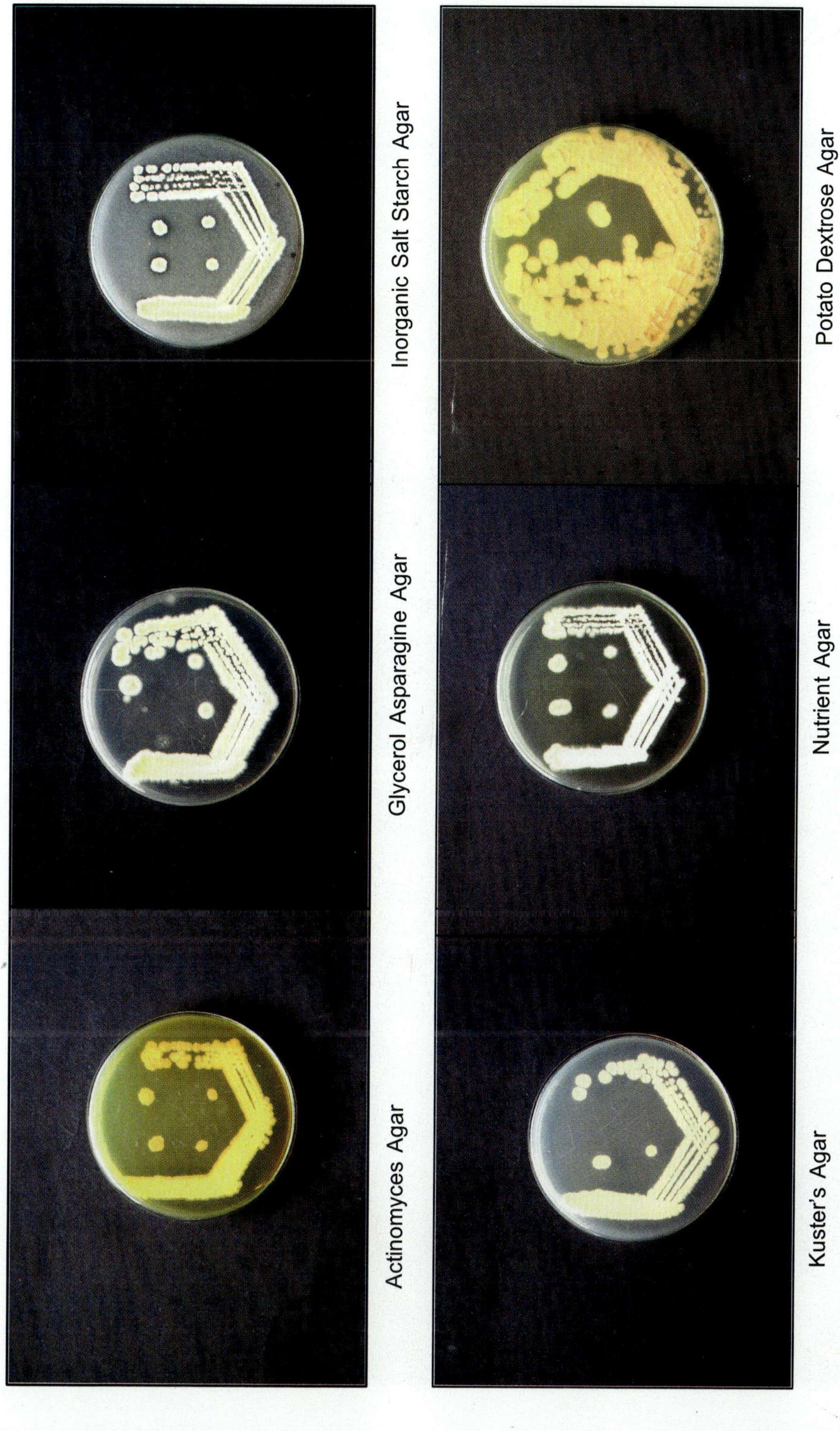

Figure 25.17: Substrate Mycelial Colors of *Streptomyces AQB. CMC.14* in Six Different Media

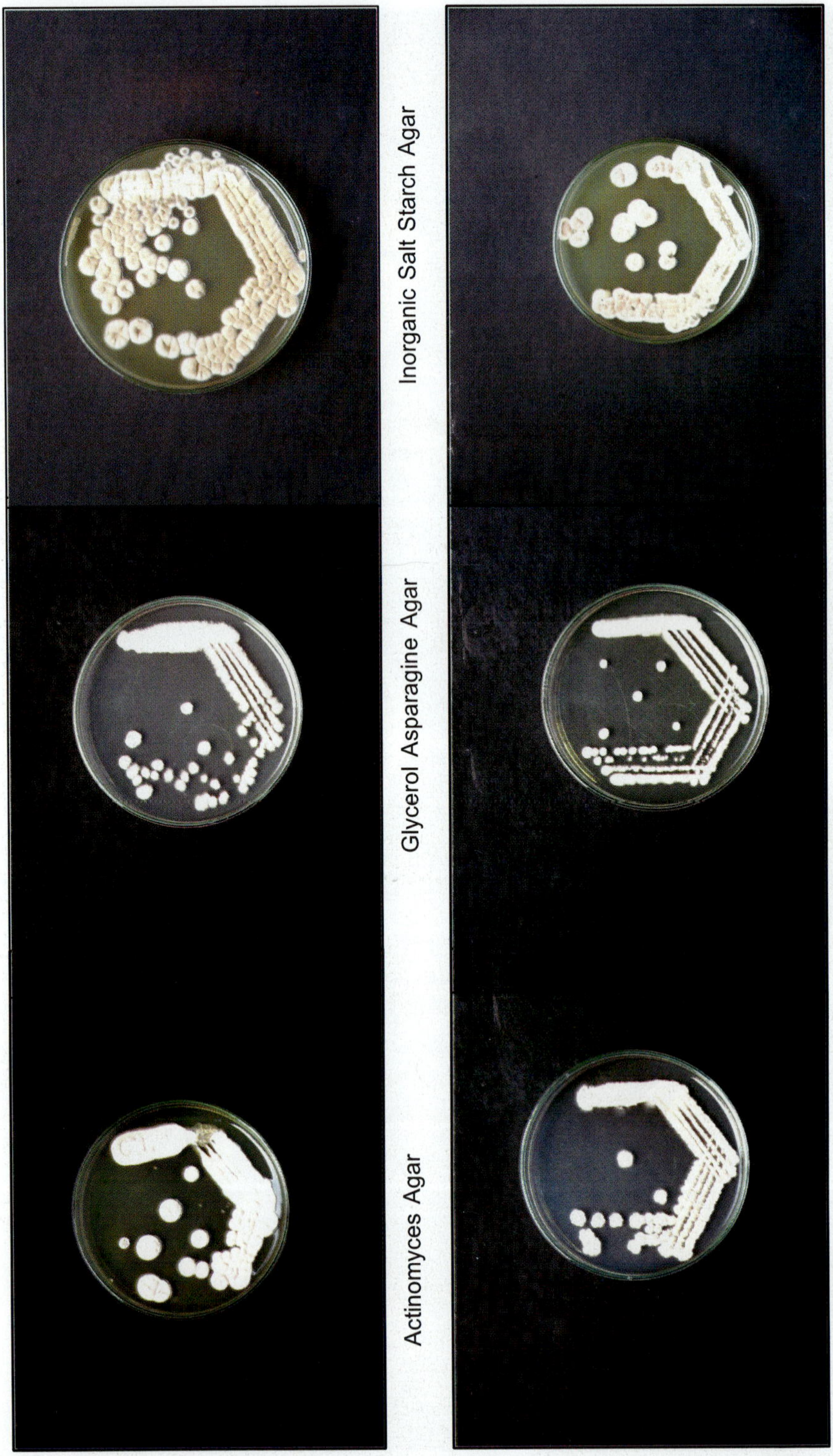

Figure 25.18: Aerial Mycelial Colors of *Streptomyces AQB.GCH.15* in Six Different Media

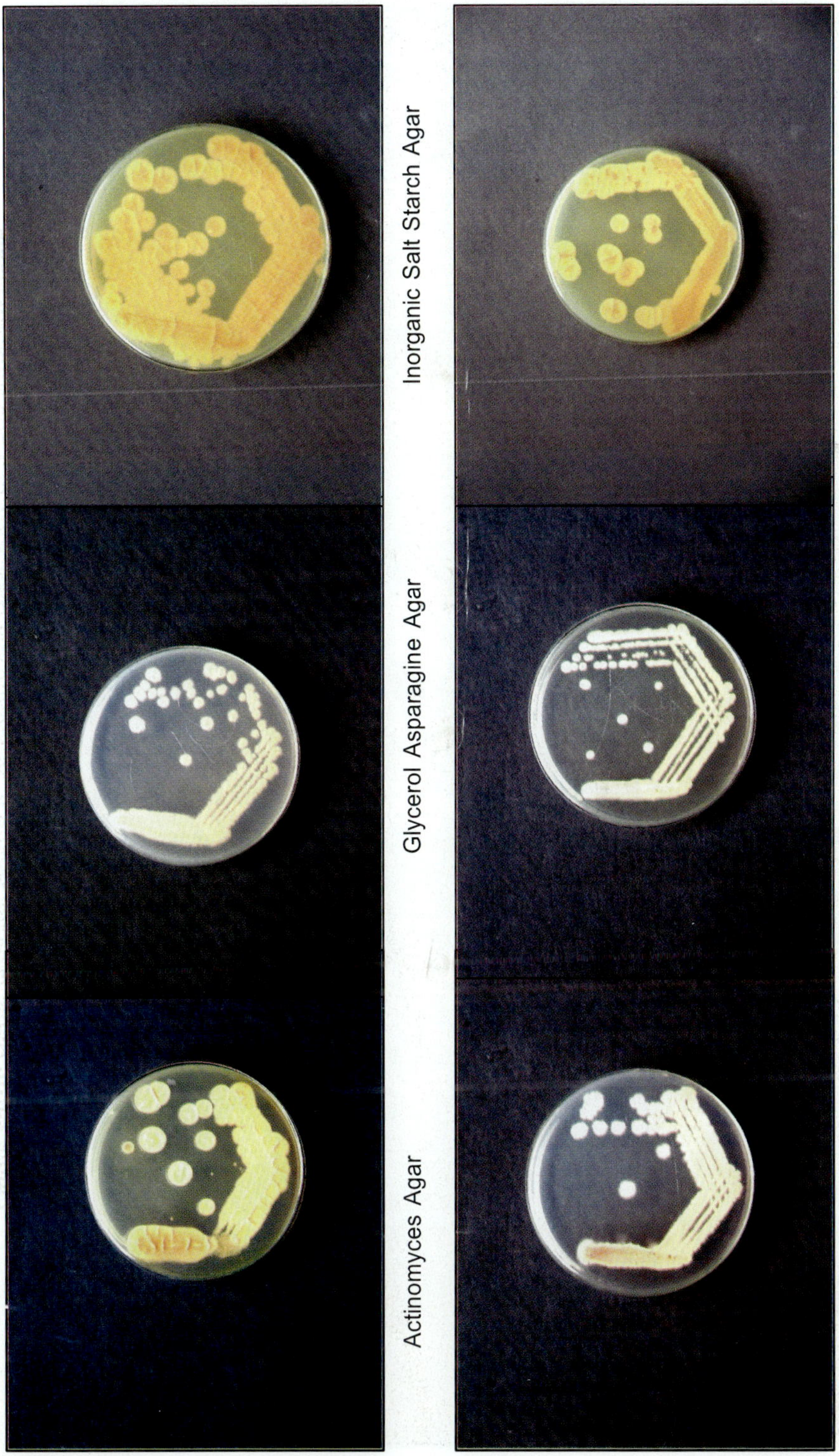

Figure 25.19: Substrate Mycelial Colors of *Streptomyces AQB. GCH.15* in Six Different Media

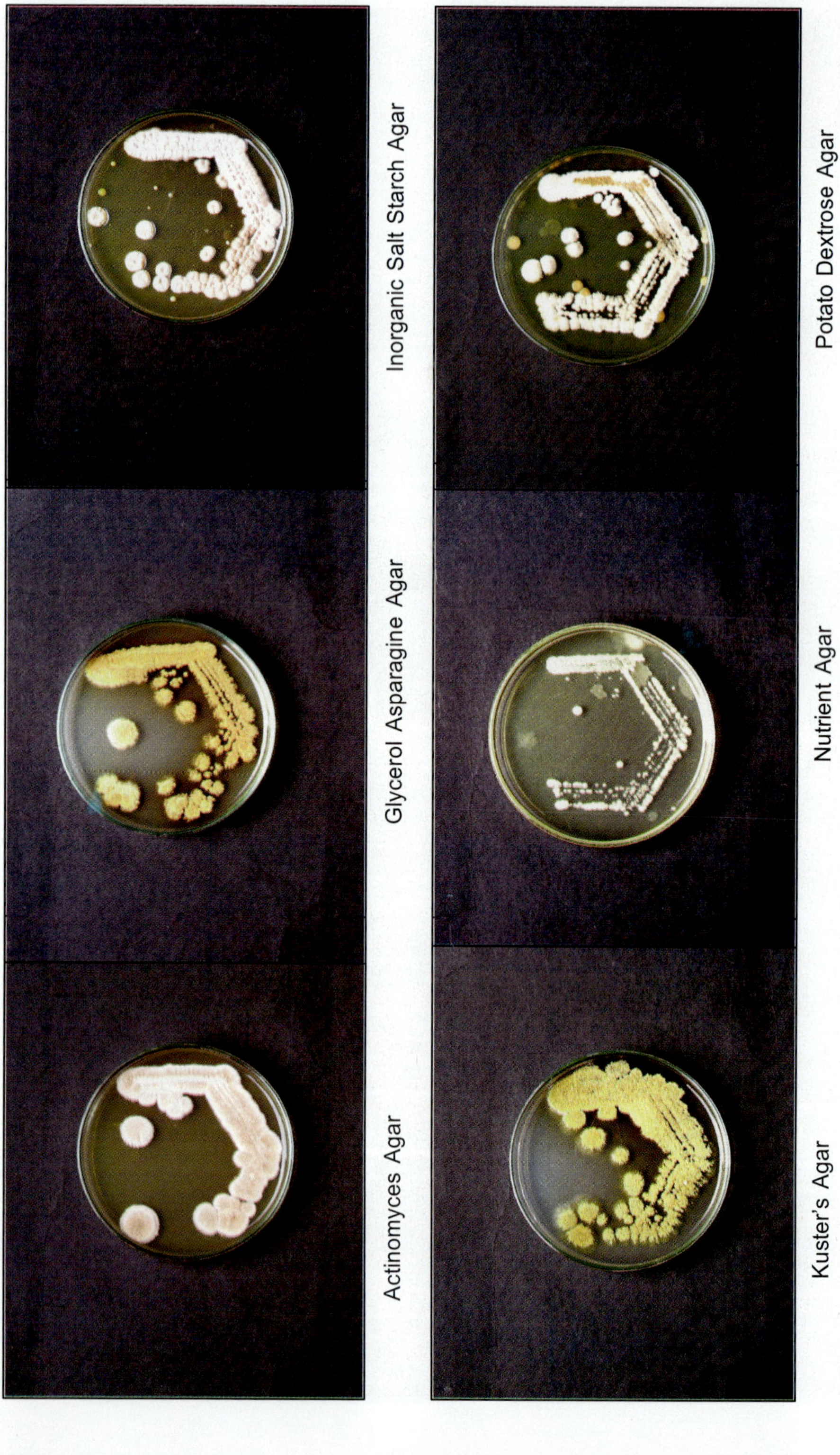

Figure 25.20: Aerial Mycelial Colors of *Streptomyces AQB.HVA.16* in Six Different Media

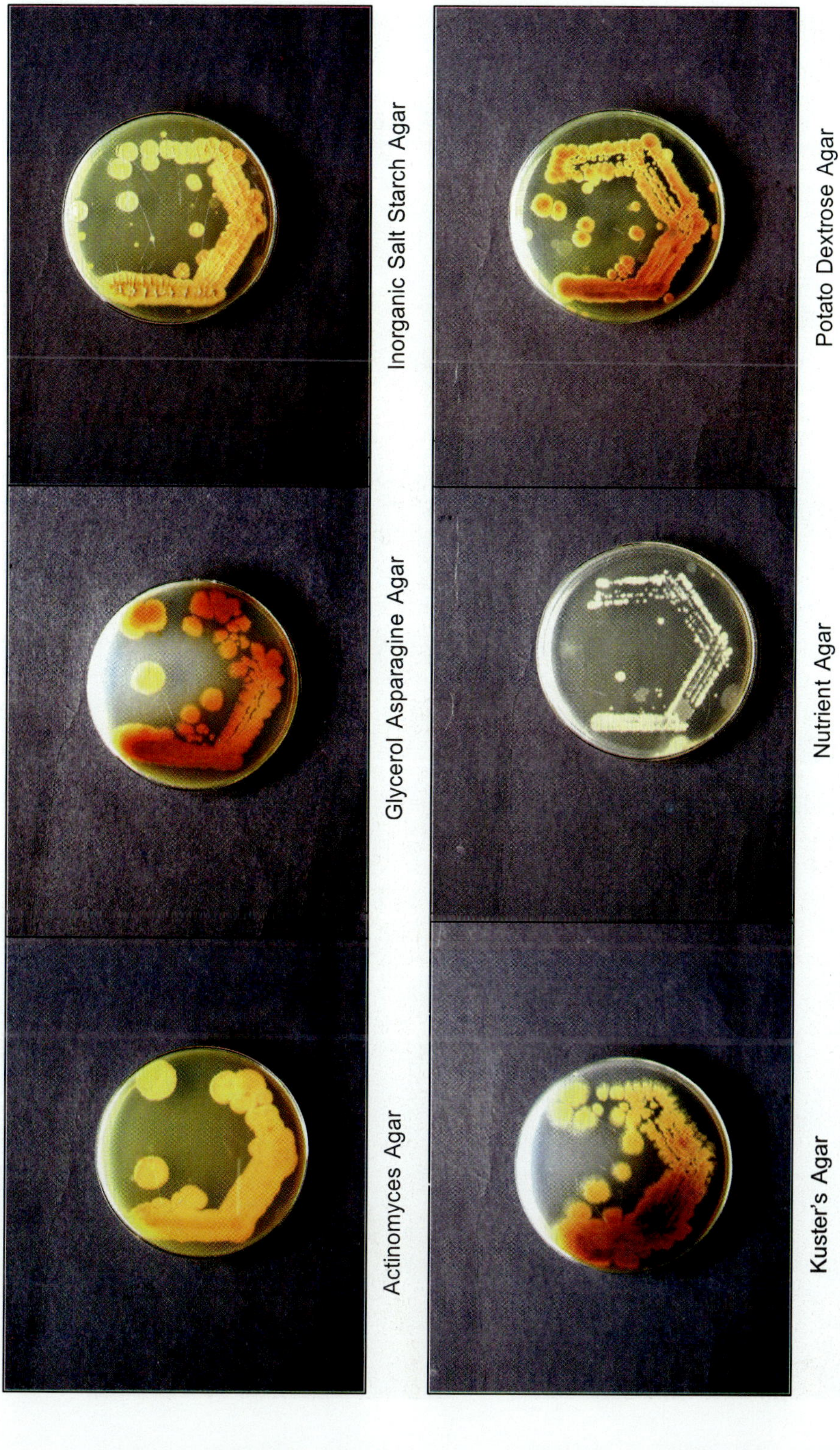

Figure 25.21: Substrate Mycelial Colors of *Streptomyces AQB.HVA.16* in Six Different Media

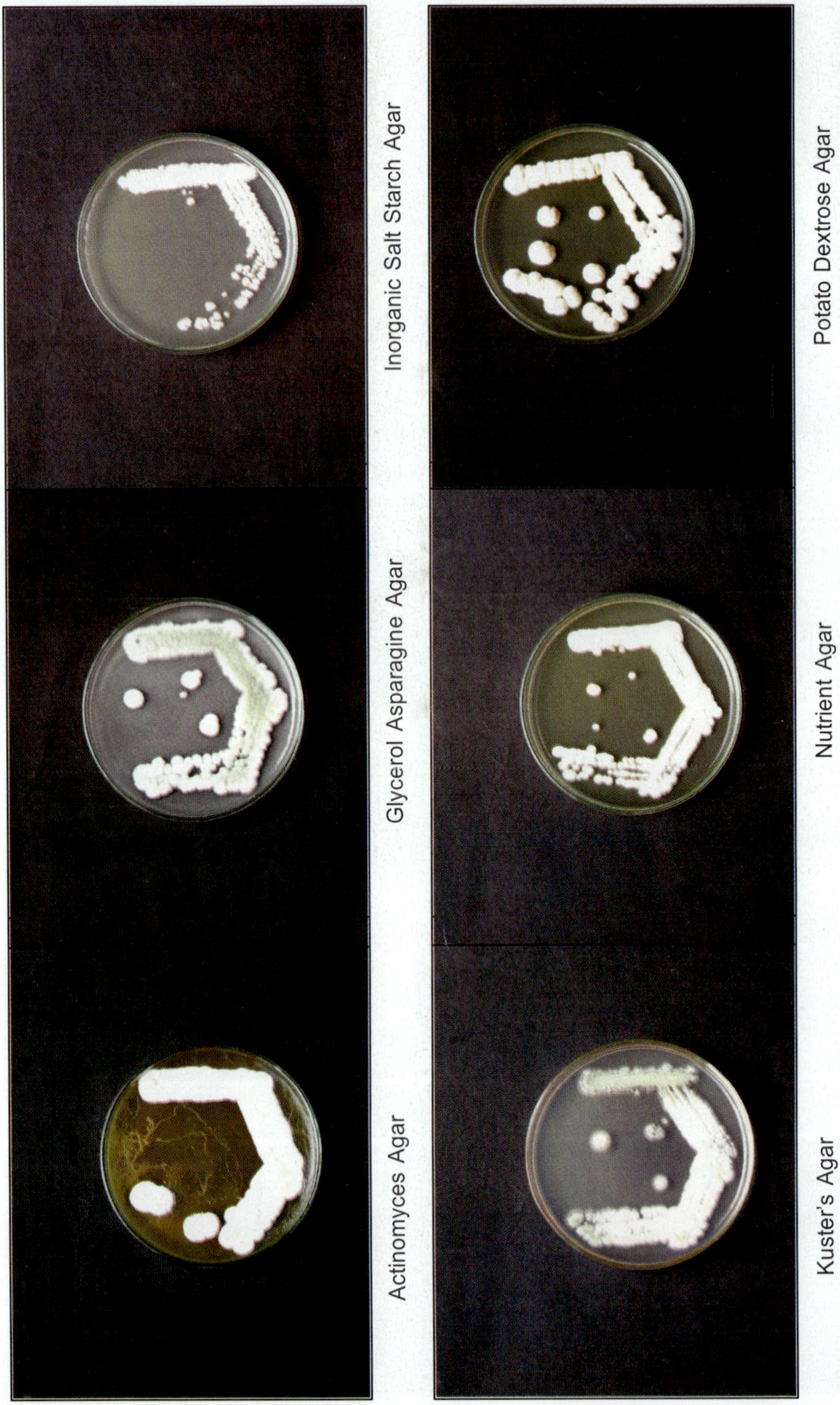

Figure 25.22: Aerial Mycelial Colors of *Streptomyces AQB.PTN.17* in Six Different Media

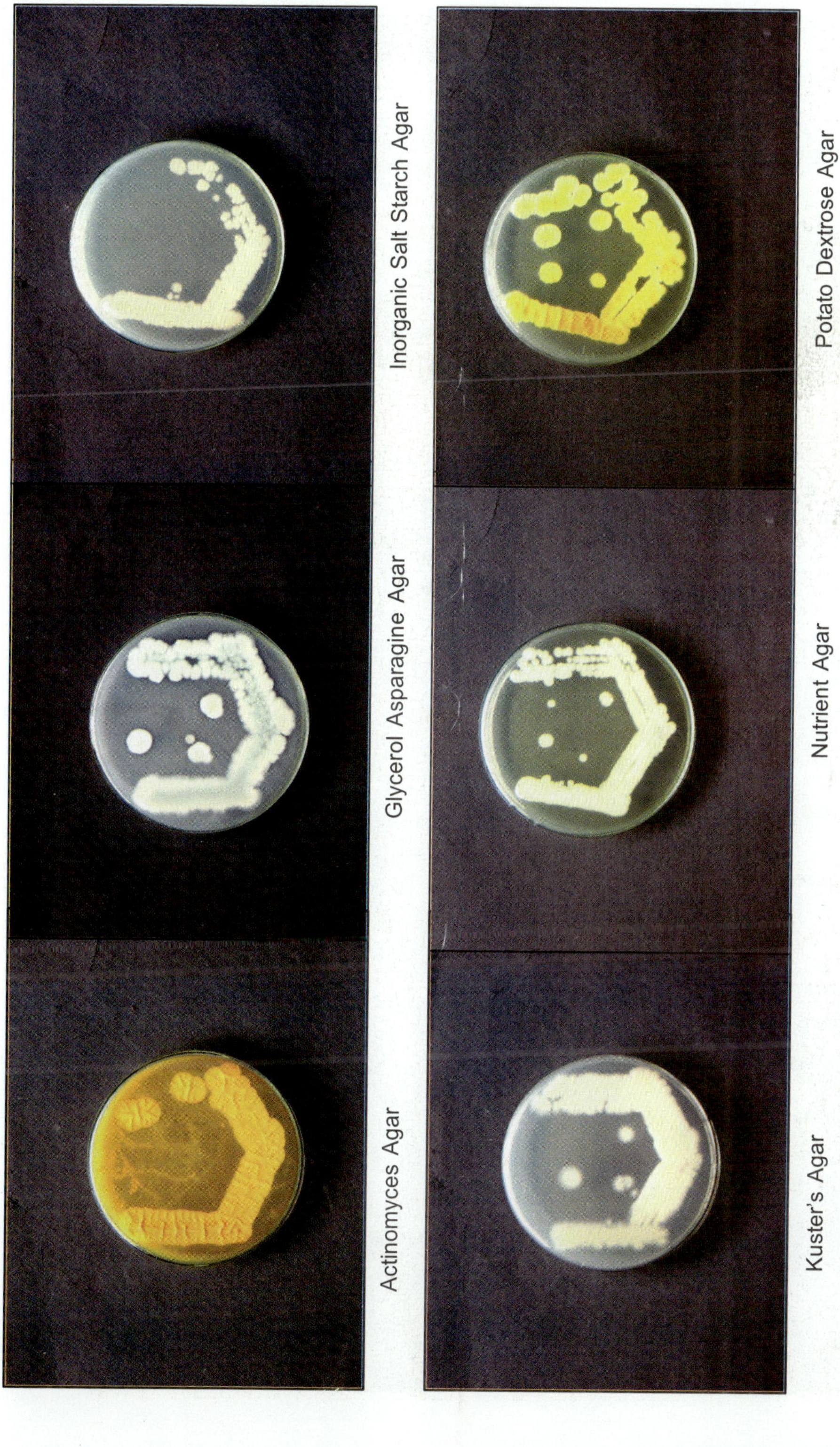

Figure 25.23: Substrate Mycelial Colors of *Streptomyces AQB.PTN.17* in Six Different Media

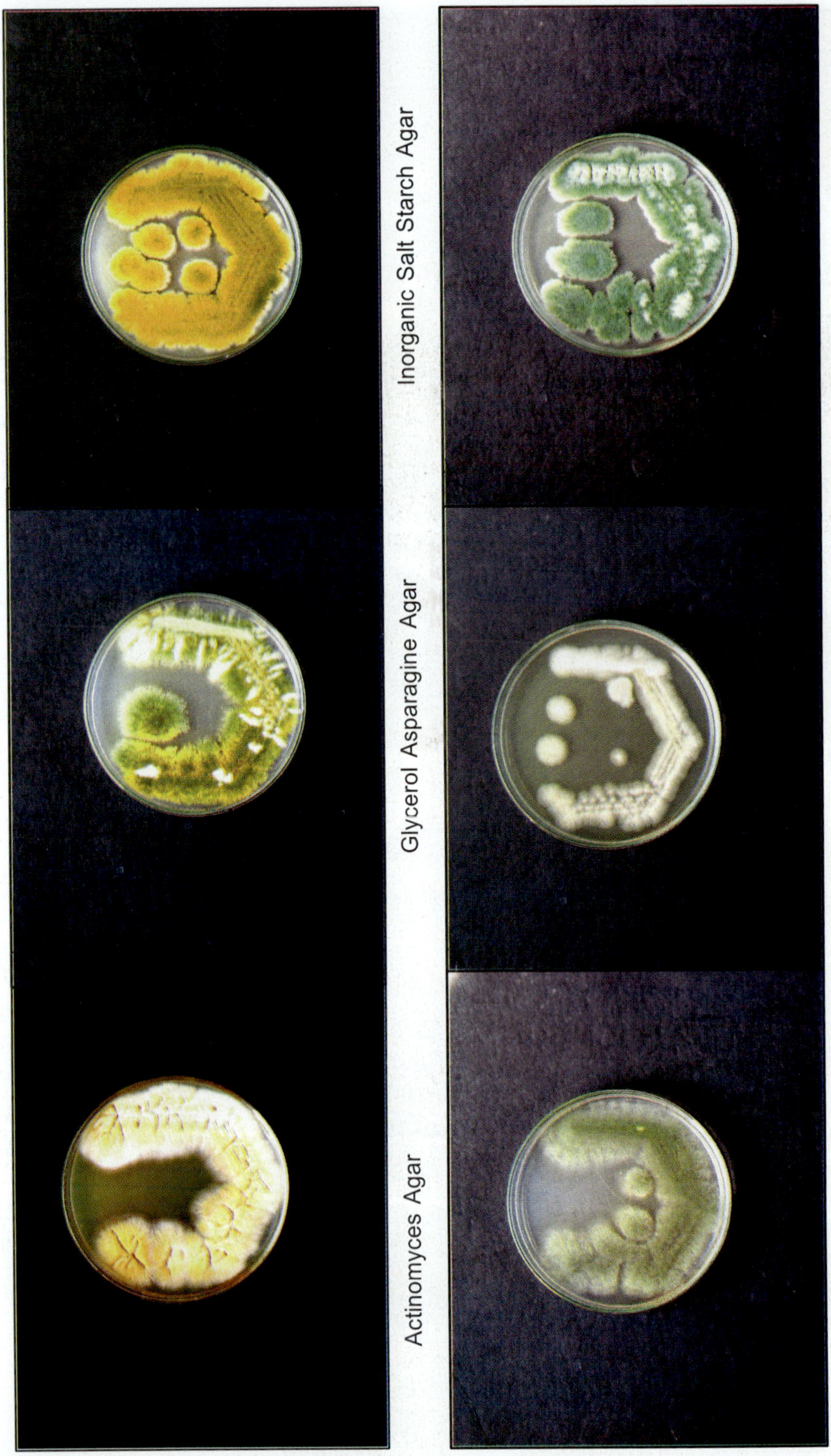

Figure 25.24: Aerial Mycelial Colors of *Streptomyces AQB.SWD.18* in Six Different Media

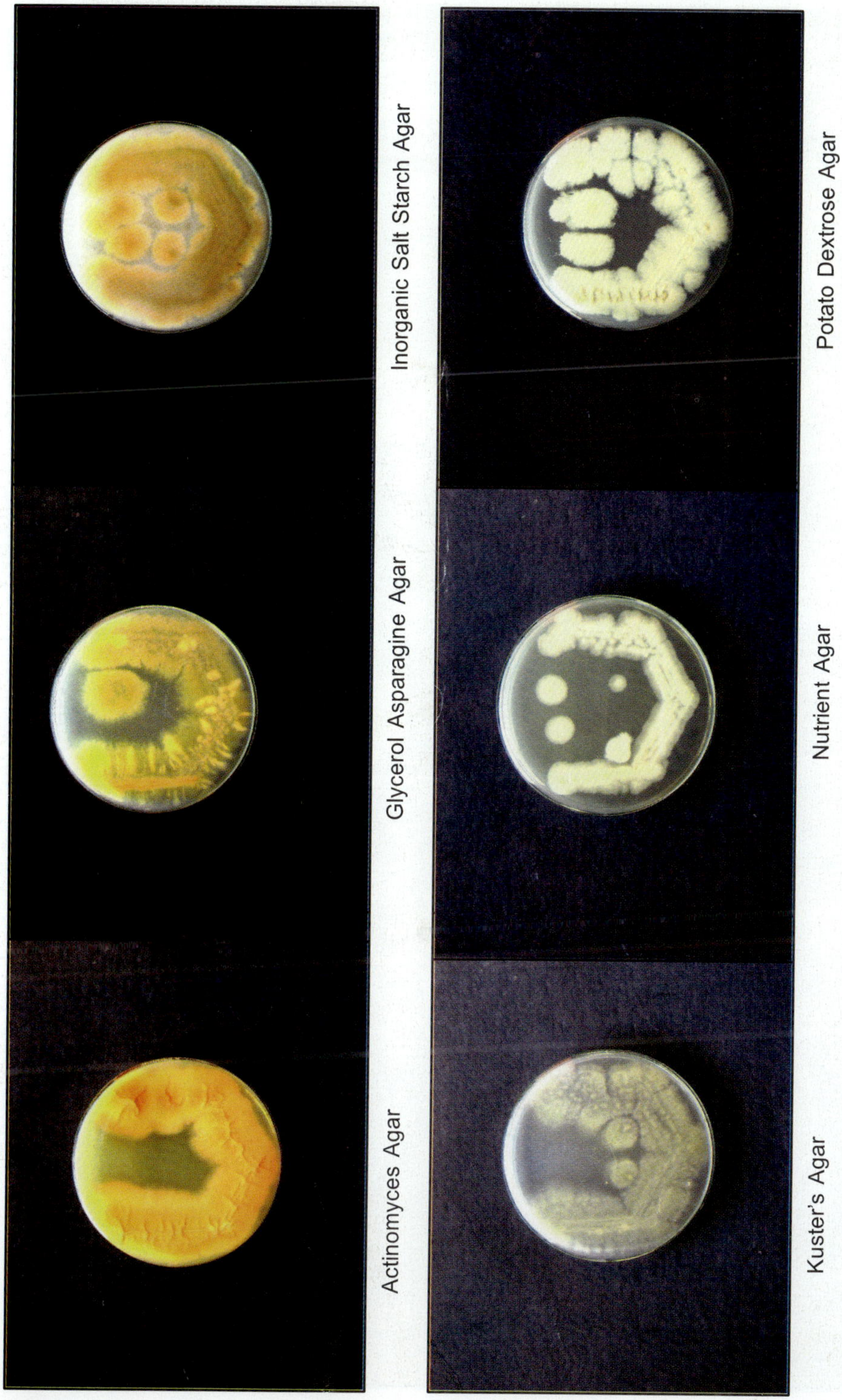

Figure 25.25: Substrate Mycelial Colors of *Streptomyces AQB.SWD.18* in Six Different Media

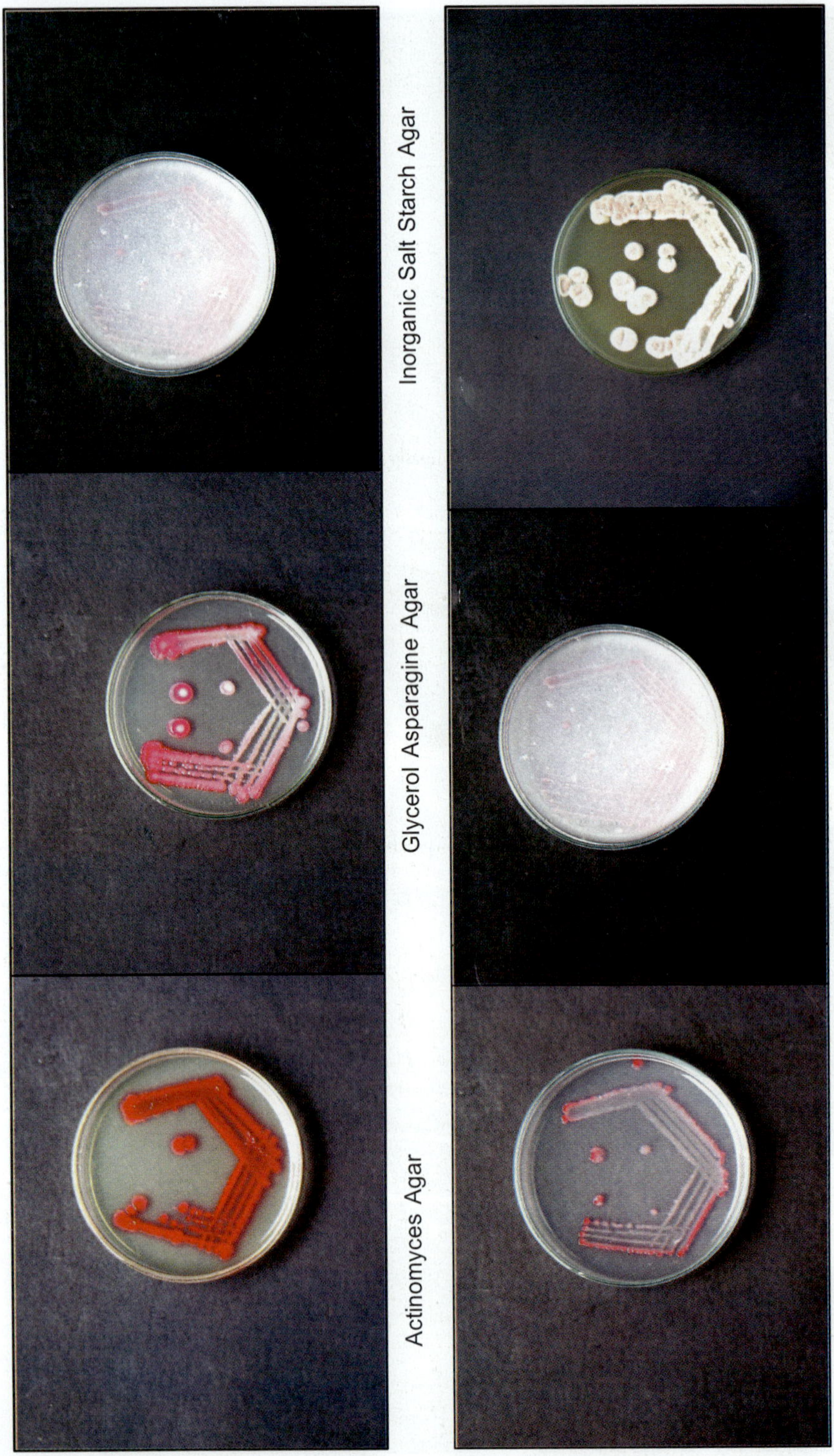

Figure 25.26: Aerial Mycelial Colors of *Streptomyces AQB.SIR.19* in Six Different Media

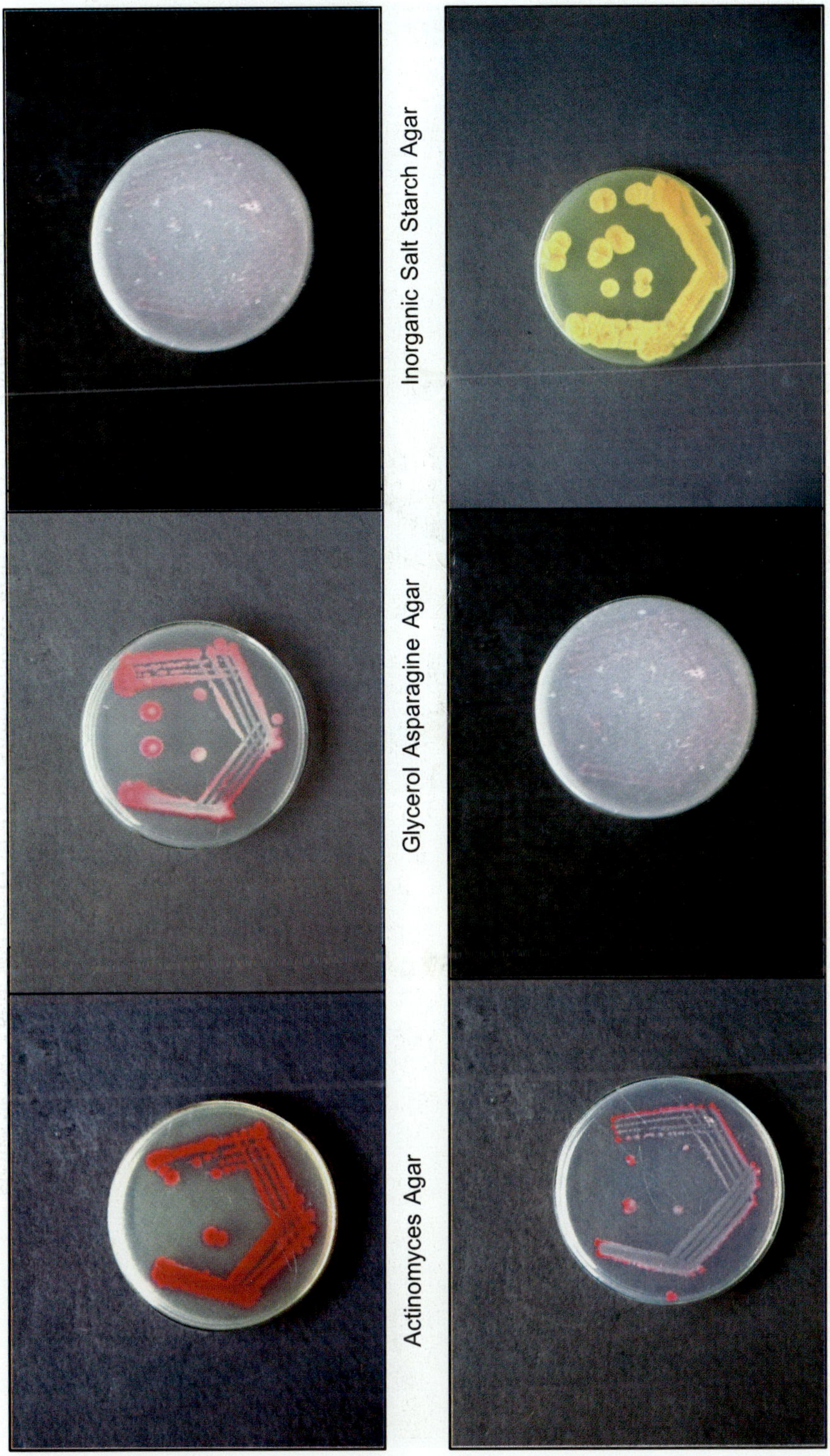

Figure 25.27: Substrate Mycelial Colors of *Streptomyces AQB.SIR.19* in Six Different Media

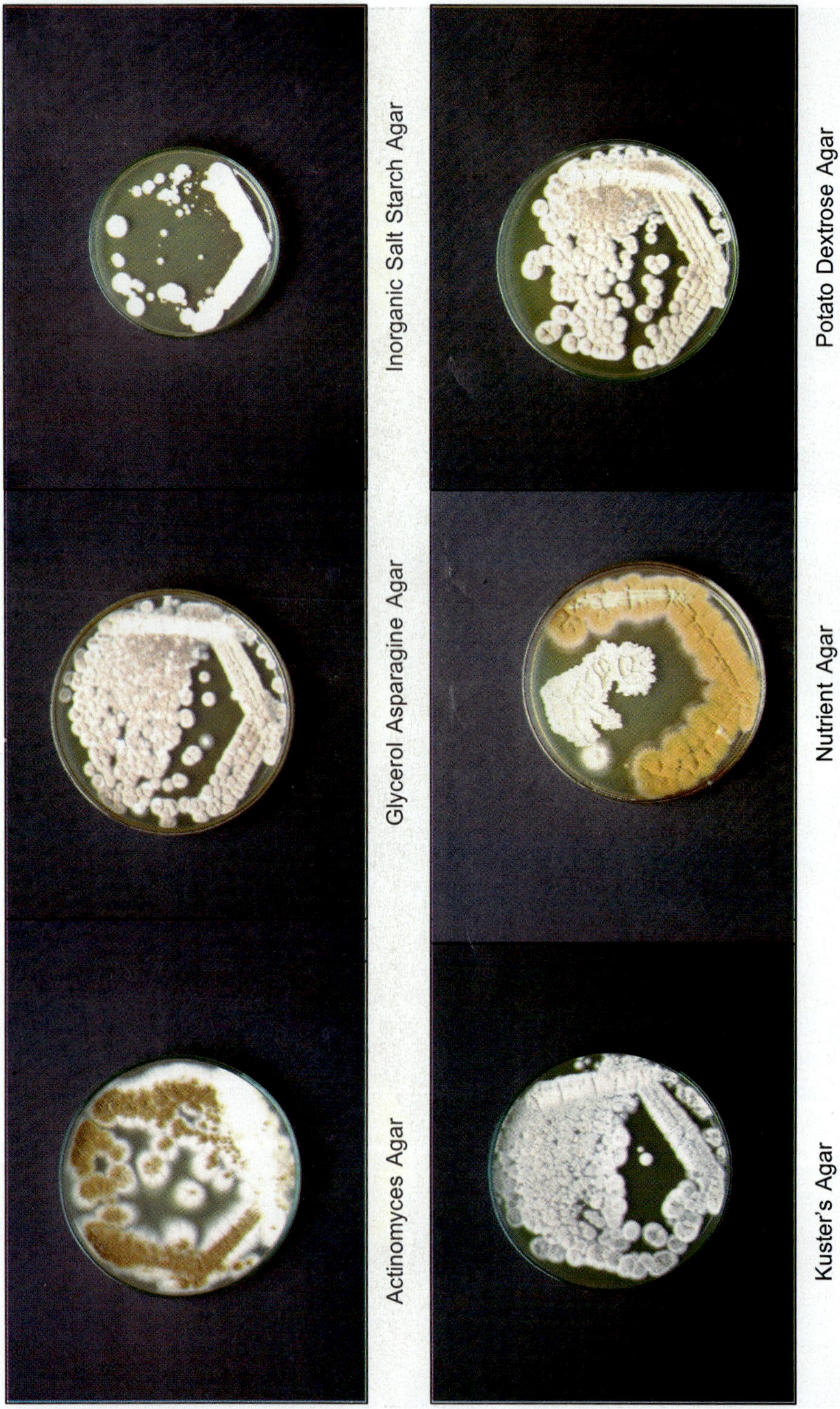

Figure 25.28: Aerial Mycelial Colors of *Streptomyces AQB.UFA.20* in Six Different Media

Figure 25.29: Substrate Mycelial Colors of *Streptomyces AQB.UFA.20* in Six Different Media

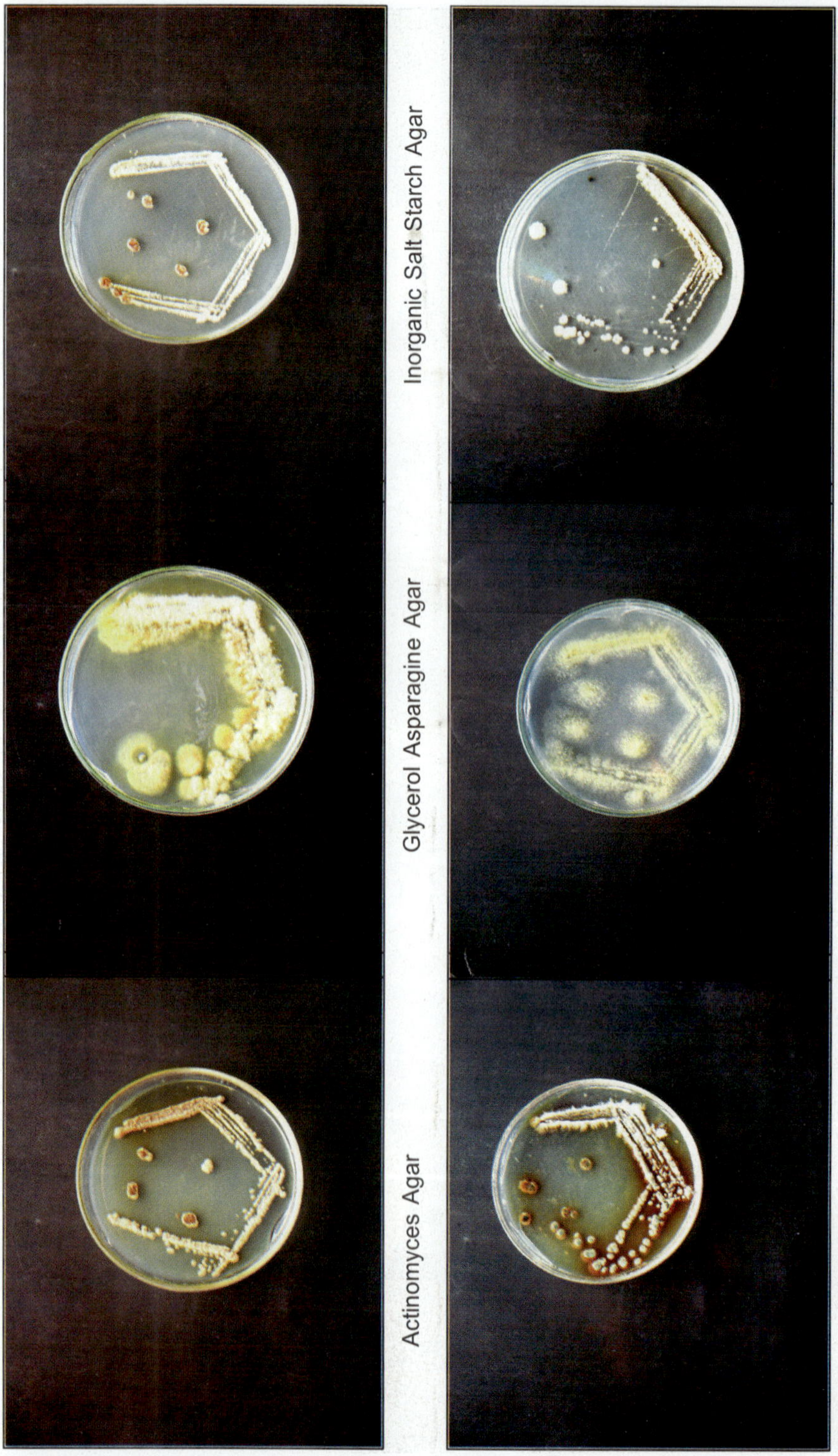

Figure 25.30: Aerial Mycelial Colors of *Streptomyces AQB.VPN.21* in Six Different Media

Figure 25.31: Substrate Mycelial Colors of *Streptomyces AQB.VPN.21* in Six Different Media

Figure 25.32: Photosynthetic Bacteria

Mass Culture of Photosynthetic Bacteria

Purple sulfur bacteria

Green sulfur bacteria

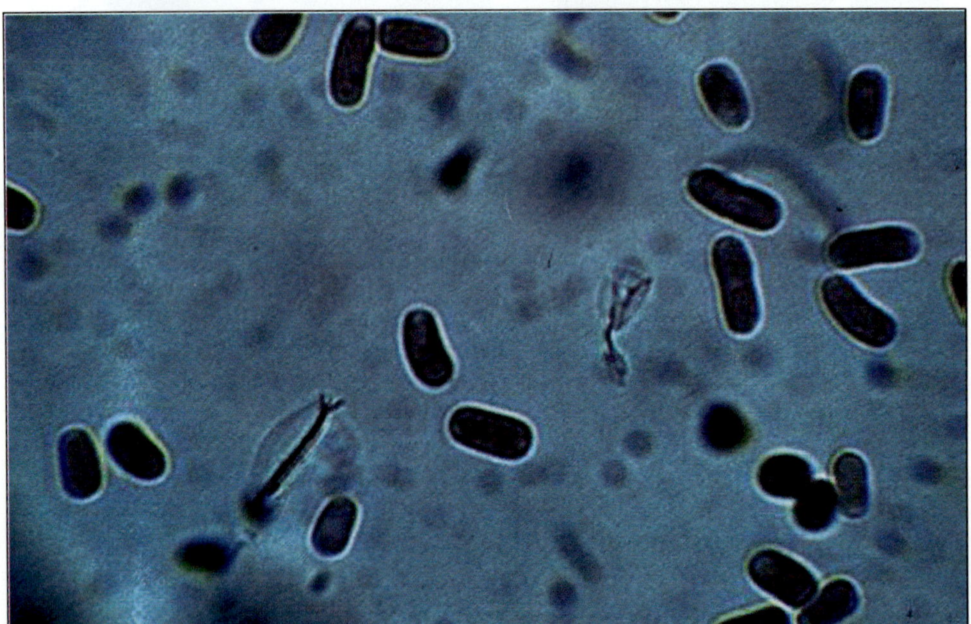

Purple Sulphur Bacteria (*Chromatium* spp), Cell size – 4.5µm

Figure 25.32–*Contd...*

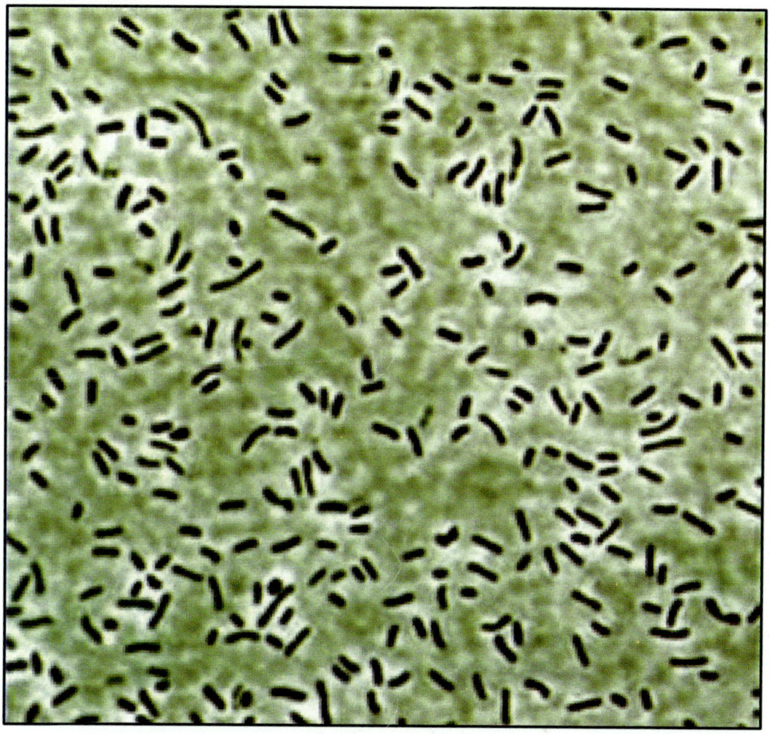

Green Sulphur Bacteria (*Chlorobium* spp), Cell size – 0.8 μm

Chapter 26
Microscope

Introduction

Types of Microscope

1. Light (or optical) microscope
2. Electron microscope

Light Microscope Includes

1. Brightfield
2. Darkfield
3. Fluorescence and
4. Phase contrast (Figure 26.4)

In light microscope magnification is obtained by a system of optical lenses using light waves.In electron microscope, use of beam of electrons instead of light to produce images. There are two types:

1. Transmission electron microscope
2. Scanning electron microscope and images (Figures 26.7 and 26.8)

Bright Field

In this microscopy, the microscopic field is brightly lighted and the microorganisms appear dark (Figure 26.1).

Once microbes stained, they absorb much light, resulting great contrast and colour differentiation. The optical part of bright field microscope and the path of the light rays follow are to produce enlargement or magnification. Generally the

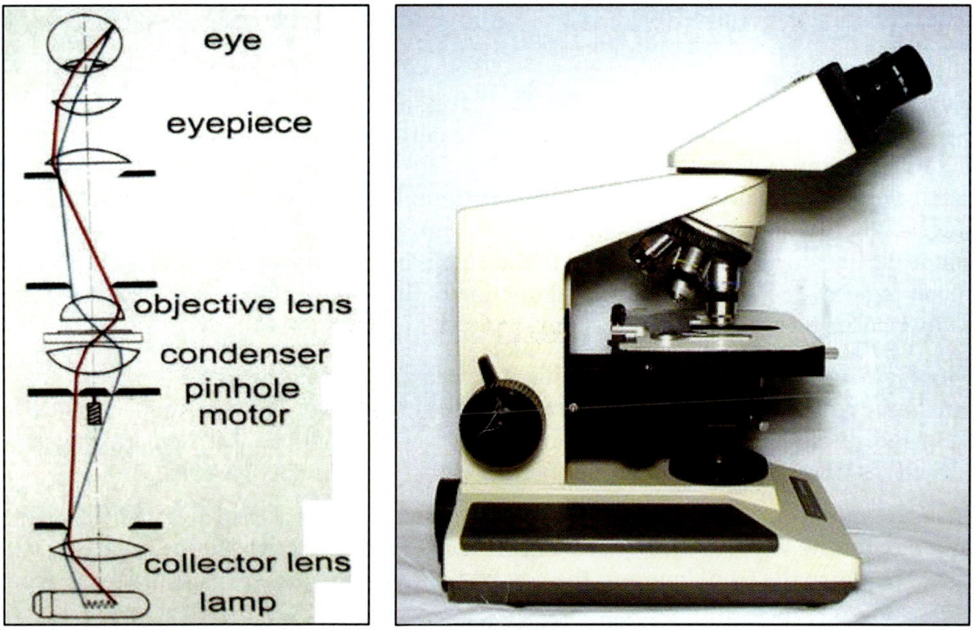

Bright Field Microscope

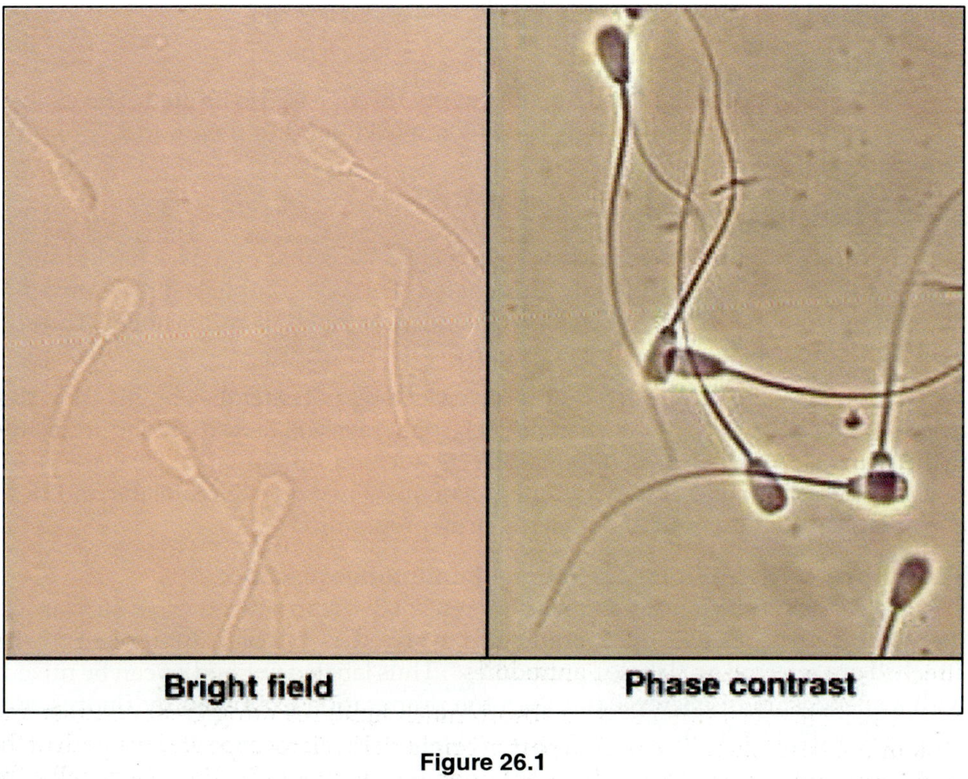

Figure 26.1

magnification is about x1000 to x2000. Above this level the image becomes fuzzy (unclear).

Resolving Power

The basic limitation of the bright-field microscope is one not of magnification but of resolving power, the ability to distinguish two adjacent points as distinct and separate. Without any distinction of two different structures, mere magnification is not beneficial. The more lines or dots per unit area that can be seen distinctly as separate lines or dots, the greater is the resolving power of the microscope. The resolving power of a microscope is a function of the wavelength of light used and the numerical aperture of the lens system.

The Relationship between Angular Aperture and Resolution

A narrow cone of light enters the low-power objective. The total angle is 640; θ is 32 –Half angle aperture and the numerical aperture is 0.6. With oil- immersion objectives the size of the cone light increased to 1160; θ is 580. N.A. is 1.3.

The magnitude of the angle is expressed as a sine value of the half aperture angle. The sine value is multiplied by refractive index 'n' of the medium filling the space between the front lens and the cover slip gives the numerical aperture (NA).

$$NA = n \sin \theta$$

With dry objective the value of 'n' is 1, since 1 is the refractive index if air. When immersion oil is used as the medium, 'n' is 1.56 and if θ is 580.

$$NA = n \sin \theta = 1.56 \times \sin 58 = 1.56 \times 0.05 = 1.33$$

The wavelength of light used in optical microscope is also limited. The visible range is between 400nm (blue light) cod700nm (red light) or 0.4μm to 0.7μm nm-ronometer and is equal to 0.001μm or 10-9m.

Dark Field Microscope

The effect produced by the dark field technique is that of a dark background against which objects are brilliantly illuminated. This is accomplished by equipping the light microscope with a special kind of condenser that transmits a hollow cone of light from source of illumination (Figure 26.2).

Most of the light directed through condenser does not enter the objective and the field is essentially dark. However some of the light rays will be scattered (diffracted) if the transparent medium contains objects such as microbial cells. Dark field microscope is valuable for the examination of unstained microorganisms (motile – live microbes) in fluid mart and hanging drop preparation

The Fluorescent Antibody Technique–Immunofluorescence

It is possible to chemically combine fluorescent dye with antibodies. *i.e.* substances that combine with specific microorganisms. Antibodies to which fluorescent dye is attached are referred as "labeled antibodies". Thus labeled antibodies can be mixed with a suspension of bacteria and then the preparation examined by fluorescent microscopy (Figure 26.3).

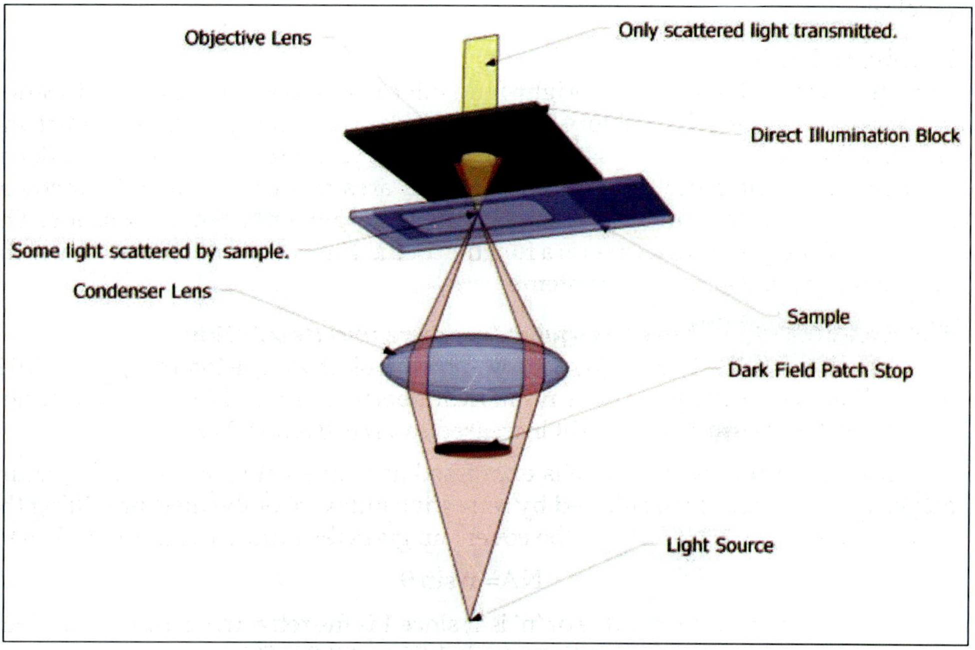

Figure 26.2

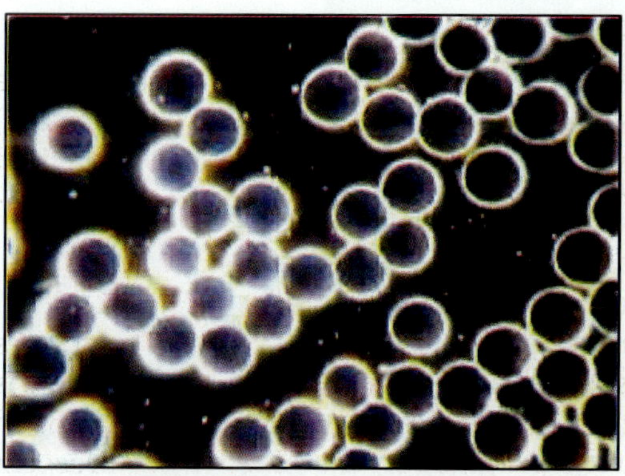

Human Blood Cell Bacteria

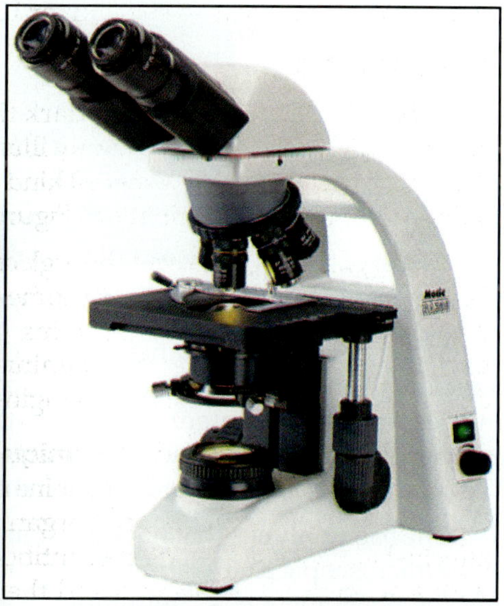

Dark Field Microscope

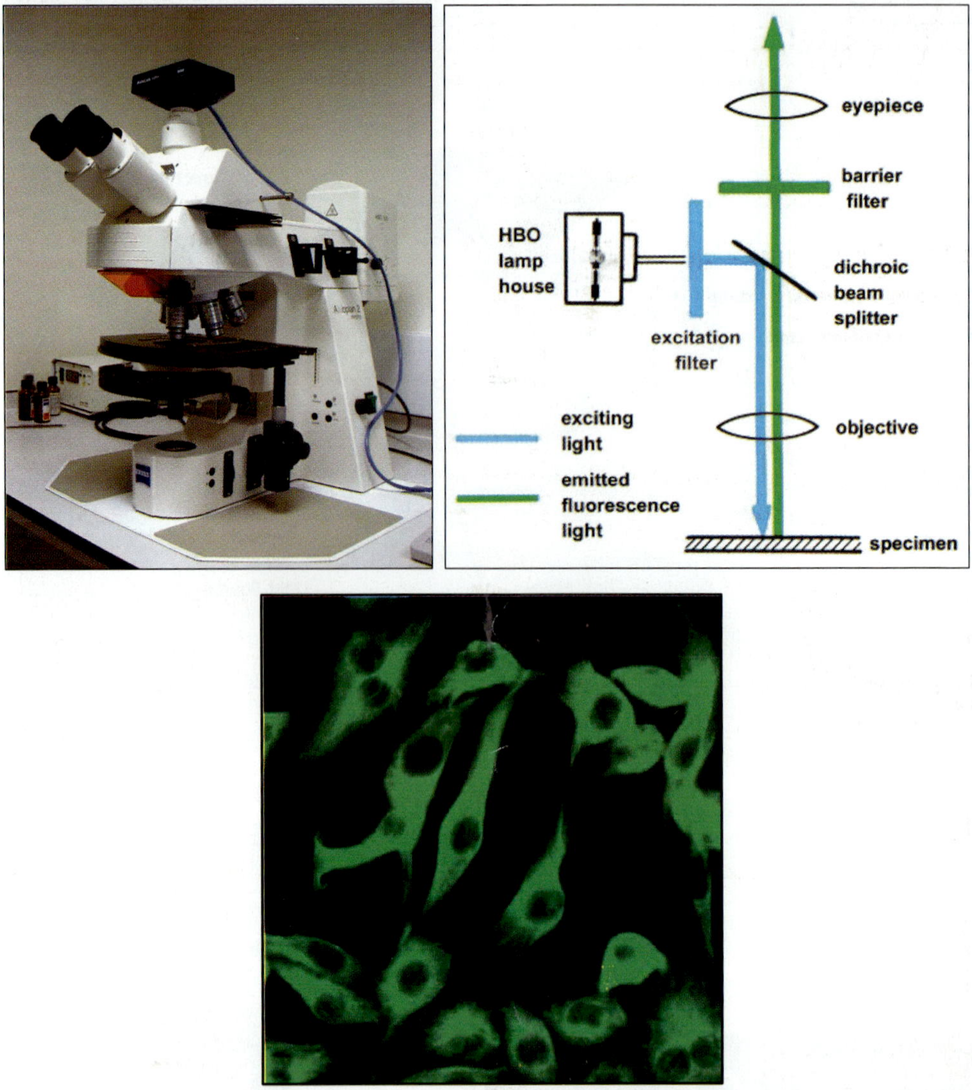

Figure 26.3: Fluorescent Microscope

Many chemical substances absorb light. After absorbing light of a particular wavelength and energy, some substance will then emit light of a longer wave length and lesser energy content. Such substances are called fluorescent. In practice, microorganisms are stained with fluorescent dye and illuminated with a blue light. The blue light is absorbed and green light emitted by the dye.

The bacterial cells that have combined with the labeled antibody will be visible in the microscopic preparation. This procedure is "fluorescent antibody technique" and the phenomenon is "immunofluorescence". It is possible to identify a "single microbialcell".

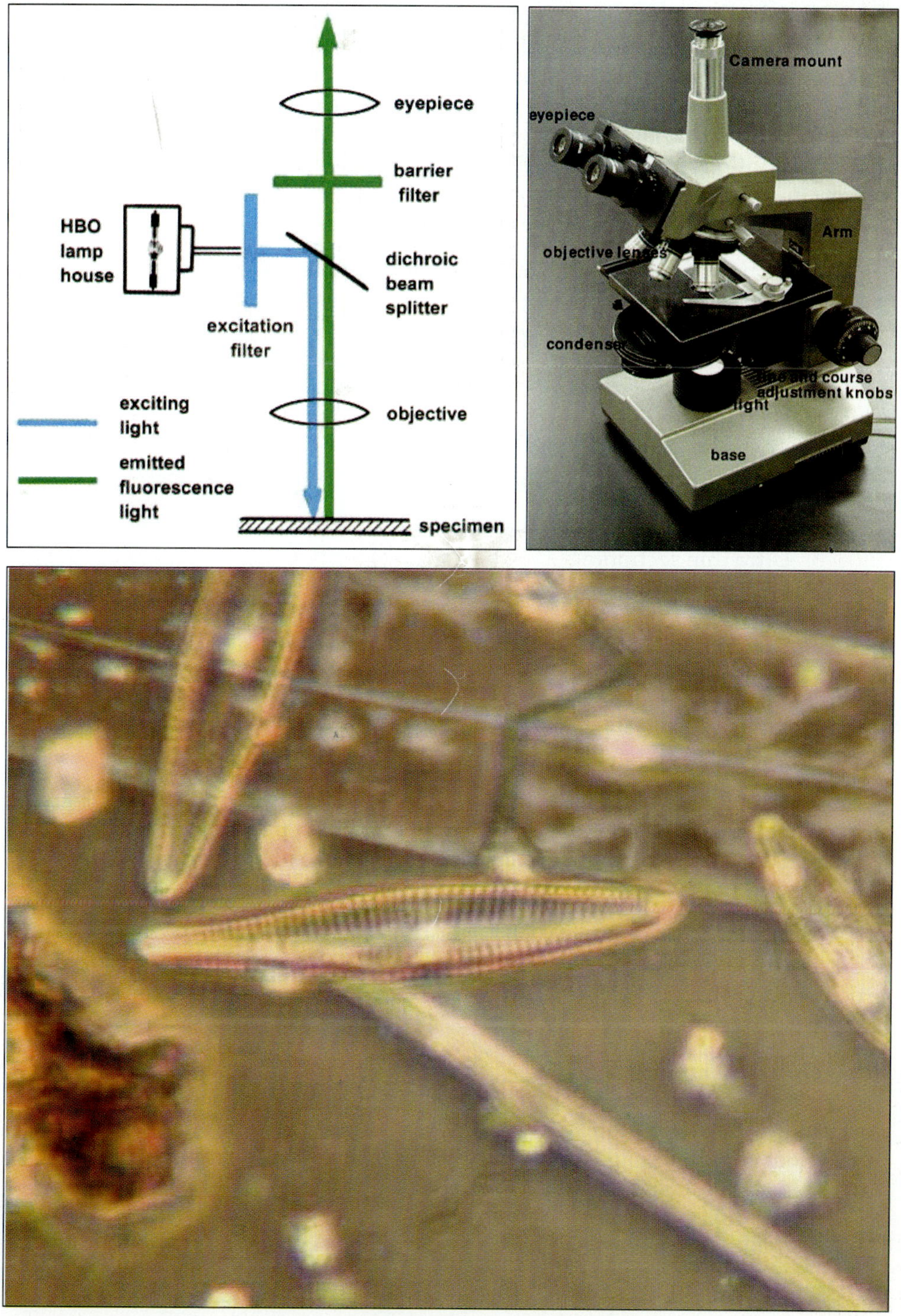

Figure 26.4: Phase-contrast Microscope

It is valuable for unstained living cells and within cells the structural difference noticed. In principle, the technique is based on the fact that light passing through one material and into another material of a slightly different refractive index and thickness will undergo a change in phase. These differences in phase or wave front irregularities are translated into variations in brightness of the structure and hence are detectable by the eye. It is possible to reveal differences in cells and their structures by phase contrast microscope.

Electron Microscope

Electron microscope (Figure 26.5) uses electron beams and magnetic fields to produce the image, whereas the light microscope uses light waves and glass lenses. 60 to 80kv electrons having the wavelength of 0.05A. A-Arogstrom is 1/10000µm or 1/10-8µm. The resolution power of electron microscope is 100times more than that of the light microscope, having magnification up to x400000. Numerous techniques are available for use with electron microscopy, useful for characterization of cellular structure. They are:

1. Shadow casting (Figure 26.6B)
2. Negative staining (Figure 26.6A)
3. Ultra thin sectioning
4. Freeze etching
5. Localization of cell constituents
6. Localization of enzymes in thin-sections.

Autoradiography

Shadow Casting

This technique involves depositing an extremely thin layer of metal (e.g. Platinum) at an oblige angle on the organism so that the organism produces a shadow on the uncoated side. It gives the topographical representation of the surface of the specimen.

Negative Staining

An electron dense material such as phosphotungstic acid can be used as a 'stain' to out line the objects. The electron opaque phosphotingstate does not penetrate structure but forms thick deposits in crevices. Fine structure of viruses and flagella of bacteria can be seen.

Ultra Thin Sectioning

For observing intracellular structure the material must be thin. Bacterial cells can be embedded in a plastic material and this can be made into ultra thin slices to level of 60nm. These slices can be used for microscopic examination. Electron microscope stain such as uranium and lanthanum salts.

Freeze-Etching

Freeze-etching was developed to prepare sections of the specimens. The specimen is sectioned while contained in a frozen-block. Carbon replicas of these exposed surfaces are then prepared which reveal internal structure of the cells.

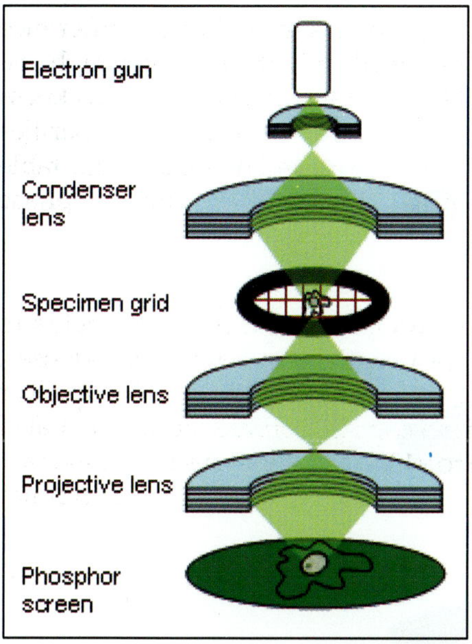

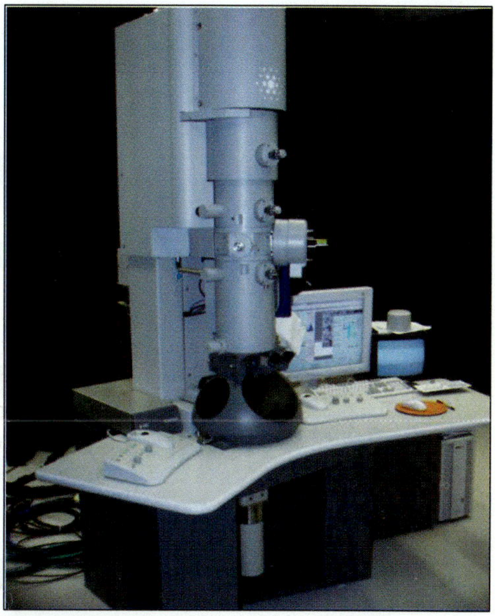

Figure 26.5: Electron Microscope

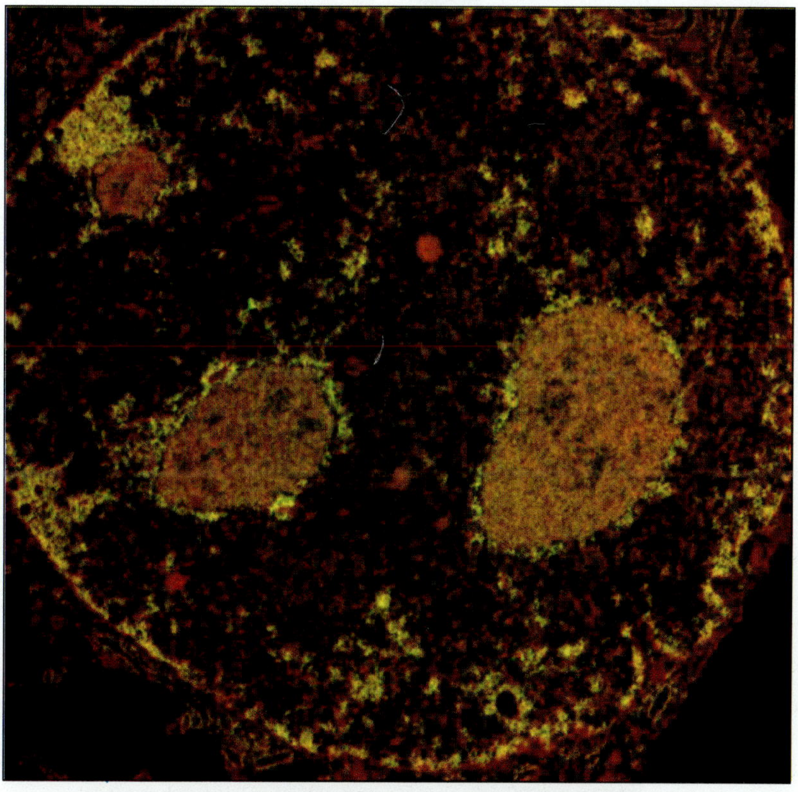

Figure 26.6A: Negative Staining Image Shadow Casting

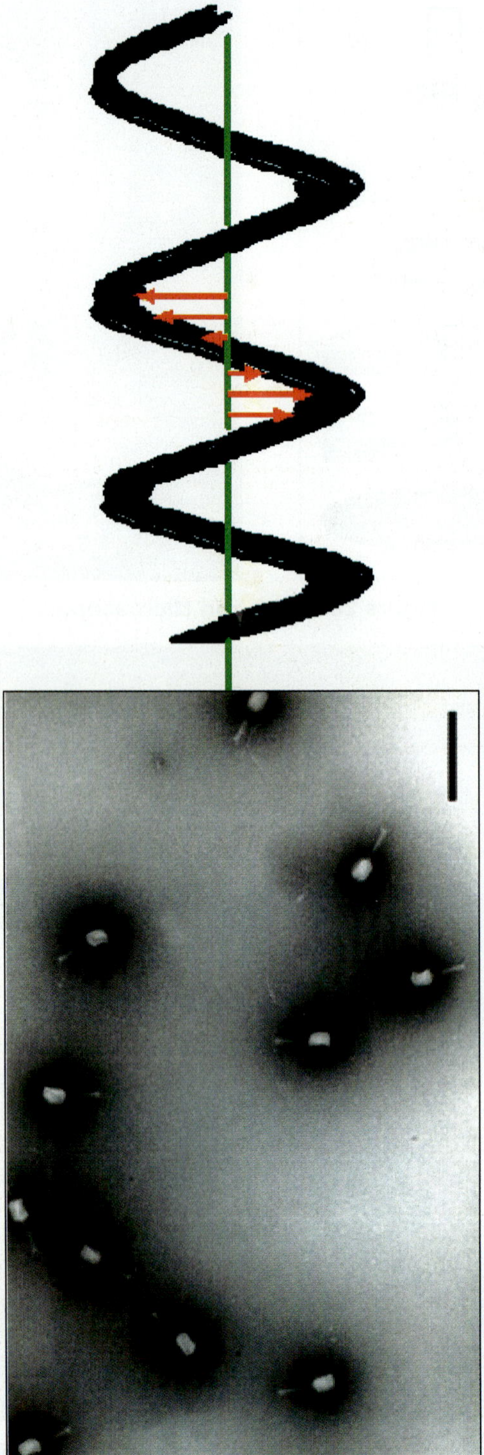

Figure 26.6B

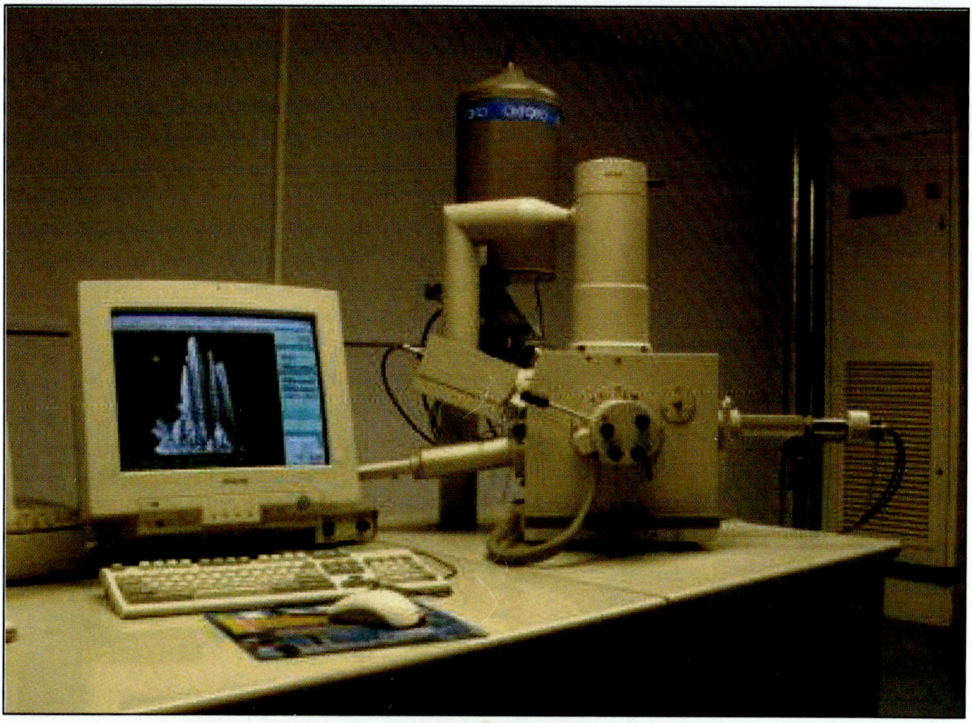

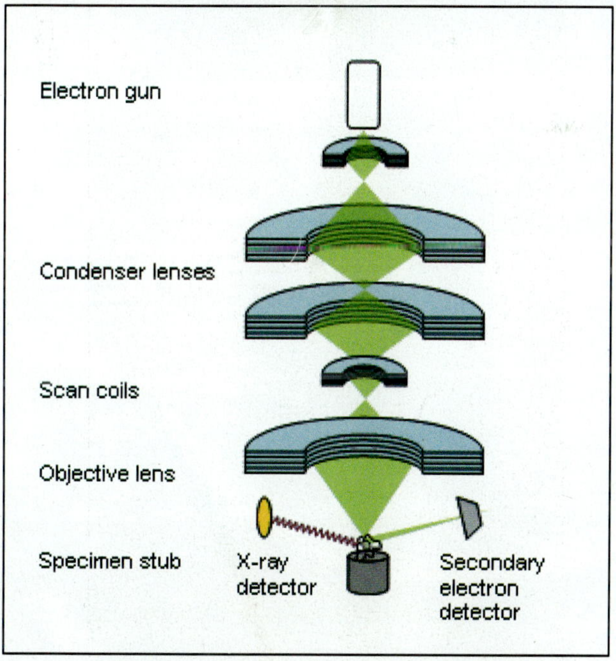

Figure 26.7: Scanning Electron Microscope

Figure 26.8: SEM Images

Localization of Cell Constituents

Thin sections of cell can be treated with ferritin-labeled antibody.Ferritin is an iron-containing substances of high density that affects markedly passage of the electron beam. The combination of ferritin-labeled antibody with antigen in the cell produce complex which manifests a higher contrast in the electron-microscope image.

Localization of Enzymes in Thin Sections

Electron microscope techniques have been developed to locate the position of enzymes within the cells.The intracellular localization of the enzyme isocitrate dehydrogenase of *Escherichia coli*. This was accomplished by first preparing ultra thin section of the bacterial cells, followed by an imno-chemical technique by which colloidal gold is affixed specifically to the isocitrate dehydrogenase enzyme.

Autoradiography

It is a cyto-chemical method in which the location of a particular chemical constituent in a specimen is determined by observing the site at which radioactive materials become positioned. The cells are first exposed to the radioactive substance to permit its uptake. In practice, the specimens prepared for microscopic examination is covered with a layer of photographic emulsion and stored in the dark for a period of time. The ionising radiation emitted during the decay of radioactive substance produces latent images in the emulsion and after photographic processing, the developed images is seen as grains of silver in the preparation (Figure 26.6).

Scanning Electron Microscopy

In this the specimen is subjected to a narrow beam which rapidly moves over (scan) the surface of the specimen (Figure 26.8). This causes the release of a shower of secondary electrons and other types of radiation from the specimen surface. The secondary electrons are collected by a detector which generates an electron signal.

Limitations of Electron Microscopy

1. Specimens examined under high vacuum and not in a living condition.
2. Drying procedures alter the morphology of organism.
3. Low penetration power of electron beam.

Types of Microscopy	*Maximum Magnification*	*Appearance of Specimen*	*Applications*
Bright -field	X1000-2000	Bacteria generally stained	Morphological features of bacteria, fungi, yeast, algae and protozoa
Dark-field	X1000-2000	Generally unstained	For microorganisms in living conditions
Fluorescent	X1000-2000	Bright and coloured: colour of the fluorescent dye	Diagnostic technique to identify organism
Phase contrast	X1000-2000	Varying degree of darkness	For examination of cellular structure of protozoa, yeast, algae and some bacteria
Electron	X200000-400000	Viewed on fluorescent screen	Examination of virus and ultra structure of microbial cells

Chapter 27
Immunostimulatory Effect on Fishes

Introduction

Aquatic ecosystems are well utilized for aquaculture in recent decades. The use of unutilized resources and modern technology hiked aquaculture production. However, the number of outbreaks of bacterial diseases in cultured fish has also increased and it causes decline in aquaculture production. Fish health is closely associated with characteristics of aquatic environment and the microbes present. The intimate contact with water which, eases the movement of chemicals into and through the mucus coated skin, and other external layers become a disadvantage to the fish when nefarious chemicals, pollutants, and contaminants enter the aquatic ecosystem (Adams, 1990, Wester *et al.*, 1994). These chemicals can have adverse effects on the fish's physiological pathways including the important mechanisms that help to protect the fish against diseases *viz* the nonspecific and specific immune responses. Without complete immune protection and barriers against viral, bacterial, fungal and protozoan pathogens that are ubiquitous in the environment, the animal is disadvantaged and become susceptible to disease-causing agents. In addition, during their whole life, farmed fishes are inevitably subjected to various stresses like handling, transportation, crowding, infections, exposure to pollutants and physiological changes that may lead to immuno-suppression and consequent infections. Even though vaccines are available in developed countries against a few diseases, aquaculture still experiences high loss of fish stocks due to outbreak of diseases. Part of this is because even efficient vaccines lose their effect one year after vaccination, and new pathogens are constantly gaining territory (Robertsen *et al.*, 1994). Further, it seems unlikely that cultured fish can be vaccinated against all potential diseases. Hence, the significance of a suitable client in preserving the health of living organisms is widely recognized(Lygren *et al.*, 1998).

In recent years, the increasing consumer concern about the residues of antibiotics, hormones, growth promoters, and the danger of development of antibiotic resistant strains has led to the use of immunostimulants in aquaculture. By definition, immunostimulants are substances that can enhance the nonspecific defense mechanisms as well as specific immune response if the treatment is followed by infection or vaccination (Anderson, 1992a).Many natural and synthetic substances have been reported that potentiate the fish immune system and increase disease resistance (Klontz and Anderson, 1970, Amend and Flender, 1976, Munn and Trust, 1983, Siegel *et al.*, 1983, Blazer and Wolke, 1984, Oliver *et al.*, 1985, Cossarini, 1985 Azuma and Jolles, 1987, Grayson *et al.*, 1987, Kitao *et al.*, 1987, Yano *et al.*, 1993; Anderson *et al.*, 1989, Kijita *et al.*, 1990, Siwiki *et al.*, 1990, Anderson, 1992a, Logambal, 1996, Rajkumar, 1996, Jayameena, 1997, Hemapriya, 1997, Venkatalakshmi and Michael, 2000, Logambal *et al.*, 2000, and Sudhakaran *et al.*, 2006). The search for new immunostimulants continues as an attempt to improve intensive fish farming. According to Rodriguez (2004) such new products should posses two characteristics.

1. Provide general stimulation and be
2. Economically affordable.

These two characters are well fulfilled by whole microorganisms.

1. First, they are rich sources of immunostimulant substances such as β-glucans, chitin, vitamins, genetic material, etc. At the same time, they act as a source of nutrients and micronutrients that affect the general fish physiology.
2. They are cheap sources of immunostimulants. New strains can be generated by genetically manipulating strains with a high content of specific substances.

This concept of using beneficial whole microorganisms is termed as "**probiotics**"and is gaining importance, as an eco-friendly disease management tool. In addition to disease resistance, they also improve water quality and growth of farmed fish. Probiotics are defined as microbial dietary adjuvants that beneficially affect the host physiology by modulating mucosal and systemic immunity as well as improving nutritional and microbial balance in the intestinal tract (Fuller, 1989). Beneficial bacteria in the best cases could be used to substitute the use of antibiotics as preventive agents of disease (Nikoskalainen *et al.*, 2001) and as growth promoters (Byun *et al.*, 1997). Immunomodulation by probiotics have been well documented in mammals including man (Gill; 1998; Kato *et al.*, 1984; Kato *et al.*, 1999; Ogawa *et al.*, 2001; Malin *et al.*, 1996). The immunomodulatory role of microbes as far as we know, still needs elaborate investigations in fish. Recently, there are a few studies that explore the immunomodulatory role of probiotics in fish. The non specific immunostimulation and colonizing efficiency in gut, skin, mucous of *Lactobacillus rhamnosus* were studied in rainbow trout (Nikoskalainen, *et al.*, 2003) and in turbot (Villamil *et al.*, 2002). The probiotic yeast cells *Saccharomyces cerevisiae* was recorded for its immunostimulatory activities in rainbow trout (Ortuno *et al.*, 2002) and in gilt head seabream (Rodriguez *et al.*, 2004). A mutant fungal probiotic *Mucor circinelloides* was demonstrated to be more effective in immunostimulation than the wild strains in sea bream (Rodriguez *et al.*, 2004).

Immunologists are working for a long period in the field of immunomodulators, including immunosuppressors and stimulators, on the immune protection in fish, and their putative points of action(Anderson, 1996). Many of the assays for detecting the changes in the protective mechanisms of fish due to immunomodulation are derived from those used in fish disease diagnostics and immunization programs. Study of specific immune response may include measuring the levels of humoral antibody with tests such as ELISA (Arkoosh and Kaattari, 1987), complement fixation test (Ingram, 1990) and microtiter agglutination (Roberson, 1990). For these assays, an immunization regimen in required to induce the specific antibody and specific cellular responses. Teleost fishes were shown to exhibit humoral immune response to a variety of antigens (Sailendri and Muthukaruppan, 1975; Sailendri, 1973, Michael and Pricilla, 1994; Rajavarthini and Michael, 1996; Sudan and Michael, 1995; Mohan, 1997; Ali, 1988; Jayaraman *et al.*, 1979) and has been used for studying a wide variety of immunomodulators (Anderson, 1996; Rajkumar, 1996; Logambal, 1996; Venkatalaksmi and Michael, 2000; Binu Ramesh *et al.*, 2006; prabhakar *et al.*, 2006).

The assays for testing changes in the non specific defense system may require more background training and insight of the technician because they often require more detailed microscope work or sophisticated equipments to record the chemical changes. (Anderson, 1996). There are many immunological assays for analysis of non specific defense mechanisms like NBT assay (Nitroblue Tetrazolium Assay) (Anderson *et al.*, 1992b), phagocytic assay (Sovenyi Kusuda, 1987) lysozyme assays (Ellis, 1992) chemiluminescent response, pinocytosis assay and chemotaxis assay (Mathews *et al.*, 1992). These assays of macrophage function have been teste on several species of fish and the results indicate that such methods can be applied as bioindicators of fish health (Warinner *et al.*, 1998; Weeks and Warinner, 1984; Weeks *et al.*, 1986; 1987, Venkatalakshmi and Michael, 2000; and Logambal *et al.*, 2003).

An initial step in many immunological studies is to collect and examine the leukocytes (Rowley, 1990). According to Anderson, (1996) though the hematological assays are generally less sensitive as bioindicators of immunomodulation, they are easy to do and technical training and laboratory requirements are minimal.

Knittel (1981) suggested that the ultimate test for modulation of the immune system by a substance is to show increased resistance to disease by comparing the test animals to control fish. The use of probiotics to displace pathogens by competitive processes or by release of growth inhibitors is now gaining acceptance in the animal industry as a better, cheaper and more effective remedy than administering antibiotics to promote health of animals (Westerdahl *et al.*, 1991). The increasing concern about the danger of development of antibiotics resistant strains and residues of antibiotics have lead to the use of biological or probiotic feed additives in the animal feeds (Verschuere *et al.*, 2000a). For most of the applications, it is recommended to use probiotics strains that are able to survive the gastro intestinal passage and have the capacity to adhere to the host intestinal tissues in order to prolong their health effects (Fuller, 1986,1989; Havenaar *et al.*, 1992; Ouwehand *et. al.*, 1999). While most of the probiotics are completely excreted a few days after ingestion (Joborn, 1998; Marteau and Vesa, 1998; Robertson *et al.*, 2000; Nikoskalainen *et al.*, 2003), only a transient colonization (Nikoskalainen *et al.*, 2003; Joborm, 1998 and Gatesoupe, 1999) or in *vitro* adhesion in the gut (Villamil *et al.*, 2002) was shown to the best of our knowledge. Of course, the fungal probiotics *Debaryomyces hansenii* and *Rhodotorula glutinis* were shown to inhabit the gut of *Onchorynchus mykiss* juveniles for 30 days and 65 days respectively (Andlid *et al.*, 1995). The bacterial probiotics, which are well studied and widely used in aquaculture, belong to the genus *Lactobacillus* has good records of so many probiotic characters. However, no observations are available for prolonged gut colonization of *Lactobacillus*.

Probiotics are described as live microbial feed supplements (Fuller, 1989) or as water additives (Moriarity, 1998). Gatesoupe (1999) insisted that microbial cells that are administered in such a way to enter the gastro intestinal tract and to be kept alive with the aim of improving health, alone can be designated as probiotics. It is argued that even if a probiotic is not able to colonize the gut, it should be able to adhere to the aquatic animal associated tissues like mucous or skin. If a probiotic fails to colonize, a regular administration of the same is required to obtain its effects (Marteau and Vesa, 1998). To overcome this problem, in the present study, an attempt is made to utilize all the benefits of probiotics without intestinal colonization using immobilization technology. Recent researches on immobilized cell (IC) technology applied to probiotic cultures have emphasized the importance of, and interest in this new technology. In dairy industry IC technology has been shown to offer many advantages for biomass and metabolite production compared to free cell systems such as high cell density and very high volumetric productivity (Doleyres *et al.*, 2004a), reuse of biocatalysts, high process stability (physical and biological) over long fermentation periods (Lamboley *et al.*, 1997) and retention of plasmid bearing cells (D-Angio *et al*, 1994; Huang *et al.*, 1996). Improved resistance to contamination, uncoupling of biomass and metabolite production, stimulation of production and secretion of secondary metabolites and physical and chemical protection of cells are also recorded (Doleyres *et al.*, 2004; Doleyres and Lacroix, 2004 and Lacroix *et al.*, 2004). However, these benefits have not yet been exploited for the aquaculture industry.

A simpler indicator for assessing the pressure of any chemical or biological agent on fish is the growth. Growth and food utilization have significant relationship with fish health and aquaculture production. The various dietary components like Protein (Akiyama *et al.*, 1981) Lipid (Bromely, 1980) Fatty acids (Paul *et al.*, 1996) Carbohydrates (Bergot, 1979; Beamish *et al.*, 1986) as well as quality of diet (Ebanasar and Jayaprakas, 1995) have been found to contribute significantly to growth and biochemical composition of fishes. In addition to this, feed additives like vitamins, (Naik *et al.*, 1999; Murthy and Naik, 2000, 2002; Bhanja *et al.*, 2001; Muruganandham *et al.*, 2003, 2004a and b), hormones (Nirmala and Pandian, 1983; Arul, 1986; Jayaprakas and Sindhu 1994) and growth promoters (Unnikrishnan, 1995; Sambhu and Jayaprakas, 2001; Abraham *et al.*, 2001) and herbal products (Eswari, 2004) also have significant role on growth and food utilization of fishes. An understanding of the influence of any water additives or feed supplements to be used in aquaculture, on growth and food utilization of fish is essential for obtaining high production potential.

Aeromonas hydrophila, is a causative agent of one of the major fish bacterial diseases in India. A wide antigenic variety of motile *Aeromonas* markedly limits the development of vaccines (Newmann, 1993, Stevenson, 1998). However vaccines to protect farmed fish against A. *hydrophila* are not commercially available to date (Anbarasu *et al.*, 1998). Pretreatment with glucan before vaccination increased the antibody production (Selvaraj *et al.*, 2005). Similarly Venkatalakshmi and Michael (2001) and Logambal *et al.* (2000) showed the potential of *Ocimum sanctum* leaf extracts in stimulating the immune response of vaccinated tilapia against *A. hydrophila*. Five candidate probiotics isolated from clown fish was shown to compete with *A.hydrophila* and *V.alginolyticus* reducing pathogen attachment (Vine *et al.*, 2004). Probiotics, along with herbal treatment significantly boost recovery from *A. hydrophila* infection (Harikrishnan and Balasundram, 2005). Residual problems and development of resistant genes arising with the use of antibiotics, threaten the environment. In fact, *Aeromonas hydrophila* is reported to be resistant to almost all antibiotics. In this scenario, probiotics seems to be the better alternative.

Fish health affects many facets of our life. The production of high quality fish for food and wild stock enhancement, and the fact that fish health reflects the quality of our aquatic environments are some of the practical concerns in the field of fishery science research. Also, fish serves as a good model for studying the vertebrate immune system because of its relative simplicity and it has both wings of the immune response *i.e.* humoral and cell mediated responses (Iwana and Nakanishi, 1996). The increased intensification of aquaculture has led to a high number of disease outbreaks with an increasing range of pathogens. Consequently the extensive use of antibiotics has led to drug resistance problems. In order to improve health and welfare in the rearing of these animals, several alternatives such as improved husbandry, nutrition, and water quality, lower stocking densities, and use of vaccines, non specific immunostimulants and bacterial probiotics such as lactic acid bacteria have been proposed. In fish, as in other aquatic organisms, the probiotics were administered directly to the water or as feed additives or via live food such as rotifers and artemia. Such probiotics have been shown to increase the health status of fish, seemingly by colonizing the gut and acting as antagonistic to pathogens. Recently emphasis has been given to studies on the immunostimulatory effects of probiotics in fish. However, though it was proved that administration of bacteria improves the survival of certain fish and shell fish after challenge with pathogens, question still remains whether there is a direct stimulation of the immune system (Tannock 1997; Ringo and Vadstein 1998; Powels *et al.*, 1998; Gatesoupe, 1999).

In the present study a pioneering attempt was made to utilize the advantages of probiotics without and or with gut colonization. The effectiveness of the probiotics was assessed in terms of both specific and nonspecific immunity. *Lactobacillus sporogenes*, a widely used probiotic has been shown to induce antibody production in humans (Malin *et al.*, 1996; Ogawa *et al.*, 2001). However stimulation of specific antibody production by probiotics against a particular pathogen has not been reported in fish so far. Work on the effect of probiotics on specific immune response of fish is still lacking.

Specific Antibody Response

Specific antibody level is used as an index to measure the acquired immunity in *Oreochromis mossambicus*. Three experimental groups (n=6 in each group) of fish were treated with the probiotic bacterium, *Lactobacillus sporogenes* in three different concentrations *vis a vis* - 2.5 x10^4(T1), 5 x10 ^{4}(T2) and 1 x 10^5(T3) CFU/ml. A separate control group was maintained similar to the experimental groups but with out probiotic exposure. All four fish groups (control, T1, T2, T3) were immunized with heat killed *Aeromonas hydrophila*. Anti- *Aeromonas* antibody titer was assayed on 7th, 14th, 23rd, and 28th days of post immunization.

The present study clearly shows a significant enhancement of specific anti-*Aeromonas* antibodies in all three treatments groups that too in a dose dependent manner. This observation is the first of its kind. The results of the present study are in agreement with earlier observations. Nikoskalaine and his associates (2003) reported for serum total immunoglobulin production in trouts, and for immune responses to specific antigen in man (Malin *et al.*, 1996; Ogawa *et al.*, 2001).The specific antibody response are also known to be modulated by other immunomodulators in fish which include both immunosuppressors (Sudhan and Michael, 1995; Prabhakar *et al.*, 2006, Binu Ramesh *et al.*, 2006) andimmunostimulators like synthetic(Rajkumar, 1997; Gopalakrishnanand Arul, 2006) and herbal products (Venkatalakshmi and Michael, 2000; Logambal *et al.*, 200; Sudhar *et al.*, 2006). Though fish depend more heavily on the non-specific defense mechanisms, specific immunity also plays an important role (Anderson, 1992). Hence a study on the impact of probiotics on specific immune response is needed to assay the role of immunostimulants against a specific pathogen. While nonspecific immunity provides a quick and immediate response, specific immune mechanisms impart memory which lowers the risk or recurrent infection. The advances of piscine humoral immunity rapid and possess practical applications in the field of vaccinology (Kaattari and Pignanelli, 1996). In addition to the traditional synthetic and herbal immunostimulants, the use of probiotics forms a novel approach/method to stimulate fish immune response.

The kinetics of the antibody response was observed to be similar in both control and experimental groups enhanced by the probiotics bacteria. In the present study, the lower concentrations 2.5x10^4 and 5x10^4 and 1x10^5 cells/ml stimulated the specific antibody response significantly when gives as a "water additive". These probiotic concentrations seem to be much far less than used in earlier studies. Nikoskaleinen *et al.* (2003) have reported a stimulation of the total immunoglobulin production, by only two doses of probiotics *i.e* 7.9 x 10^4 and 2.8 x 10^8 CFU/g feed stimulated the response when given as feed supplement. The differences could be attributed to the route of administration or the species of the probiotics. In the present study the magnitude of the antibody response was maximum in fish groups subjected to the water additive method of administration. The reason could be that vegetative cells used were in the log phase of their growth. Hence, it took minimum time for proliferation and much more cells are in the ready state to secrete extra cellular products that have immunostimulatory effect or antimicrobial effect or stress relieving effect. Among fish groups that were exposed to probiotic supplemented feed, the lowest concentration (T1- $2.5x10^4$ CFU/gm) stimulated the antibody response to the least compared to control group fish. But the peak day of antibody response was advanced to 7th day of post immunization when compared to control group fish. When compared to control group, delay in peak day was observed in the T2 group. Since spores of *L. sporogenes* were used for feed supplementation, the delay in the kinetics and magnitude of the response could be attributed to the time taken for germination of spores and growth of the probiotic cells. The route method and conditions of exposure, were known to have direct impacts on the degree of immunity elicited and regionality of the response and such aspects of immunity are essential for designing delivery systems and the timing of immunization (Kaattari and Pignanelli, 1996). Recent studies on the various molecular metabolites of these probiotics support the immunomodulatory role of *Lactobacillus sp*. The oligonucleotides of *L.gasseri* were shown to induce B lymphocytes by enhancing the expression of the immune receptor CD86 (Kitazawa *et al.*, 2001). The purified chromosomal DNA from 12 strains of *Lactobacillus acidophilus* was shown to induce proliferation of splenic B-lymphocytes. Cloned and amplified DNA from *L.gassori* JCM 131 induced B lymphocyte mitogenic activities (Kitazawa *et al.*, 2001). Similarly dietary intake of *Lactobacillus rhamnosus* HN001 enhanced the production of both Th 1 and Th2 cytokines in antigen primed mice (Cross *et al.*, 2002). Enhanced antibody production in the present study could be correlated with its ability to stimulate the number of activated macrophages and there by enhancing antigen presentation (Logambal *et al.*, 2000) and the direct effect on the immunopoietic cells themselves. Based on the above observations, the potentiality of *Lactobacillus sp.*, in stimulating antibody response in the present study might be attributed to its mutagenic properties on T cells and B cells. Further, significant correlation was observed between antibody response and the number of lactobacilli colonized in gut, which was verified by mathematical modeling. Colonization of *Lactobacillus sporogenes* in gut could be correlated with increased antibody production. The present study also shows similar enhancement in both water and feed supplemented groups of fish.

The present study also shows similar enhancement in antibody responses even if the probiotic cells were not allowed either to leak into the aquatic environment or to enter the gut. *L. sporogenes* enhanced antibody response. The observed enhancement in the antibody response, in the absence of gut colonization could be attributed to that of the extra cellular products of these probiotics that were released/leaked into the aquatic environment and entered in to the body of the fish. Products from probiotics such as antibodies, organic acids, hydrogen peroxide (Ringo and Gatesoupe, 1998), siderophores (Gram and Melchiorsen, 1996) and bacteriocins (tenBrink *et al.*, 1994) are well known to be released into the aquatic environment. These extra cellular products from the probiotics also have profound influence on other aquatic organisms. However, studies are required to show, whether the probiotic cells as such have any direct/physical role to play. It remains to be answered whether the colonized probiotics act through these extra cellular products or they play a synergistic role. The present observation provides a scope for such studie in future. Based on earlier reports and present investigations, a probable sequence of events that lead to the stimulation of immune responses in fish by the probiotic is proposed.

Chapter 28

Streptomyces as Immunostimulant on Ornamental Fish, *Vieja synspilus* (Flower Horn)

Introduction

Immune-stimulants are chemical compounds that activate white blood cells (Leukocytes) and hence may render animals more resistant to infections by viruses, bacteria, fungi and parasites. Most immune-stimulants are chemical compounds which exist as structural elements of bacteria, mycelial fungi and yeasts. However, there are also a couple of purely synthetic compounds, originally made for other purposes, which have incidentally been found to possess immune-stimulating properties. Haemagglutination assay was done to asses the activity of antigen with antibody. Reactive oxygen species is done to evaluate if any free radical release when affected by pathogen. *Streptomyces* and photosynthetic bacteria were chosen because of high antibacterial activity and pigment production which act as antioxidant.

During evolution of animals their immune system has developed mechanisms to detect chemical structures which are typical for potentially dangerous micro-organisms and use those structures as «alarm signals» to switch on the defense against infections. In the presence of such chemical signals the immune system will respond as if challenged by a pathogenic microbe. Hence, administration of an immune-stimulant prior to an infection may elevate the defense barriers of the animal and thus provide protection against an otherwise severe or lethal infection.

Immune-stimulants may improve health and performance of farm animals, including fish and shrimp in aquaculture, if used prior to:

1. Situations known to result in stress and impaired general performance of animals (*e.g.* handling, change of temperature and environment, weaning of larvae to artificial feeds).
2. Expected increased exposure to pathogenic micro-organisms and parasites(*e.g.* spring and autumn blooms in the marine environment, high stocking density.
3. Developmental phases when animals are particularly susceptible to infectious agents (*e.g.* the larvae phase of shrimp and marine fish, smoltification in salmon and sexual maturation).

Immune-stimulants may act in synergy with antibiotics and thus strengthen the effect of curative medication. However, immune-stimulants are primarily prophylactic agents which should be used to elevate the general defense barrier of the organism and hence to reduce the risk of disease, not as a curative medicine. If used at an advanced stage of a disease development, an immune-stimulant may be recognized by the organism as an "apparent infection" in addition to an existing one, and it may therefore possibly aggravate disease symptoms, at least for a short while.

Fry and larval fish are particularly susceptible to microbial infections and are often subject to high mortality due to opportunistic pathogens in hatcheries. 50 young fish do not possess a mature specific immune system (Ellis, 1988) and depend on the nonspecific cellular defense mechanisms to be resistant to microbial infections (Trust, 1986). The use of immune-stimulants to reduce mortality of larval and juvenile fish in hatcheries is therefore another important application.

It has been shown that immune-stimulants may act in synergy with antibiotics in preventing infections in human patients (Smets *et al.*, 1988; de Felippe, *et al.*, 1993) and in fish (Thompson *et al.*, 1993). There is a sound biological basis for such a synergy; the aggressiveness of the infectious bacterium is lowered by the anti-microbial agent while the antibacterial mechanisms of the body itself are being stimulated. It is therefore worth consideration to use immune-stimulants in combination with curative anti-microbial agents at an early phase of disease development, or prior to anticipated disease outbreak.

The development of parasites which attach to skin and mucosal membranes are inhibited by complement factors, enzymes and enzyme inhibitors. These chemical defense weapons are produced by specialized cells in the surface tissues, and are continuously secreted. Certain immune-stimulants, *e.g.* b-1,3/1,6-glucans, stimulate these cells and thereby the defense against surface parasites. It has been reported that immune-stimulants are indeed effective against sea-lice on farmed salmon. It is not generally recognised that the immune system is one of the largest "organs" in the body of higher animals; in man it constitutes 3 per cent of the body weight. This organ is distributed all over the body and represents its "armed forces". The largest division of this defence is localized in the gut tissues, where specialized lymphoid organs (*Peyer's patches*), lymphoid nodules, and the whole range of different white blood cells are found. The gut is also equipped with special gateways (*M-cells*) for entrance of small particles and bacteria. This major lymphoid organ of the body may be activated or down-regulated by food and feed ingredients, by micro-organisms and by immune stimulants. Since signals are transmitted from the lymphoid organs in the gut to the rest of the body, the well-being of an animal, or human, is very sensitive to changes in the processes which go on in the digestive tract.

Fishes, on the other hand, have not developed histologically distinct and specialised lymphoid tissues such as the Peyer's patches of warm blooded animals, and special gateways for absorption of particles, like the M-cells. However, all the cell types responsible for a local immune response are present in the intestinal wall of fish (*e.g.* carp and tilapia), including immunoglobulin positive lymphocytes (B-cells), immunoglobulin negative lymphocytes (T-cells), and antigen presenting macrophages (Davina *et al.* 1980; Doggett and Harris,1987; Rombout *et al.*,1989and 1993). Like in warm blooded animals, the gut associated immune system of fishes seems to function in the absorption of macro-molecules from the intestine and in the production of antibodies in response to antigens (Hart *et al.*, 1988). In other words, immune cells in the gut of fish may function like mammalian M-cells in sampling macromolecules. By this assumption one may explain the fact that b-1.3/1.6-glucans, also when administered in the feed of fish, enhance their disease resistance, growth performance and the efficacy of injected vaccines.

By definition, an immunostimulant is a chemical, drug, stressor, or action that elevates the non-specific defense mechanism or the specific immune response (Anderson, 1992). This definition describes the complexity of the interaction between the immune system and its environment. The immune system can be found in any possible state between the normal state and the high "alert" state. Each individual is also likely to present a different state. Any environmental stress (Fevolden *et al.*, 1994), any substance present in the food will have an effect. This also explains the wide variety of immunostimulants in use today (Anderson, 1992; Secombes, 1994). The sole injection of a phosphate buffer will already trigger some defense mechanism. On the other extreme, the injection of a vaccine or bacterin together with an immunostimulant will probably give the highest response. A number of studies have dealt with the evaluation of well known immunostimulants such as polysaccharides, lipoproteins, lipopolysaccharides, etc. mainly derived from the cell wall of microorganisms.

Carotenoids stimulate the immune system, prevent cancer, and act as free radical scavengers (Ames 1983; Bendich and Olson 1989; Di Mascio *et al.*, 1991; Diplock 1991). These physiological functions had not been considered in the ecological literature and led to the alternative hypothesis that carotenoid-dependent ornaments actually indicate the bearer's immune condition and health status (Lozano 1994).

Carotenoid pigments are known to have antioxidant properties, scavenging potentially harmful free radicals and protecting cells and tissues (including the immune system) from oxidative damage (Lozano, 1994; Møller *et al.*, 2000; Krinsky 2001). Recent experimental studies have demonstrated the immunomodulatory action of carotenoids in several species of colourful birds and fish (McGraw and Ardia 2003; Saino *et al.*, 2003). Thus, carotenoid coloration stands as a model system in which to study the honesty reinforcing mechanisms of sexual trait expression; those individuals that accumulate the most carotenoids achieve superior health and communicate their carotenoid and health status to females directly with their bright colors.

Observations

Figure 28.1: Experimental Setup of Microbes (*Streptomyces* and Photosynthetic Bacteria) Incorporated Feed as the Immunostimulant

Figure 28.2–*Contd...*

Figure 28.2–*Contd...*

Figure 28.2–*Contd...*

Figure28. 2: Diseased Flower Horn Fish after Injected with *Aeromonas hydrophila*

Figure 28.3: Recovered Flower Horn After Fed with *Streptomyces* and Photosynthetic Bacteria Incorporated Feed

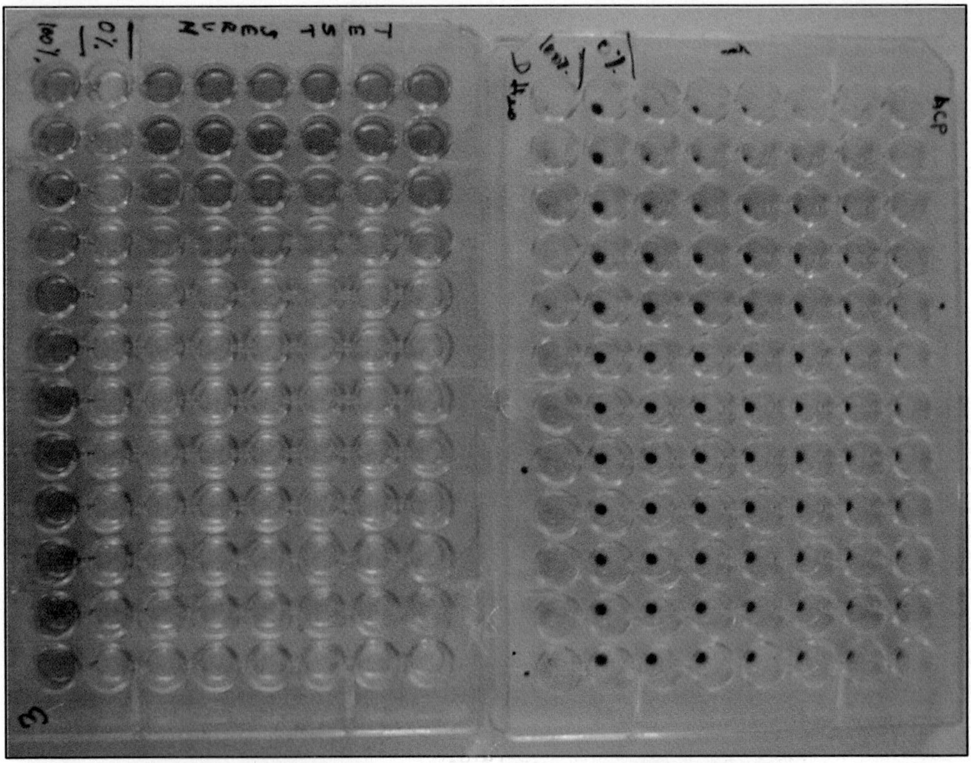

No Mat Formation (Treated blood)

Mat Formation (After treatment)

Figure 28.4: Haemagglutination Assay

Low ROS concentration after treatment

High ROS concentration in treated sample

Figure 28.5: Reactive Oxygen Species

The experimental set up for the Flower horn fish (*Vieja synspilus*) was stocked in six glass tank each carrying five fishes (Figure 28.1). Initially the fishes were injected with virulent *Aeromonas hydrophila* ($1x10^8$ cells/fish).After treatment with stress leads to loss of weight and coloration (Figure 28.2). Fish infected with *Aeromonas hydrophila* may have many different symptoms. These range from sudden death; otherwise healthy fish for lack of appetite, swimming abnormalities, pale gills, bloated appearance, and also the skin ulcerations were noticed. The skin ulcers may occur at any site on the fish body surface and often are surrounded by a bright rim of red tissue. Other organs commonly affected with this disease include the gills, kidneys, liver, spleen, pancreas, and skeletal muscle. The symptoms vary since they are dependent upon a number of factors including the virulence of the organism, the resistance of the fish to infection, the presence or absence of bacteremia or septicemia, and stress factors associated with the fish. Because of the variability of these symptoms, the diagnosis of this disease based only upon symptoms which is highly unreliable and may be economically disastrous to the fish producer. Obviously the best prevention against *Aeromonas hydrophila* infection is to never have this disease. This may sound absurd, but by minimizing the stress factors of fish through proper handling, stocking levels, nutrition, transportation, and water mishandled, overcrowded, transported under poor quality, fish are much less susceptible to this disease. Excellent sanitation and filtration procedures are an absolute must in order to minimize the chances of this disease ever occurring. Once the diagnosis of *Aeromonas hydrophila* infection of fish has been established (which should be left to those trained in infectious fish diseases and supported by proper laboratory diagnostic tests), treatment should begin immediately [Swann and Randy White (1989)].

Table 28.1: ROS Concentration in *Aeromonas hydrophila* Treated Fish Samples and Microbes Incorporated Feed given Fish Samples

ROS Concentration in 655 nm	*Control*	*Feed 1*	*Feed 2*	*Feed 3*	*Feed 4*	*Feed 5*	*Feed 6*
ROS concentration in *Aeromonas hydrophila* Treated Fish samples	2.365	1.602	1.780	1.928	2.008	1.900	1.618
ROS concentration in Microbe incorporated feed Fish samples	1.469	1.223	1.259	1.244	1.228	1.358	1.340

The Microbe (*Streptomyces* and Photosynthetic bacteria) incorporated feed when supplemented to the Flower horn fishes (*Vieja synspilus*) which already treated with *Aeromonas hyrophila,* it results in recovery of the fish from the stress as well as the recovery of pigmentation which is lost during treatment (Figure 28.3). The Haemagglutination assay clearly proves the presence of Antigen before treated (no mat formation) and after supplemented with microbe incorporated feed (mat formed) (Figure 28.4). As far as Reactive Oxygen Species studies concerned the concentration of ROS increased during the fish treated with *Aeromonas hyrophila* and it get decreased with the fishes fed with microbe incorporated feed (Figure 28.5). When it is read at 655 nm, the values for the treated ranges from 1.602 – 2.365 whereas for the Microbial feed supplemneted it ranges from 1.223 – 1.469 (Table 28.1), when compared to control feed the microbe incorporated feed showed better results. These prove that microbes (*Streptomyces* and Photosynthetic bacteria) supplemented feed acts as an Immunostimulant to the fish which results in recovery from stress as well as from recovery of pigmentation.

Conclusion

Streptomyces sp., and photosynthetic bacteria are widely recognized as industrially important microorganisms because of their ability to produce many kinds of novel secondary metabolites, including antibiotics. In the course of screening for new metabolites, several studies were carried out in order to isolate new *Streptomyces* species from different habitats. Recent studies are focusing on the response of antioxidant system of bacteria, which is important in terms of biotechnology, such as *Streptomyces* growth in various oxidative stress conditions. The production of reactive oxygen species (ROS) in bacteria, attack almost, all cell components, including DNA, protein and lipid membrane (Toyokuni *et al.*, 1995; Ames *et al.*, 1993). This ROS, which includes the super oxide anion, hydrogen peroxide and the hydroxyl radical in bacteria, is mainly related to respiratory chain activity (Gonzalez-Flecha and Demple 1995; Lushchak 2001). Organisms defend themselves against destructive ROS by non-enzymatic (Vitamins E and C or flavonoids and metal chelators) and enzymatic means (antioxidant enzymes).The immnostimulants from microbial source act very efficiently against harmful pathogen by decreasing the Reactive oxygen species level of the fish. The free radicals which are released during pathogenesis like super oxide, catalyze, peroxides may be reduced on supplementation of microbial cells incorporated feed which act as immunostimulant.

Bibliography

Alex, E., Maya, K., Dhevendaran, K. and Natarajan, P. (1993). Phosphatase producing microbes associated with fish and shellfish in Trivandrum coast. *Fish. Technol.*, 30: 62-66.

Alexander, J.B., Bowere, D.A. and Shamshoom, S.M. (1981). Hyperosmotic infiltration of bacteria into trout: Route of entry and the fate of the infiltrated bacteria. *Develop. Biol. Standard.*, 49, 441-445.

Amend, D.F. (1981). Potency testing of fish vaccines. *Dev. Biol. Standard.*, 49, 447-454.

Amend, D.F. and Fender, D.C. (1976). Uptake of bovine serum albumin by rainbow trout hyperosmotic solutions. A model for vaccinating fish. *Science*, 192, 793-794.

Amend, D.F. and Johnson, K.A. (1981). Current status and future needs of *Vibrio anguillarum* bacterins. *Develop. Biol. Standard.* 49, 403-417.

Ames BN, Shigenaga MK and Hagen TM. (1993) Oxidants, antioxidants, and the degenerative diseases of aging. *Proc Natl Acad Sci USA*;90: 7915–22.

Ames, B. N. (1983). Dietary carcinogens and anticarcinogens. *Science* (Washington, D.C.) 221: 1256–1264

Anderson, D.P. (1992). Immunostimulants, adjuvants and vaccine carriers in fish: application to aquaculture. *Annual Rev. of Fish Diseases*, 1992: 281 -307.

Anderson, J.W. and Conroy, D.A. (1970). *Vibrio* diseases in fishes. In: *A Symposium on Diseases of Fishes and Shellfishes*, ed. S.F. Snieszko, Washington, DC: *American Fisheries Society, Special Publication* No. 5, 226-272.

Annie Mathew (1995). *Studies on antagonistic streptomycetes associated with fish and shell fish of Veli Lake, Kerala* – Ph.D Thesis, University of Kerala.

Annie Mathew, Dhevendaran, K., Georgekutty, M.I. and Natarajan, P. (1994). L-asparaginase activity in antagonistic streptomycetes associated with clam *Villorita cyprinoides* (Hanley). *Indian J. Mar. Sci.*, 23: 204-208.

Antipa, R. and Amend, D.F. (1977). Immunization of Pacific salmon: Comparison of intraperitoneal injection and hyperosmotic infiltration of *Vibrio anguillarum* and *Aeromonas salmonicida* bacterins. *J. Fish. Res. Board Can.*, 34, 203-208.

Antipa, R., Gould, R. and Amend, D.F. (1980). *Vibrio anguillarum* vaccination of sockeye salmon, *Oncorhynchus nerka* (Walbaum) by direct and hyperosmotic immersion. *J. Fish.Dis.*, 3, 161-165.

Antoine, T., Carraro, S., Micha, J.C. and Van Hove, C. (1986), Comparative aooetency for *Azolla* of *Cichlasoma* and *Oreochromis* (Tilapia). *Aquaculture*, 53: 95-99.

Aoki, T., Egusa, S. and Arai, T. (1974). Detection of R factor in naturally occurring *Vibrio anguillarum* strains. *Antimicrobial Agents* and *Chemotherapy*, 6, 534-538.

Aoki, T., Kitao, T., Itabashi, T.wada, Y. and Sakai, M. (1981). Proteins and lipopolysaccharides in the membrane of *Vibrio anguillarum. Develop. Boil. Standard.*, 49, 225-232.

Austin, B., Stuckey, L.F., W. bertson, P.A., Effendi, I., and Griffith, D.R.W. (1995). A probiotic strain of *Vibrio alginoyticus* effective in reducing diseases caused by *Aeromonas salmoniioda, Vibrio anguillarm* and *Vibrio ordalii. Fish Diseases* 18: 93-96.

Austin.B. and Austin, D.A, (1987). In: *Bacterial Fish Pathogens*.pp.268-287.

Barbier, P., Kalimanzira, C. and Micha, J.C. (1985), L'amenagement de zones marecageuses en ecosystems agro-piscicoles. FUCID: Namur, 35 pp.

Barness, LA. 1994. Dietary sources of nucleotides – from breast milk to weaning. *J. Nutr.* 124(Suppl. 1S): 128S-130S.

Baudin-Laurencin, F. and Tangtrongpiros, J. (1980). Some results of vaccination against vibriosis in Brittany. In: *Fish Diseases*. ed. W. Ahne, Berlin: Springer-Verlag. pp. 60-68.

Beena, V.S 1993. *Studies on the parasites and diseases of the bagrid catfishes of the genus Mystus(Scopoli)*, Ph.D Thesis, University of Kerala.

Bendich, A. (1989). Carotenoids and the immune responsJe.. *Nutr.*, 119: 112-11 5.

Bendich, A. (1992). Physiological role of antioxidants in the immune system. *J. Dairy Sci.*, 76: 2789-2794.

Bendich, A., and J. A. Olson. (1989). Biological actions of carotenoids. Federation of the American Societies for *Experimental Biology Journal* 3: 1927–1932.

Benenson, A.S. (1970). ed. *Control of Communicable Diseases in Man*. Washington: The American Public Health Association.

Bergman, A.M. (1909). Die rote Beulen-krankheit des Aals. *Berichtaus der Koniglichen Bayerischen Versuchsstation*, 2, 10-54.

Blazer, V.S. (1992). Nutrition and disease resistance in fish. *Annual Rev. of Fish diseases*, 309-323.

Bonaveri, G.F. (1761). Quoted by Drouin de Bouville (1907).

Boyd, C.E. 1968, Fresh water plants: A potential source of protein. *Economic Botany*, 22: 359-368.

Bricknell I.R., Burno. D.W., Bowden. T.J. and Smith. P (1996) Fat cell necrosis syndrome in Atlantic halibut, Hippoglossus hippoglossus. L. *Aquaculture*, 144, 65-69.

Brisinello, W., Doimi, M., Giorgetti, G. and Sarti, M. (1985). Vaccination trials against vibriosis in sea bass (*Dicentrarchus labrax*) fry. *Bull. Eur. Ass. Fish Pathol.*, 5, 55-56.

Burke, J. and Rodgers, L. (1981). Identification of pathogenic bacteria associated with the occurrence of 'red spot' in sea mullet. *Mugil cephalus* L., in south-eastern Queensland. *J. Fish Dis.*, 3, 153-159.

Burton, G.W. (1989). Antioxidation of carotenoids. *J. Nutr.*, 11 9: 109-1 11.

Cahu, C., Zambonino Infante, J., Escaffre, A.-M., Bergot, P. and Kaushik, S. (1998). Preliminary results on sea bass (*Dicentrarchus labrax*) larvae rearing with compound diet from first feeding. Comparison with carp (*Cyprinus carpio*) larvae. *Aquaculture* 169: 1–7.

Canestrini, G. (1893). La malattia dominate delle anguille. *Atti Institute Veneto Service*, 7, 809- 814.

Carver, J.D. and W.A. Walker. (1995). The role of nucleotides in human nutrition. *Nutr. Biochem.* 6: 58-72.

Cassani, J.R. (1981), Feeding behaviour of under yearling hybrids of the grass carp, *Ctenopharyngodon idella* female and the big head, *Hypophthalmichthys* nobilis *male, on selected species of aquatic plants. J. Fish Biol.*, 18: 127-131.

Caumette, P. (1987), Role des bacteries phototrophes et des bacteries sulfato-reductrices dans les milieux lagunaires. ORSTOM, Paris, 395 p.

Chart, H. and Trust, T.J. (1984). Characterization of the surface antigens of the marine fish pathogens, *Vibrio anguillarum* and *Vibrio ordalii*. *Can. J. Microbiol.*, 30, 703-710.

Chew, B.P. (1995). Antioxidant vitamins affect food animal immunity and health. *J. Nutr, 125:* 1804s-1808s.

Cisar, J.O. and Fryer, J.L. (1969). An epizootic of vibriosis in chinook salmon *Bull. Wildl. Dis. Assoc.*, 5, 73-76.

Coe, W.B. and Turk, M. (1973), In: *Symposium: Processing Agricultural and Municipal wastes* (G.E. Inglett, ed), AVI Publ., West Port, Connecticut, 29 p.

Collins.C.H., Lyne, P.M and Grange, J.M, 2001, Collins and Lynes Microbiological Methods- 7th Edition. Arnold London, 493pp.

Colorni, A., Paperna, I. and Gordin, H. (1981). Bacterial infections in gilthead sea bream *Sparus aurata* cultured at Elat. *Aquaculture*, 23, 257-267.

Costantini D, Dell'Omo G, Casagrande S, Fabiani A, Carosi M, Bertacche V, Marquez C, Snell H, Snell H and Tapia W, Gentile G (2005) Inter-population variation of carotenoids in Galápagos land iguanas (*Conolophus subcristatus*). *Comp Biochem Physiol* B 142: 239–244.

Coutteau, P. and Lavens, P. (1989) The use of yeast as a single-cell protein in aquacultural diets. *Artemia* Reference Center, Med. Fac. Landbouww. *Rijksuniv. Gent*. 54: 1583–1592.

Croy, T.R. and Amend, D.F. (1977). Immunization of sockeye salmon (*Oncorhynchus nerka*) against vibriosis using the hyperosmotic infiltration technique. *Aquaculture.* 12, 317-325.

Daniel, S.G. (1994). Natural feed for pond fishes. *Science*, Dec. 31.

Das, S.K., Manissery, J.K. and Varghese, T.J. (1994), Growth response of Indian major carps *Catla catla* and *Labeo rohita* to formulated diets containing leaf powders as protein source. *Fish. Technol.*, 3192): 102-107.

Davina, J. H. M., Rijkers, G. T., Rombout, J. H. W. M., Timmermans, L. P. M.and van Muiswinkel, W. B. (1980). Lymphoid and non-lymphoid cells in the intestine of cyprinid fish, p. 129-140. In: *Development and differentiation of vertebrate lymphocytes,* (J. D. Horton, ed.). Elsevier/North-Holland Biomedical Press, Amsterdam.

de Felippe, J., da Rocha e Silva, M., Maciel, F. M. B., de Macedo Soares, A.and Mendes, N. F. (1993). Infection prevention in patients with severe multiple trauma with the immunomodulator beta 1-3 polyglucose (Glucan). *Surgery, Gynecology and Obstetrics,* 177, 383-388.

Devresse, B. (2000). Nucleotides: a key nutrient for the immune system of shrimp. *Feed Mix* 8(3): 20-22.

Dhevendaran, K., Santhanam, R and Selvakumar, N. (1978). Bacteria associated with plankton at Porto Novo. *Curr. Sci.* 45., 77-79.

Dhevendaran, K., Maya, K. and Natarajan, P. (1992), Studies on microbial enzymes and cellulolytic activity in *Aeromonas* of retting ground. *Proc. Nat. Acad. Sci.*, 62: 547-556

Dhevendaran, K. and Maya, K. (1998). Arylsulfatase in fish and shellfish of Veli lake, South India. *J. Mar. Biotech.*, 6: 71-75.

Dhevendaran, K and Mathew, A.K. (1999 a). Systematics of *Streptomyces* associated with the gut of alimentary canal of estuarine fish and shell fish of India. *Fish. Technol*, 36: 90-95.

Dhevendaran, K and Mathew, A. K. (1999 b) Antibiotic and L. asparaginase activity of Streptomyces isolated from fish, shell fish and sediment of *Veli* estuarine Lake along the Kerala coast. *Indian. J.Mar. Sci;* 28: 335-337

Dhevendaran, K. and Salley, V. (2000). Distribution of coliforms and bacterial pathogens in the backwater system of Poonthura region.12th *Keral Sci. Cong.* Kozhikode, Kerala.

Dhevendaran, K., Praseetha, P.K.and Annithakumari, Y.K. (2004). Microbial association with seaweeds of Kovalam Coast. *J. Seaweeds Utli.,* 1: 12–18.

Di Mascio, P., M. E. Murphy, and H. Seis. (1991). Antioxidant defense systems: the role of carotenoids, tocopherols, and thiols. *American Journal of Clinical Nutrition* 53: 194S–200S.

Diplock, A. T. (1991). Antioxidant nutrients and disease prevention: an overview. *American Journal of Clinical Nutrition* 53: 189S–193S.

Doggett, T. A.and Harris, J. E. (1987). The ontogeny of gut-associated lymphoid tissue in *Orechromis mossambicus. J. Fish Biol.* 31, suppl. A, 23-27.

Duff, D.C. B. (1942). The oral immunization of trout against *Bacterium salmonicida. J. Immunol.,* 44, 87-94.

Dykstra.M.J., Noga.E.J., Levine, J.F., Moye, D.W and Hawkins.J.H. (1986); Characterization of the *Aphanomyces* species involved with ulcervative mycosis(UM) in Menhaden. *Mycologia,* 78: 664-672.

Egidius, E. and Andersen, K. (1978). Host-specific pathogenicity of strains of *Vibrio anguillarum* isolated from rainbow trout *Salmo gairdneri* (Richardson) and Saithe *Pollachius viriens* (L). *J. Fish. Dis.,* 1, 45-50.

Egidius, E. and Andersen, K. (1979). Bath immunization –a practical and non-stressing method of vaccinating farmed sea rainbow trout (*Salmo gairdneri* Richardson) against vibriosis. *J. Fish Dis.* 2, 405-410.

Egidius, E., Wiik, R., Andersen, K., Hoff, K.A. ans Hjeltner, B. (1986). *Vibrio salmonicida* sp. nov; a new fish pathogen. *Int. J. System. Bacteriol.,* 36, 518-520.

Ellis, A. E. (1988). Ontogeny of the immune system in teleost fish. pp. 20-31. In: Fish vaccination, (A. E. Ellis, ed.). Academic Press, London.

Evelyn, T.P.T. (1971). First records of vibriosis in Pacific salmon cultured in Canada, and taxonomic studies of the response bacterium, *Vibrio anguillarum. J. Fish. Res. Board Can.,* 28, 517-525.

Evelyn, T.P.T. and Ketcheson, J.E. (1980). Laboratory and field observations on antivibriosis vaccines. In. *Fish Diseases,* ed. W. Ahne, Berlin: Springer-Verlag, pp. 45-54.

Evelyn. T.P.T. (1984). Immunization against pathogenic vibriosis. In. *Sympossium on Fish Vaccination,* ed. P. de Kinkelin, Paris: Off. Int. Epiz. Pp. 121-150.

Ezura, Y., Tajima, K., Yoshimizu, M. and Kimura, T. (1980).Studies on the taxonomy and serology of causative organisms of fish vibriosis. *Fish Path.,* 14, 167-179.

Fathima, E.J. (1973). *Studies on the bacterial flora of the alimentary tract of the Indian mackerel.* M.Sc. Dissertation, Annamalai University.

Fenchel, T. (1971). Aspects of decomposer food chains in marine benthos. *Verh. Dt. Zool. Ges., Bd.,* 65(5): 14-23.

Fevolden, S.E., Roed, K.H. and Gjerde, B. (1994). Genetic component of post-stress cortisol and lysozyme activity in Atlantic salmon - correlations to disease ressitance. *Fish* and *Shellfish immunology,* 4: 507-519.

Fletcher, T.C. and White, A. (1973). Antibody production in the plaice after oral and parenteral immunization with *Vibrio anguillarum* antigens. *Aquaculture*. 1, 417-428.

Fryer, J. L., Nelson, J.S. and Garrison, R.L. (1972). Vibriosis is fish. In: *Progress in Fishery and Food Science*, ed. R.W. Moore, Seattle: University of Washington, Publications in Fisheries, pp. 129-133.

Fryer, J. L., Rohovec, J. S., Tebbit, G.L., McMichael, J.S. and Pilcher, K.S. (1976). Vaccination for control of infectious diseases in Pacific salmon. *Fish. Path.*, 10, 155-164.

Fryer, J.L., Amend, D.F., Harrell, L.W., Novotony, A.S., Plumb, J.A., Rohovec, J.S. and Tebbit, G.L. (1977). *Development of Bacterins and Vaccines for Control of Infectious Diseases of Fish*. Oregon State University. Sea Grant College Program. Publ. No. ORESU-T-77-012.

Fryer, J.L., Rohovec, J.S. and Garrison, R.L. (1978). Immunization of salmonids for control of vibriosis. *Mar. Fish. Review*. 40, 20-23.

Fuller, R. (1978). Epitheial attachment and other factors controlling the colonization of the intestine of the gnotobiotic chicken by *Lactobacilli. Appl. Bacteriol.*, 46: 335-342.

Fuller, R., and Turvey, A. (1971). Bacteria associated with the intestinal wall of the fowl (*Gallus domesticus*). *Appl. Bacteriol.*, 34: 617-622.

Gatesoupe, F.T. (1999). The use of probiotics in aquaculture. *Aquaculture*, 180: 147-165.

Gerhard Sandmann (2002) Molecular evolution of carotenoid biosynthesis from bacteria to plants *Physiologia Plantarum* 116: 431–440.

Gerking, S.D. (1984), Assimilation and maintenance ration of a herbivorous fish, *Sarpa salpa*, feeding on a green alga. *Trans. American Fish. Soc.*, 113: 378-387.

Ghittino, P., Andruetto, S. and Vigliani, E. (1972). "Redmouth" enzootic in hatchery rainbow trout caused by *Vibrio anguillarum*. Riv. Ital. Piscic. *Ittiopathologia*, 7(2), 41-45.

Gibbons, N. E. (1936). Studies on salt fish. 1. Bacteria associated with the reddening of salt fish. *J. Biol. Board Can.*, 3, 70-76.

Gibson, G.R. and M.B. Roberfroid. (1995). Dietary modulation of the human colonic microbiota: introducing the concept of prebiotics. *J. Nutr.* 125(6): 1401-1412.

Giorgetti, G., Tomasin, A.B. and Ceschia, G. (1981). First Italian anti-vibriosis vaccination experiments of freshwater farmed rainbow trout. *Develop. Biol. Standard.*, 49, 455-459.

Gonzalez-Flecha B.and Demple, B. (1995). Metabolic sources of hydrogen peroxide in aerobically growing *Escherichia coli*. *J Biol Chem*;270: 13681–7.

Goodwin T.W. (1984). (Ed.) The biochemistry of the carotenoids. Vol.II Animals. Chapman and hall Ltd., New York, USA., VAS., 160-72.

Gould, R.W., O'Leary, P.J., Garrison, R.L., Rohovec, J.S. and Fryer, J.L. (1978). Spray vaccination: a method for the immunization of fish. *Fish. Path.*, 13, 63-68.

Grafen, A. (1990). *J. Theor. Biol.* 144, 517.

Grether GF, Hudon J.and Millie DF (1999) Carotenoid limitation of sexual coloration along an environmental gradient in guppies. *Proc R Soc Lond* B 266: 1317–1322.

Grey, W.D. (1970), The use of fungi as food and in feed processing. Buttrworth and Co., USA, p. 60.

Grimes, D.J., Stemmler, J., Hada, H., May, E.B., Maneval, D., Hetrick, F.M., Jones, R.T., Stoskopf, M. and Colwell, R.R. (1984). *Vibrio* species associated with mortality of sharks held in captivity. *Microbial Ecol.*, 10, 271-282.

Hacking, M. A. and Budd, J. (1971). *Vibrio* infection in tropical fish in a freshwater aquarium. *J. Wildl. Dis.*, 7, 273-280.

Hanlon, R.D.G. (1982).The breakdown and decomposition of allochthonous and autochthonus plant litter in an oligotrophic lake (Llyn Frongoch). *Hydrobiologia,* 88. 281-288.

Harrell, L.W., Etlinger, H.M. and Hodgins, H.O. (1975). Humoral factors important in resistance of salmonid fish to bacterial diseases. I. Serum antibody protection of rainbow trout (*Salmo gairdneri*) against vibriosis. *Aquaculture.* 6, 211-220.

Harrell, L.W., Etlinger, H.M. and Hodgins, H.O. (1976). Humoral factors important in resistance of salmonid fish to bacterial disease. II. Anti-*Vibrio anguillarum* activity in mucus and observations on complement. *Aquaculture*, 7, 363-370.

Harrison, F. C., and Kennedy, M. E. (1922) The red discoloration of cured cod fish. *Trans. Roy. Soc. Can., Sect.* 5, 16, 101-152.

Hart, S., Wrathmell, A. B., Harris, J. E.and Grayson, T. H. (1988). Gut immunology in fish: A review. *Dev. Comp. Immunol.*, 12, 453-480.

Hastein, T. and Smith J.E. (1977). A study of *Vibrio anguillarum* from farms and wild fish using principal components analysis. *J. Fish Biol.*, 11, 69-75.

Hastein, T., and Holt G. (1972). The occurrence of vibrio disease in wild Norwegian fish. *J. Fish Biol.* 4, 33-37.

Hastein, T., and Refsti, T. (1986). Vaccination of Rainbow Trout against vibriosis by injection, dip and bath. *Bull. Eur. Ass. Fish Pathol.* 6, 45-49.

Hayashi, K., Kobayashi, S., Jamata, T., and Ozaki, H. (1964). Studies on the vibrio disease of rainbow trout (*Salmo gairdneri irideus*). II. Prophylactic vaccination against the vibrio-disease. *J. Fac. Prefect. Univ. Mie.* 6, 181-191.

Holm, K.O. and Jorgensen, T. (1987). A successful vaccination of Atlantic salmon, *Salmo salar* L. against 'Hitra disease' or cold water vibriosis. *J. Fish Dis.*, 10, 85-90.

Holt, G. J. (1993) Feeding larval red drum on microparticulate diets in a closed recirculating water system. *J. World Aquacult. Soc.* 24: 225–230.

Hood, M.A. and Meyers, S.P. (1973), Microbial aspects of penaeid shrimp digestion. *Proc. Gulf Caribb. Fish. Inst.*, 26: 81-92.

Horne, M.T., Tatner, M., McDerment, S. and Agius, C. (1982). Vaccination of rainbow trout, *Salmo gairdneri* Richardson, at low temperatures and the long-term persistence of protection. *J. Fish Dis.*, 5, 343-345.

Hossain, M.A. and Jauncey, K. (1989), *Aquacult. Fish. Manage.* 20: 255

Howard-Williams, C. and Davies, B.R. (1979), The rates of dry matter and nutrient loss from decomposing *Potamogeton pectinatus* L. in brackish, South-temperate lake. *Freshwat. Biol.*, 9: 13-21.

Howard-Williams, C., Davies, B.R. and Cross, R.H.M. (1978), The influence of periphyton on the surface structure of a *Potamogeton pectinatus* L.leaf. (An hypothesis). *Aquat. Biol.*, 5: 87-91.

IFIC. (2004).Backgroundonfunctionalfoods.From*http: //www.ific.org/nutrition/functional/index.cfm*. International Food Information Council, 2004.

Itami, T. and Kusada, R. (1980). Studies on spray vaccination against vibriosis in cultured ayu. II. Duration of vaccination efficacy and effect of different vaccine preparations. *Bull. Jap. Soc. Sci. Fish.*, 46, 699-703.

Itoh, K. (1973), Food requirements of copepods estimated from their metabolic rates. *Bull. Plankt. Soc. Japan.*, 20: 78-83

Jauncey, K. (1982), The effects of varying dietary protein level on the growth, food conversion, protein utilization and body composition of juvenile tilapias (*Sarotherodon mossambicus*). *Aquaculture*, 27: 43-45.

Johnson, K.A. and Amend, D.F. (1983). Efficacy of *Vibrio anguillarum* and *Yersinia ruckeri* bacterins applied by oral and anal intubation of salmonids. *J. Fish Dis.* 6, 473-476.

Johnson, K.A., Flynn, J.K. and Amend, D.F. (1982). Onset of immunity in salmonid fry vaccination by direct immersion in *Vibrio anguillarum* and *Yersinia ruckeri* bacterins. *J. Fish Dis.*, 5, 197-205.

Johnson, K.A., Flynn, J.K. and Amend, D.F. (1982). Duration o immunity in salmonids vaccinated by direct immersion with *Yersinia ruckeri* and *Vibrio anguillarum* bacterins. *J. Fish Dis.* 5, 207-213.

Johnson.E. and W. Schroeder (1995) Microbial carotenoids. *Ad. Bioch. Eng. Biotechnol.*, 53: 119-178.

Joseph, j.D. and Williams, J.E. (1975), Shrimp-head oil: A potential feed additive for mariculture. *Proc. World Maricult. Soc.*, 6.

Kaushik, S.J. and Luquet, P. (1980), Influence of bacterial protein incorporation and of sulphur amino acid supplementation to such diets on growth of rainbow trout. *Salmogairdnerii* Richardson. *Aquaculture*, 19: 163-17.

Kawai, K., Kusada, R. and Itami, T. (1981). Mechanisms of protection in ayu orally vaccinated for vibriosis. *Fish Pathol.*, 15, 257-262.

KihlBerg, R. (1972). The microbe as a source food. *Ann. Rev. Microbial.*, 26: 427-466.

Kobayashi, M. Haque, M.Z. (1971), Contribution to nitrogen fixation and soil fertility by photosynthetic bacteria. *Pant. and soil*. Special Vol., 443-456.

Kobayashi, M., Kobayashi, M. and Nakanishi, H. (1971) *J. Ferment. Technol.*, 49; 817.

Krinsky, N. I. (2001). Carotenoids as antioxidants. Nutrition, 17, 815–817.

Kusuda, R., Kawai, K., Jo. Y., Akizuki, T., Fukunaga, M. and Kotake, N. (1978). Efficacy of oral vaccination for vibriosis in cultured ayu. *Bull. Jap. Soc. Sci. Fish.*, 44, 21-25.

LaDon Swann and M. Randy White, D.V.M. (*1989*). Diagnosis and Treatment of "Aeromonas hydrophila" Infection of Fish *A Guide to Approved Chemicals in Fish Production and Fishery Resource Management*, , University of Arkansas Cooperative Extension Service.

Lakshmanaperumalsamy, P. (1978) *Studies on actinomycetes with special reference to antagonistic Streptomycetes from sediment of Porto Novo coastal zones*. Ph.D. Thesis, Annamalai University, India.

Lakshminarayana, D. (1988). *Myxosporidian parasites of fresh water fishes of Andhra Pradesh*. M.Phil. Dissertation, Andhra University Visakhapatnam.

Lall, S.P. and Olivier, G. (1993). Role of micronutrients in immune response and disease resistance in fish. In *Fish Nutrition in Practice*, INRA (ed). Les Colloques, n"61, Paris, pp. 1 O1 -1 18.

Latscha, T. (1991) Carotenoids in aquatic animal nutrition. In proceedings of Aquaculture, feed processing and nutrition workshop, 68-79.

Levine, M.A., Wolke, R.E. and Cabelli, U.J. (1972). *Vibrio anguillarum* as a cause of disease in winter flounder (*Pseudopleuronectes americanus*). *Can. J. Microbiol.*, 18, 1585-1892.

Lewis, William, M. and Michael Bender, (1960). Heavy mortality of golden shiners during harvest due to bacterium of the genus *Aeromonus. Prog.Fish; Cult.*, 22(1): 11-14.

Lewis, William, M. and Michael Bender, (1961). Free Living Ablity of a warm water fish pathogen of the genus *Aeromonus* and factors contributing to its infection of the golden shiner, *Prog.Fish.Cult.*, 23(3): 124-126.

Li, P. and Gatlin, D. M., III (2003) Evaluation of brewers yeast (*Saccharomyces cerevisiae*) as a feed supplement for hybrid striped bass (*Morone chrysops _ M. saxatilis*). *Aquaculture* 219: 681–692.

Lochhead, A. G. (1934) Bacteriological studies on the red discoloration of salted hides. *Can. J. Research*, 10, 275-286.

Lochhead, A. G. (1952). Soil Microbiology. *Ann. Rev. Microbiol*. 6: 185-206.

Lochhead, A. G., and Chase, F. E. (1943). Qualitative studies of soil microorganisms. V. Nutritional requirements of the predominant bacterial flora. *Soil Sci.*, 55, 185-195.

Lom. J. (1970a). Protozoa causing diseases in marine fishes. In: A symposium on diseases of fishes and shellfishes (editor) Sniezsco. S.F. *Amer. Fish. Soc. Spl. Publ.* pp. 101-12.

Lom. J. (1970b). Some aspects of host-parasite relation in myxosporodians of the genus Henneguya. *J. Parasitol.*, 56: 442-445.

Lom. J., Dykova, I. (Ed.) (1992). *Protozoan Parasites of Fishes*. Elsevier Science publishers. Amesterdam. pp. 315.

Lozano, G. A. (1994). Carotenoids, parasites, and sexual selection. *Oikos*, 70, 309–311.

Lushchak VI. (2001). Oxidative stress and mechanisms of protection against it in bacteria. *Biochemistry* (Mosc);66: 476–89.

M. Andersson (1994) (Ed.). *Sexual Selection* Princeton Univ. Press, Princeton, NJ.

Maeda, M., Nogami, K., and Ishibash, N. (1992). Utility of microbial food assemblages for culturing a crab. *Portunus trituberculatus. Aquaculture* 21: 31-38.

Mahesh kumar.S., (1985). *The epizootiology of fin rot in hatchery reared Atlantic Salmon* (*Salmo salar*), Masters thesis, University of Maine, Orono, ME.

Mandal, A.K. (1984). A study on the haematozoa of fishes of commercial importance from India. *Tech. Monograph*. 9. Z.S.I. of India. pp 49.

Manju, K.G.and K. Dhevendaran(1997), Effect of bacteria and actinomycetes as single cell protein feed on growth of juveniles of *Macrobrachium idella* (Hilgendorf). *Indian J. Exp. Biol.*, 35: 53-55.

Manju, K.G. and Dhevendaran, K. 2002.Influence of macrophytes and *Azolla* spp., on the growth conversion efficiency and gut flora of fish, *Oreochromis mossambicus*(Peters) *J.Aquacul. Tropics.*, 17: 221-230.

Margolis.L. (1953), The effect of fasting on the bacterial flora of the intestine of fish. *J.Fish.Res.Bd.Canada*, 10: 62-6.

Mary, P.P., Chandramohan, D. and Natarajan, R (1975), The gut microflora of some commercially important fishes from Poroto Novo waters.*Bull.Dept.Mar.Sci.*, 7: 185 – 199.

Mateo, C.D. (2005). *Aspects of nucleotide nutrition in pigs*. Ph.D dissertation, Department of Animal and Range Sciences, South Dakota State University, USA, pp. 171.

Mattheis, T. (1964). Das Vorkommen von *Vibrio anguillarum* in Ostseefischen. *Zentralblatt fur Fischerei N. F.* 12, 259-263.

Maya, R., Dhevendaran, K., Annie Mathew, M.I., Georgekutty and Natarajan, P. (1995), Seasonal variations of bacteria in fish *Etroplus suratensis* and *Etroplus maculatus* (Pisces. Cichlidae). *Indian J. Mar. Sci.* 24: 225-228.

Maya, K., Dhevendaran, K and Natarajan, P. (1990). Arylsulfatase bacteria associated with fish, *Therapon jarbua. Int. conf. Microb. Poikilotherms. Paris, France.*

Maya, K., Dhevendharan, K. and Natarajan, P. (1990), Arylsulfatase producing bacteria in the gut of *Therapon Jarbua.* International colloquium on Microbiology in poikilotherm, (R.Lessel, Ed.) Elsevier Science Publishers Biochemical Division.PP 203-210.

McCarthy, D.H. (1976). *Vibrio* disease in eels. *J.Fish. Biol.*, 8, 317-320.

McEvoy, L. A., Navarro, J. C., Hontoria, F., Amat, F. and Sargent, J. R. (1996) Two novel *Artemia* enrichment diets containing polar lipid. *Aquaculture* 134: 339–352.

McGraw, K. J. and Ardia, D. R. (2003). Carotenoids, immunocompetence, and the information content of sexual colors: an experimental test. *American Naturalist*, 162, 704–712.

Mercy Abraham (1989). *Studies on the nutritional grouping of bacteria in the gut of a culturable freshwater fish, Labeo rohita (Hamilton).* M.Phil. Dissertation, university of Kerala.

Meyer, F.P. (1970). Seasonal fluctuations in the incidence of disease on fish farms. In: *A Symposium on disease of fishes and shell fishes* (ed; S.F.Snieszko) Washington; *Am.Fish. Soc.*, pp;51-59.

Møller, A. P., Kimball, R. T. and Erritzøe, J. (1996). Sexual ornamentation, condition, and immune defence in house sparrow Passer domesticus. Behavioral Ecology and Sociobiology, 39, 317–322.

Montencourt, B.S. (1998) *Trichoderma resei* Cellulases Trends. *Biotechnol.*, 1: 156-160.

Moriarty, D.J. W. (1978), Quantitative studies on bacteria and algae in the food of the mullet *Mugil cephalus* L. and the prawn *Metapenaeus bennettae* (Racek and Dall). *J. Exp. Mar. Biol. Ecol.*, 22: 131-143.

Muller A.P (2000).*Avian Poult. Biol. Rev.* 11, 137.

Muroga, K., Jo., Y. and Nishibuchi, M. (1976). Pathogenic *Vibrio* isolated from cultured eels. I. Characteristics and taxonomic status. *Fish Path.,* 12, 141-145.

Muroga, K., Nishibuchi, M. and Jo, Y. (1976). Pathogenic *Vibrio* isolated from cultured eels. II. Physiological characteristics and pathogenicity. *Fish Path.*, 11, 147-151.

Muroga, K., Takahashi, S. and Yamanoi, H. (1979). Non- cholera *Vibrio* isolated from diseased aya. *Bull. Jap. Soc. Sci. Fish.*, 45, 829-834.

Natarajan, P. and James, (1977). Bibliography of parasites and diseases of marine and freshwater fishes of India.*J.Fish.Biology*.

National Research Council (1993). *Nutrient requirements of* fish. Committee on Animal Nutrition. Board on Agriculture. National Research Council. National Academy Press, Washington, D.C. 1993.114 p.

Negro, J.J., Tella, J.L., Blanco, G., Forero, M.G. and Garrido-Fernández, J. (2000). Diet explains interpopulation variation of plasma carotenoids and skin pigmentation in nestling white storks. *Physiol Biochem Zool* 73: 97–101.

Newell, R.C. (1970). Biology of Intertidal animals. *Logos*, London, 555pp.

Nishino. H. (1999). Cancer Prevention by carotenoids Pure and *Applied Chemistry* 71: 2273-2278.

Odum, W.E. (1970). Utilization of the dorect grazing and plant detritus food chains by the striped mullet, *Mugil cephalus*. In: *Marine Food Chains*, J.H. Steele (Ed), University of California Press, Berikeley, pp. 222-240.

Oki, A. and Sercarz, E. (1985). T-cell tolerance studied at the level of antigenic determinants. *J. Exp. Med.*, 161, 897-911.

Olson, V.A. and Owens, I.P.F. (1998).*Trends Ecol. Evol.* 13: 510.

Olvera-Novoa, M. A., Martý´nez-Palacios, C. A. and Olivera-Castillo, L. (2002) Utilization of torula yeast (*Candida utilis*) as a protein source in diets for tilapia (*Oreochromis mossambicus* Peters) fry. *Aquacult. Nutr.* 8: 257–264.

Oppenheimer, C.H. (1962) (Ed.). Marine fish diseases. In: *Fish as Food*, Vol. 2, New York: Academic Press, pp. 541-572.

Pacha, R.E. and Kiehn, E.D. (1969). Characterization and relatedness of marine vibrios pathogenic to fish: physiology, serology and epidemiology. *J. Bact.*, 100, 1242-1247.

Parker, R.B. (1974). Probiotics. The other half of antibiotic story. *Anim. Nutr. Health*, 29: 4-8.

Person Le Ruyet, J., Alexandre, J. C., The´baud, L. and Mugnier, C. (1993) Marine fish larvae feeding: formulated diets or live prey? *J. World Aquacult. Soc.* 24: 211–224.

Petit.H., Sance.S., Negre-Sadargues, G. Castello R. and Trilles J.P. (1991). Ontogeny of carotenoid metabolism in the prawn *Penaeus japonicus* Bate (1888). A qualitative approach. *Comp. Biochem., Physiol.*, 99B, 667-671.

Planas, M., and Cunha., I. (1999). Larviculture of marine fish: problems and perspectives. *Aquaculture*, 177: 171-190.

Prejs, A. and Blaszczyk, M. (1977), Relationship between food and cellulose activity in freshwater fishes. *J. Fish Biol.*, 11: 447-452.

Provasoli, L., Conklin, D.E. and D'Agostino, A.S. (1970), Factors inducing fertility in aseptic Crustacea. *Helgolander wiss. Meeresunters.* 20: 443-454.

Quan, R. (1992). Dietary nucleotides: potential for immune enhancement. In: *Foods, Nutrition and Immunity* (M. Paubert-Braquet, C. Dupont and R. Paoletti, eds). *Dyn. Nutr. Res.* 1. Karger, Basel, pp. 13-21.

Ransom, D. P. (1978). Bacteriologic, immunologic and pathologic studies of *Vibrio* sp. pathogenic to salmonids. Ph.D. thesis, Oregon State University, Corvallis.

Ransom, D. P., Lannan, C. N., Rohovec, J. S. and Fryer, J. L. (1984). Comparison of histopathology caused by *Vibrio anguillarum* and *Vibrio ordalii* and three species of Pacific salmon. *J. Fish Dis.*, 7, 107-115.

Reed, L.J. and Muench, H. (1938). A simple method for estimating 50 per cent end points. *Am.J. Hygiene*, 27, 493-497.

Rekha, R., Maya, K., Dhevendaran, K. and Natarajan, P. (1992). Studies on the influence of artificial feeds on the gut microflora in *Mystus gulio. In*: *Current Trends in Fish and Fishery Biology* and *Aquatic Ecology* (Yousuf, A.R., Raina, M.K. and Qadri, M.Y., eds.). 153-162.

Rieper, M. (1978). Bacteria as food for marine harpacticoid copepods. *Mar. Biol.*, 45: 337-345.

Rieper, M. (1982), Feeding preferences of marine harpacticoid copepods for various species of bacteria. *Mar. Ecol. Prog. Ser.*, 7(3): 303-307.

Roach, S., and Tannock, G.W. (1980). Indigenous bacteria that influence the number of *Salmonella typhimurium* in the spleen of intravenously challenged mice. *Cana. Microbiol.*, 26: 408-411.

Roberfroid, M.B. (2000). Prebiotics and probiotics: are they functional foods? *Amer. J. Clinic. Nutr.* 71(6): 1682S-1687S.

Roberts, R.J., Wotten, R., Macrae, I., Millar, S., and Struthers, W. (1989). Ulcerative disease survey, Bangladesh. Final report to the government of Bangladesh and the overseas Development Administration institute of Aquaculture, Stiriling University, Scotland, P 104.

Robinson, J. (1950). *A possible explanation of microbial halophilism.* Thesis, McGill University.

Rombout, J. I I. W. M., Bot. H.E.and Taverne-Thiele, J.J. (1989.) Immunologic importance of the second gut segment of carp. II. Characterization of mucosal leucocytes. *J. Fish Biol.*, 35, 167-178.

Rombout, J. H. W. M., Taverne-Thiele, J.J.and Villena, M.I. (1993). The gut-associated lymphoid tissue (GALT) of carp (*Cyprinus carpio* L.): An immunocytochemical analysis. *Dev. Comp. Immunol.* 17, 55-66.

Rombout, J.H.W.M., Lamers, C.M.J., Heffrich, M.H., Dekker, A. and Taverne-thiele, J.J. (1985). Uptake and transport of intact macromolecules in the intestinal epithelium of carp (*Cyprinus carpio*) and the possible immunological implications. *Cell and Tissue Research*, 239, 519-530.

Rosenkvist-Jensen, L. (1982). Results from vaccination of Danish rainbow trout against vibriosis *Devel. Comp. Immunol.* Suppl. 2, pp.187-191.

Ross, A. J., Martin, J. E. and Bressler, V. (1968). *Vibrio anguillarum* from an epizootic in rainbow trout (*Salmo gairdeneri*) in the USA. *Bull. Off. Int. Epiz.*, 69, 1139-1148.

Rouatt J.W (1968). Nutritional classification of soil bacteria and their value in ecological studies. In: *The Ecology of Soil Bacteria*, T.R.G.Gray and D.Parkinson (eds) PP 360–370.

Rucker, R. R. (1959). Vibrio infections among marine and fresh-water fish. *Progr. Fish. Cult.*, 21, 22-25.

Saino, N., Ferrari, R., Romano, M., Martinelli, R. and Møller, A. P. (2003). Experimental manipulation of egg carotenoids affects immunity of barn swallow nestlings. *Proceedings of the Royal Society of London*, Series B, 270, 2485–2489.

Sandifer, P.A. and Joseph, J.D. (1976). Growth responses and fatty acid composition of juvenile prawns (*Macrobrachium rosenbergii*) fed a prepared ration augmented with shrimp head oil. *Aquaculture*, 8: 129-138.

Sarada, P. (1996). *Influence of Azolla and microbial feed on the growth and conversion efficiency of the fish, Oreochromis mossambicus* (Peters). M.Phil Dessertation, University of Kerala, 65.

Schneider, R and Nicholson, B.L., (1980). Bacteria associated with finrot disease in hatchery reared Atlantic salmon (*Salmo salar*). *Can.J.Fish and Aquat.Sci.*, 37(10): 1505-1513.

Schoop, G. (1935). Obligat halophile Mikroben. *Zentr. Bakt. Parasitenk.*, Abt. I, 134, 14-26.

Secombes, C.J. (1994). Enhancement of fish phagocyte activity. *Fish* and *Shellfish Immunology*, 4: 421 -436.

Seki, H. (1989). Marine microorganisms associated with the food of young salmon. *Appl. Microbiol.*, 17: 252-255.

Sera, H., Ishida, Y. and Kadota, H. (1972), Bacterial flora in the digestive tracts of marine fish. III. Effects of H^+concentration and gastric juices on survival of the indigenous bacteria. *Bull. Japan. Soc. Sci. Fish.*, 38: 859-863.

Sera, H., Ishida, Y. and Kadota, H. (1974). Bacterial flora in the digestive tracts of marine fish. In: *Effects of the ocean environment on microbial activities* (R.R. Colwell and R.Y. Moitra, Eds.) University Park Press, Baltimore, Maryland, USA, pp. 467-490.

Sera, H and Ishida, Y. (1972). Bacterial flora in the digestive tracts of marine fish. II changes of bacterial flora with time lapse after ingestion of diet. *Bull.Jpn Soc Sci*: Fish., 38: 633.

Shiranee, P., Natarajan, P.and Dhevendaran, K. (1993). The role of gut and sediment bacterial flora in the nutrition of cultured pearlspot (*Etroplus suratensis Bloch*). *Isrl. J. Aquacul. Bamidgeh*, 45(2): 45-58.

Smets, P., Zalisz, R.and Boisnic, F. (1988). Biostim® (RU 41740), immune functions and infection. *Advances in the Biosciences*, 68, 85-102.

Smith, P.D. (1982). Analysis of the hyperosmotic and bath methods for fish vaccination-Comparison of uptake of particulate and non-particulate antigens. *Devel. Comp. Immun.*, Suppl. 2, pp. 181-186.

Smitha. G.S. (2005). *Studies on the distribution of Streptomyces in Akkulam lake and their screening of carotenoid pigments*. M.Sc. Dissertation, M.G. University.

Smoragiewiez, W., Bielecka, M., Babucgowski, A., Boutard, D., and Dubeau, H. (1993). Les probiotiques. *Cana. Microbiol.*, 39: 1089-1095.

Snieszko, S.F. (1973). Recent advances in scientific knowledge and developments pertaining to diseases of fishes.*Adv.Vet.Sci.Comp.Med.*, **17**: 291-314

Srivastava, K.P and Floodgate, G.D. (1986). Studies on the intestinal micro flora of Dab, *Limanda limanda*, *J.Mar. Biol.Assoc.India*. 8: 1-7.

Stahl, W. and Sies, H. (1999). In: *Antioxidant Food Supplements in Human Health*, L. Packer, M. Hiramatsu, T. Yoshikawa, Eds. (Academic Press, San Diego, CA,), pp. 183– 202.

Staley, T.J., Bryant, P.M and Pfenning, T.N. (1989). *Bergeys Manual of Systematic Bacteriology* 3rd Edition. William and Wilkins. Baltimore, 2298pp.

Strout, R.G., Sawyer, E.S. and Coutermarsh, B.A. (1978). Pathogenic vibriosis in confinement-reared and feral fishes of the maine-New Hampshire coast. *J. Fish. Res. Board Can.*, 35, 403-408.

Subasinghe, R. (1997). Fish health and quarantine. In: *Review of the State of the World Aquaculture. FAO Fisheries circular no.* 886. Food and Agriculture Organization of the United Nations, Rome, Italy. P. 45-49.

T. Von Schantz, S. Bensch, M. Grahn, D. Hasselquist, H. Wittzell, (1999).*Proc. R. Soc. Lond. Ser. B* 266, 1.

Tacon A.G. (1981). Speculative review of possible carotenoid function in fish. *Progr. Fish Culturist*, 43: 205-208.

Tacon, A.G.J.. (1987). The nutrition and feeding of farmed fish and shrimp. A training manual. Vol II Nutrient source and composition. *FAO*, GCP/RLA/O75/ITA. Field document 5/E, 126 p.

Tacon, A.G.J. (1994). Feed ingredients for carnivorous fish species. Alternatives to fishmeal and other fishery resources. FAO, *Fisheries Circular* N"881. FIRVC881, 35 p.

Tacon, A.G.J. and Jackson, A.J. (1985),. Utilization of conventional and unconventional protein sources in practical fish feeds. *In*: *Nutrition and Feeding in Fish* (C.B. Cowey, A.M. Mackie and J.G. Bell, eds.), Academic Press, London, Pp. 119-145.

Tacon, A.G.J. (1993). Feed ingredients for warmwater fish. Fish meal and other processed feedstuffs. FAO, *Fisheries Circular* N"856. FIRVC586, 64 p.

Tacon, A.G.J. and Kurmaly, K. (1996). Nutrition and health management in shrimp culture. In *Book of Abstracts of the 1996 Annual Meeting of the World Aquaculture Society*, Bangkok, Thailand. LeRoy Creswell, R. (ed), Ft. Pierce, Florida, pp. 397.

Tatner, M.F. and Horne, M.T. (1983). Susceptibility and immunity to *Vibrio anguillarum* in post-hatching rainbow trout fry. *Salmo gairdneri* Richardson 1836. *Dev. Comp. Immunol.* 7, 465-472.

Thompson, A., White, A., Fletcher, T. C., Houlihan, D. F., Secombes, C. J. (1993). The effect of stress on the immune response of Atlantic salmon (*Salmo salar* L.) fed diets containing different amounts of vitamin C. *Aquaculture*, 114, 1-18

Toyokuni, S., Okamoto, K. and Yodol, J. (1995). Persistent oxidative stress in cancer. *FEBS Lett.*, 358: 1–3.

Trust, T. J. (1986). Pathogenesis of infectious diseases of fish. *Ann. Rev. Microbiol.* 40: 479-502.

Trust, T.J. and Sparrow, R.A.H. (1974), The bacterial flora in the alimentary tract of freshwater salmonid fishes. *Can. J. Microbiol.*, 20(9): 1219-1228.

Uauy, R. (1989). Dietary nucleotides and requirements in early life. In: *Textbook of Gastroenterology and Nutrition in Infancy* (E. Lebenthal, ed). 2nd Ed., Raven Press, New York, NY, USA, pp. 265-280.

Uauy, R. (1994). Non-immune system responses to dietary nucleotides. *J. Nutr.* 124(Suppl. 1S): 157S-159S.

Van Buren, C.T., A. Kulkarni and F.B. Rudolph. (1994). The role of nucleotides in adult nutrition. *J. Nutr.* 124(Suppl 1S): 160S-164S.

Vanajakumar. (1981). *Studies on actinomycetes associated with molluscs from Porto Novo coastal water.* Ph.D Thesis. Annamalai University, India.

Varghese. T.J., Devaraj, K.V., Shantharam.B. and Shetty, H.P.C. (1976). In Proceedings of the Symposium on Development and Utilization of Inland Fishery Resources, Colombo, Srilanka. 1: 408.

Verlhac, V., Gabaudan, J. and Schielre, J. (1996). Influence of astaxantin on nonspecific immune response of rainbow trout. In *Book of Abstracts of the 1996 Annual Meeting of the World Aquaculture Society,* Bangkok, Thailand. LeRoy Creswell, R. (ed), Ft. Pierce, Florida, pp. 426.

Viele, P., Kerstetter, T.H. and Sullivan, J. (1980). Adoptive transfer of immunity against *Vibrio anguillarum* in rainbow trout, *Salmo gairdneri* Richardson, vaccinated by the immersion method. *J. Fish Biol.*, 17, 379-386.

Vincke, P. and micha, J.C. (1985). Fish culture in rice fields. P. 297-314. *In: Proceedings of the 16th Session of the International Rice Commission.* Los Banos, Laguna, Philippines, FAO, Rome 326 p.

Waksman, S.A and Henrici, G. (1934.) (Ed.). *The Actinomycetes.* Ronald Press Co., New York, p. 79-80.

West, P.A. and Lee, J.V. (1982). Ecology of *Vibrio* species including *Vibrio cholerae,* in natural waters of Kent, England. *J. Appl. Bacterial.*, 52, 435-448.

Zhu Xuebao (1989). Bacteria floccule as diet of filter-feeding fishes. *J. Fish. China-Shuichan Xuebao.*, 13(4): 339-345.

Zobell, C.E. and Landon, W. (1971). Bacterial feed for the filter feeder. In: *Aquatic Microbiology* (eds). G.Rheinheimer, Jon Wiley and Sons., London.

Index